CAMBRIDGE LIBRARY COLLECTION

Books of enduring scholarly value

Botany and Horticulture

Until the nineteenth century, the investigation of natural phenomena, plants and animals was considered either the preserve of elite scholars or a pastime for the leisured upper classes. As increasing academic rigour and systematisation was brought to the study of 'natural history', its subdisciplines were adopted into university curricula, and learned societies (such as the Royal Horticultural Society, founded in 1804) were established to support research in these areas. A related development was strong enthusiasm for exotic garden plants, which resulted in plant collecting expeditions to every corner of the globe, sometimes with tragic consequences. This series includes accounts of some of those expeditions, detailed reference works on the flora of different regions, and practical advice for amateur and professional gardeners.

Planting and Rural Ornament

William Marshall (1745–1818), an experienced farmer and land agent, published this work in 1795, and early in 1796 produced a second edition (reissued here), 'with large additions'. The two-volume work was intended as a practical guide for the owners or managers of large estates on how to establish and maintain timber plantations, both for their financial value and also as important decorative elements in the landscaping of the surroundings of the owner's house. The work covers the practical issues of planting, propagating and transplanting, discusses the choice of trees for different commercial purposes, and the planning and maintenance of hedgerows, as well as ornamental buildings. Volume 2 begins with an account of the Linnaean system of plant classification and its sexual basis, and supplies both an alphabetical list of trees and shrubs in their Latin Linnaean classes, and an index of plants under their English names.

Cambridge University Press has long been a pioneer in the reissuing of out-of-print titles from its own backlist, producing digital reprints of books that are still sought after by scholars and students but could not be reprinted economically using traditional technology. The Cambridge Library Collection extends this activity to a wider range of books which are still of importance to researchers and professionals, either for the source material they contain, or as landmarks in the history of their academic discipline.

Drawing from the world-renowned collections in the Cambridge University Library and other partner libraries, and guided by the advice of experts in each subject area, Cambridge University Press is using state-of-the-art scanning machines in its own Printing House to capture the content of each book selected for inclusion. The files are processed to give a consistently clear, crisp image, and the books finished to the high quality standard for which the Press is recognised around the world. The latest print-on-demand technology ensures that the books will remain available indefinitely, and that orders for single or multiple copies can quickly be supplied.

The Cambridge Library Collection brings back to life books of enduring scholarly value (including out-of-copyright works originally issued by other publishers) across a wide range of disciplines in the humanities and social sciences and in science and technology.

Planting
and
Rural Ornament

Being a Second Edition, with Large Additions,
of Planting and Ornamental Gardening:
A Practical Treatise

VOLUME 2

WILLIAM MARSHALL

CAMBRIDGE
UNIVERSITY PRESS

CAMBRIDGE
UNIVERSITY PRESS

University Printing House, Cambridge, CB2 8BS, United Kingdom

Cambridge University Press is part of the University of Cambridge.
It furthers the University's mission by disseminating knowledge in the pursuit of education, learning and research at the highest international levels of excellence.

www.cambridge.org
Information on this title: www.cambridge.org/9781108075916

This edition first published 1796
This digitally printed version 2015

ISBN 978-1-108-07591-6 Paperback

PLANTING

AND

RURAL ORNAMENT.

VOLUME THE SECOND.

PLANTING

AND

RURAL ORNAMENT.

BEING

A SECOND EDITION,

WITH

LARGE ADDITIONS,

OF

PLANTING AND ORNAMENTAL GARDENING,

A PRACTICAL TREATISE.

VOLUME THE SECOND.

LONDON:

Printed for G. NICOL, Bookseller to his Majesty, Pall-Mall; G. G. and J. ROBINSON, in Paternoster Row; and J. DEBRETT, Piccadilly.

M,DCC,XCVI.

CONTENTS

OF THE

SECOND VOLUME.

INTRODUCTION.

FOR a Book of Reference, the DICTIONARY form is the moſt convenient. This part of our Work is entirely of that Nature; we have therefore adopted an alphabetical arrangement. Our reaſons for making uſe of the Linnean names, as the ground-work of this arrangement, are manifold: a great number of the plants here treated of, have no Engliſh generic name belonging to them: yet it was neceſſary, to that conciſeneſs and ſimplicity which is the baſis of our plan, to arrange them agreeably to their reſpective genera; becauſe, in general, the individuals of the ſame genus have ſimilar appearances and ſimilar propenſities, which being placed together, in one point of view, their deſcription and mode of culture are rendered infinitely more eaſy and compendious, than they could poſſibly be, if treated of, ſeparately, under diſtinct and detached ſpecies. Beſides, even many of the ſpecies, now common in our ornamental grounds and ſhrubberies, have not yet had any Engliſh name given to them; and there are many more, whoſe Engliſh names are local and unſettled; whereas the

Linnean names are the ſame every where, and are known to the whole world *.

We do not mean to enter into the diſpute about the Sexual Syſtem of Linneus: it is enough for our purpoſe, that it is, at preſent, the prevailing ſyſtem; and that, being founded in nature, its principles can never be overturned: we are, nevertheleſs, ſo far from thinking it a *perfect* ſyſtem, that we believe it capable of very great improvement: at preſent, however, it is our buſineſs to take it as we find it; and for the uſe of ſuch of our readers as are unacquainted with its principles, it is proper that we ſhould here give its outline.

Every PERFECT FLOWER has four principal parts, which, in general, are obvious to the naked eye; namely, the CALYX, or outer guard; the COROLLA, or coloured leaves; the STAMINA; and the PISTILLUM. The calyx is evident in the moſs roſe; being thoſe elegant rough leaves which incloſe the bluſhing beauties of the flower: it is alſo conſpicuous in the primroſe; being the angular tube out of which the more delicate parts of the flower iſſue. The corolla of the primroſe is the yellow ornament which, by unbotanical obſervers, is itſelf conſidered as the flower. The ſtamina are conſpicuous in

* An Alphabet of Engliſh Names will be given at the end of this Volume.

moſt

moſt flowers, and are diſtinguiſhed by the farina or duſt, with which they are covered. In the primroſe, they proceed from the inſide of the tube of the corolla; and, when matured, form themſelves into a circle round the top of the tube. The piſtillum, in the primroſe, is the delicate white pillar, which, riſing from the bottom of the tube of the corolla, ſhews its flatted top, in the middle of the ſtamina, and in the center of the flower. According to Linneus, the calyx is an expanſion of the outer bark; the corolla, of the inner bark; the ſtamina, of the wood; and the piſtillum, of the pith of the plant; and, according to his Sexual Syſtem, the ſtamen is the male, and the piſtillum the female, part of generation.

Whether this laſt is or is not a fact, in nature, has been the ſubject of much diſpute. But, to the FLORAL SYSTEM, it is a matter of no great import. The parts themſelves, and not their functions, are the baſis of the Linnean ſyſtem; and we are clear in our opinion, that if that great man had conſidered his Syſtem, as being what it in reality is, merely FLORAL, without having unfortunately clogged it with the idea of SEXUAL, he would have ſaved himſelf a hoſt of enemies, and would, beyond a doubt, have rendered his Syſtem infinitely more ſimple and ſcientific, and conſequently more uſeful, than it really is. But it is now too late to regret: his Syſtem is eſtabliſhed; and himſelf no more.

Having, however, ſaid thus much, it would be unpardonable in us not to add, that whether we conſider his genius, his perſeverance, or the Syſtem he has formed, notwithſtanding its imperfections, he died one of the greateſt characters the world has known.

The VEGETABLE KINGDOM is divided, by LINNEUS, into twentyfour CLASSES: theſe Claſſes are ſubdivided into ORDERS; the Orders into GENERA; the Genera into SPECIES; and the Species into VARIETIES.

His principle of Claſſification is ſeen in the following

"KEY

"KEY OF THE SEXUAL SYSTEM.

MARRIAGES of PLANTS.
Florescence.

PUBLIC MARRIAGES.
Flowers visible to every one.

In ONE BED.
Husband and wife have the same bed.
All the flowers hermaphrodite: stamens and pistils in the same flower.

WITHOUT AFFINITY.
Husbands not related to each other.
Stamens not joined together in any part.

WITH EQUALITY.
All the males of equal rank.
Stamens have no determinate proportion of length.

1. ONE MALE.
2. TWO MALES.
3. THREE MALES.
4. FOUR MALES.
5. FIVE MALES.
6. SIX MALES.
7. SEVEN MALES.
8. EIGHT MALES.
9. NINE MALES.
10. TEN MALES.
11. TWELVE MALES.
12. TWENTY MALES.
13. MANY MALES.

WITH SUBORDINATION.
Some males above others.
Two stamens are always lower than the others.

14. TWO POWERS. | 15. FOUR POWERS.

WITH AFFINITY,
Husbands related to each other.
Stamens cohere with each other, or with the pistil.

16. ONE BROTHERHOOD.
17. TWO BROTHERHOODS.
18. MANY BROTHERHOODS.
19. CONFEDERATE MALES.
20. FEMININE MALES.

IN TWO BEDS.
Husband and wife have separate beds.
Male flowers and female flowers in the same species.

21. ONE HOUSE.
22. TWO HOUSES.
23. POLYGAMIES.

CLANDESTINE MARRIAGES.
Flowers scarce visible to the naked eye.

24. CLANDESTINE MARRIAGES

His CLASSES are:

I. ONE MALE. *(Monándria).*

One husband in marriage.

One stamen in an hermaphrodite flower.

II. TWO MALES. *(Diándria.)*

Two husbands in the same marriage.

Two stamens in an hermaphrodite flower.

III. THREE MALES. *(Triándria.)*

Three husbands in the same marriage.

Three stamens in an hermaphrodite flower.

IV. FOUR MALES. *(Tetrándria.)*

Four husbands in the same marriage.

Four-stamens in the same flower with the fruit.

(If the two nearest stamens are shorter, it is referred to Class 14.)

V. FIVE MALES. *(Pentándria.)*

Five husbands in the same marriage.

Five stamens in an hermaphrodite flower.

VI. SIX MALES. *(Hexándria.)*

Six husbands in the same marriage.

Six stamens in an hermaphrodite flower.

(If the two opposite stamens are shorter, it belongs to Class 15.)

VII. SEVEN MALES. *(Heptándria).*

Seven husbands in the same marriage.

Seven stamens in the same flower with the pistil.

VIII. EIGHT MALES. *(Octándria.)*

Eight husbands in the same marriage.

Eight stamens in the same flower with the pistil.

IX. NINE MALES. *(Enneándria.)*

Nine husbands in the same marriage.

Nine stamens in an hermaphredite flower

X. TEN MALES. *(Decándria.)*

Ten husbands in the same marriage.

Ten stamens in an hermophrodite flower.

XI. TWELVE

XI. TWELVE MALES. *(Dodeca'ndria.)*
Twelve husbands in the same marriage.
Twelve stamens to nineteen *in an hermaphrodite flower.*

XII. TWENTY MALES. *(Icosa'ndria.)*
Generally twenty husbands, often more.
Stamens inserted on the calyx (not on the receptacle *) *in an hermaphrodite flower.*

XIII. MANY MALES. *(Polya'ndria.)*
Twenty males or more in the same marriage.
Stamens inserted on the receptacle, *from* 20 *to* 1000 *in the same flower with the pistil.*

XIV. TWO POWERS. *(Didyna'mia.)*
Four husbands, two taller than the other two.
Four stamens: of which the two nearest are largest.

XV. FOUR POWERS. *(Tetradyna'mia).*
Six husbands, of which four are taller.
Six stamens, of which four are longer, and the two opposite ones shorter.

XVI. ONE BROTHERHOOD. *(Monade'lphia.*
Husbands, like brothers, arise from one base.
Stamens are united by their filaments † *into one body.*

XVII. TWO BROTHERHOODS. *(Diade'lphia).*
Husbands arise from two bases, as if from two mothers.
Stamens are united by their filaments into two bodies.

XVIII. MANY BROTHERHOODS. *(Polyade'lphia.)*
Husbands arise from more than two mothers.
Stamens are united by their filaments into three or more bodies.

XIX. CONFEDERATE MALES. *(Syngene'sia.)*
Husbands joined together at the top.
Stamens are connected by the anthers ‡ *forming a cylinder (seldom by the filaments).*

* "The base by which the parts of the fructification are connected."

† The *thread* or body of the stamen.

‡ The *tips* or heads of the stamen.

XX. FEMININE

XX. FEMININE MALES. (*Gyna'ndria.*)
Hufbands and wives growing together.
Stamens are inferted in the piftils (not on the receptacle).

XXI. ONE HOUSE. (*Monœ'cia*).
Hufbands live with their wives in the fame houfe, but have different beds.
Male flowers and female flowers are on the fame plant.

XXII. TWO HOUSES. (*Diœ'cia.*)
Hufbands and wives have different houfes.
Male flowers and female flowers are on different plants.

XXIII. POLYGAMIES. (*Polyga'mia.*)
Hufbands live with wives and concubines.
Hermaphrodite flowers, and male ones, or female ones in the fame fpecies.

XXIV. CLANDESTINE MARRIAGES. (*Cryptoga'mia.*)
Nuptials are celebrated privately.
Flowers concealed within the fruit, or in fome irregular manner."

His ORDERS are diftinguifhed by different parts of the flowers, according to the Claffes. Thofe of the firft thirteen Claffes are taken from the number of females or piftils (reckoning "from the bafe of the ftyle *; but if there is "no ftyle, the calculation is made from the "number of ftigmas" †); as ONE FEMALE (*Monogynia*), TWO FEMALES (*Digy'nia*), THREE FEMALES (*Trigy'nia*), &c. Thofe of the fixteenth, feventeenth, eighteenth, twentieth, twenty-firft, and twenty-fecond Claffes, are taken from the number of males, or ftamens. Thofe of the fourteenth, are diftinguifhed by SEEDS

* The *fhaft* or body of the piftil.

† The *fummits* or heads of the piftil.

NAKED

NAKED (*Gymnospe'rmia*), and SEEDS CLOATHED (*Angiospe'rmia*). Those of the fifteenth, by the formation of the seed-vessel, or pod ; as, WITH SILICLE (*Siliculo'sa*), and, WITH SILIQUE (*Siliquo'sa*). Those of the twenty-third are ONE HOUSE (*Monœ'cia*) ; TWO HOUSES (*Diœ'cia*) ; and THREE HOUSES (*Triœ'cia*). Those of the twenty-fourth are FERNS, MOSSES, FLAGS, and FUNGUSSES. Those of the nineteenth Class (consisting chiefly of plants with compound discous flowers, as the thistle, dandelion, &c.) are, EQUAL POLYGAMY (*Polyga'mia Equa'lis*) ; SUPERFLUOUS POLYGAMY (*Polyga'mia Supe'rflua*) ; FRUSTRANEOUS POLYGAMY (*Polyga'mia Frustra'nea*) ; NECESSARY POLYGAMY (*Polyga'mia Necessa'ria*) ; SEPARATE POLYGAMY (*Polygamia Segrega'ta*) ; MONOGAMY (*Monoga'mia*).

The following is Linneus's account (*literally* as it stands in the Lichfield translation) of the Orders last mentioned.

"EQUAL POLYGAMY consists of many marriages with promiscuous intercourse.

That is, *of many florets furnished with stamens and pistils.*

The flowers of these are vulgarly called Flosculous.

SPURIOUS POLYGAMY, where the beds of the married occupy the disk, and those of the concubines the circumference.

That is, *the hermaphrodite florets occupy the disk, and the female florets without stamens surround the border, and that in three manners :*

(*a*) SUPER-

(*a*) SUPERFLUOUS POLYGAMY, when the married females are fertile, and thence the concubines superfluous.

That is, *when the hermaphrodite flowers of the disk are furnished with stigmas, and produce seeds; and the female flowers also, which constitute the circumference, produce seeds likewise.*

(*b*) FRUSTRANEOUS POLYGAMY, when the married females are fertile, and the concubines barren.

That is, *when the hermaphrodite flowers of the disk are furnished with a stigma, and produce seeds; but the florets which constitute the circumference having no stigma, produce no seeds.*

(*c*) NECESSARY POLYGAMY, when the married females are barren, and the concubines fertile.

That is, *when the hermaphrodite flowers, from the defect of the stigma of the pistil, produce no seed; but the female flowers in the circumference produce perfect seeds.*

(*e*) SEPARATE POLYGAMY, when many beds are so united that they constitute one common bed.

That is, *when many flower-bearing calyxes are contained in one common calyx, so as to constitute one flower.*"

His GENERA are taken from the construction of the parts of fructification. All plants, whether herbs, shrubs, or trees, whose flowers and seeds correspond, as to figure and disposition, are of the same GENUS.

His SPECIES are distinguished by the leaves, and other more permanent parts of the plant. Or, it may be said of trees and shrubs, the SPECIES is determined by the natural properties of the seed: for, let the exterior of a plant, or tribe of plants, be what it may, if the seed do not produce near

near reſemblances of the parent ſtock, but plants whoſe appearances or properties are different from it (as in the caſe of apples, pears, &c.); ſuch plants are not conſidered as forming a diſtinct SPECIES, but are deemed VARIETIES.

It now only remains to offer, to the *Engliſh* reader, a few remarks concerning the due PRONUNCIATION of the LINNEAN TERMS, which we have thought it right to accent, in this SECOND EDITION.

A *vowel* when accented, in the ſecond place of ſyllables from the termination of a word, and followed by a ſingle conſonant, is long; as in *A'cer*, *bacca'ta*, *commu'nis*, *gla'ber*, *Ilex*, *ni'gra*, *orienta'lis*, *ſempervi'rens*, *villo'ſa*, *&c.* *&c.* but, when accented, in the third place, it varies with the quality of the word in which it occurs. In ſubſtantives, uſed as GENERIC TERMS, it is, in this ſituation, generally ſhort; as in *Be'tula*, *Ce'raſus*, *Cle'matis*, *He'dera*, *Juni'perus*, *Lycium*, *Peri'ploca*, *Pla'tanus*, *Po'pulus*, *Robi'nia*, *Ti'lia*:—*Meze'reum* is an exception. On the contrary, in adjectives, uſed as SPECIFIC TERMS, the vowel, accented in the third place of ſyllables, is generally long; as in the ordinary terminations, *fo'lius* and *o'ides*, in *arbo'rea*, *auſtri'acus*, *cœru'lea*, *herba'cea*, *hu'mile*, *lu'tea*, *mono'ica*, *purpu'rea*, *Syri'aca*, *&c.*—*Balſami'fera*, *ladani'fera*, *&c.* *ſylva'tica*, *tre'mula*, *vi'ride*, are exceptions.

The

The *final e* is pronounced as a ſyllable, is never *ſilent*, as in the Engliſh language. Thus *Ane'mone*, *campe'ſtre*, *canade'nſe*, *Da'phne*, *ga'le*, *hu'mile*, *officina'le*, *vulga're*, are rendered Anemony, campeſtry, canadenſy, Daphny, galey, humily, officinaley, vulgarey. If joined with *s*, the *e* in termination has a ſimilar power: as in *A'-bi-es*, *Ke'r-mes*.

Ea are ever pronounced ſeparately, as two ſyllables; whether in the Subſtantives *Alce'-a*, *Colute'-a*, *Ite'-a*, *Phillyre'-a*, *Staphyle'-a*; or in the Adjectives *acu-le-a'tus*, *arbo're-a*, *ca'pre-a*, *cœru'le-a*, *lu'te-a*, *purpu're-a*, &c. &c.

The *i*, in the termination of a word, is long; as in *crus-galli*.

Oi are ſeparately pronounced, in the termination *oides*, whether the accent be laid on the former or the latter; alſo in *Be'nzoin*,—*Ben-zo-in*.

The *y*, when accented in the third place of ſyllables, is ſhort; as in *Cy'tiſus*, *Ly'cia*; rendered Cittiſus, Lyſhia. But different, in the ſecond place; as *Py'rus*, *Sty'rax*, in which the *y* is long.

The letter *c*, in ſome particular combinations, has the power of the Engliſh *ſh*; as in *diœ'cia*, *her-*

herba'cea, Ly'cia, monœ'cia, Pista'cia, which correſpond, in pronunciation, with Diœſhia, Herbaſhia, Lyſhia, Monœſhia, Piſtaſhia.

Ch has the power of *k*: thus *Andra'chne, Chionа'nthus, Dista'chya,* are equivalent in articulation to Andrackny, Kyonanthus, Diſtackia.

In the ACCENTUATION of the Linnean terms, we have not been inattentive to the labors of the LICHFIELD SOCIETY. We have not, however, followed implicitly their accented catalogues; which, in ſome particulars, are unintelligible to practical men; and ours is a work intended to convey practical knowledge. We aſpire not at a place in the library alone; we are equally ambitious to enjoy the freedom of the morning room, and the Society of its fair inhabitants; and ſhall not be aſhamed if we are found on the duſty table of the planter's ſeed room.

We have, therefore, endeavoured to retain ſo much of the eſtabliſhed pronunciation of the names of the plants we have treated of, as we think will render them intelligible, in converſation and practice, without giving cauſe of offence in the cloſet.

The principal deviations we have judged it right to make from theſe catalogues, are in the terms

terms *Ane'mone*, *Arbu'tus*, *Colute'a*, *Glyci'ne*, *Hype'ricum*, *Ite'a*, *Philyre'a*; which, in the lifts alluded to, ftand *Anemo'ne*, *Arbutus*, *Colu'tea*, *Gly'cine*, *Hyperi'cum*, *I'tea*, *Philly'rea*; Innovations which, we truft, we are warranted in rejecting. Neverthelefs, we have brought the terms together, here, to give the reader a favorable opportunity of forming his own judgment, and of correcting with his pen, what he may think we have done amifs.

ALPHA-

ALPHABET

OF

PLANTS.

ACER.

LINNEAN Claſs and Order, *Polygamia Monoecia*: Male flowers containing eight ſtamens, and hermaphrodite flowers containing eight ſtamens and one piſtil, upon the ſame plant. There are Eleven SPECIES: Ten of which are natives of, or have been introduced into, this country.

1. A'CER *Pſeu'do-pla'tanus*: The SYCAMORE; *a tall deciduous tree*; native of the continent of Europe, but doubtful whether or not of this iſland.

2. A'CER *Campe'ſtre*: The COMMON MAPLE; *a low deciduous tree*; common in our woods and hedges.

3. A'CER *Negu'ndo*: the ASH-LEAVED MAPLE; *a deciduous tree*; native of Virginia and Carolina.

4. A'CER *Platanoi'des*: the NORWAY MAPLE; *a deciduous tree*; native of Norway and the north of Europe.

5. A'CER *Monſpeſula'num*: the MONTPELIER MAPLE; *a low deciduous tree*; growing common about Montpelier.

6. A'CER *Cre'ticum*: the CRETAN MAPLE; *a low deciduous tree*; native of the Eaſt.

7. A'CER *Ru'brum*: the SCARLET MAPLE; *a deciduous tree*; native of Virginia and Pennſylvania.

8. A'CER *Sacchari'num:* the SUGAR MAPLE; *a deciduous tree*; native of Pennſylvania.

9. A'CER *Tarta'ricum:* the TARTARIAN MAPLE; *a low deciduous tree*; native of Tartary.

10. A'CER *Pennſylva'nicum:* the PENNSYLVANIAN MAPLE; or the MOUNTAIN MAPLE; *a tall deciduous ſhrub*; native of Pennſylvania.

1. The SYCAMORE. This tree grows to a great height and ample ſize, throwing out a wide-ſpreading top. Its leaves are vine-ſhaped; and, on their firſt appearance, are of a pleaſant green; but their beauty ſoon goes off, being liable to be perforated and disfigured by inſects during the ſummer months, which reduces the value of the Sycamore as an *ornamental*: it has however, long been conſidered as a *timber tree* in this country, having been much uſed by the turners for wooden bowls, diſhes, trenchers, &c.; but, ſince the cuſtom of uſing earthen ware has become ſo prevalent, its value for this purpoſe is greatly decreaſed. Neverthele ſs, near the ſea coaſt it may be planted with advantage, as it is known to withſtand the attacks of the ſea air with peculiar hardineſs. HANBURY ſays, The Sycamore being wounded exudes a great quantity of liquor, of which is made good wine. There are two *Varieties* of the Sycamore: one with broad leaves and large keys; the other with variegated leaves.

The PROPAGATION of the Sycamore is very eaſy. In the autumn, when the keys are ripe, they may be gathered, and in a few days after ſown, about an inch and a half deep, in beds of common mould. In the ſpring the plants will appear, and make a ſhoot about a foot and a half by the autumn following, if the ground of the ſeminary be tolerably good, and they are kept clean from weeds. The ſpring after they come up, they ſhould be planted in the nurſery, in rows two feet and a half aſunder, and their diſtance in the rows muſt be one foot and a half. Here they may remain till they are big enough to plant out finally, with no farther trouble than taking off unſightly ſide branches, and ſuch as have a tendency to make the tree forked, except digging between the rows, which muſt always be done every winter. This tree will grow upon almoſt any ſoil.

2. The COMMON MAPLE is too well known to need a de-

a defcription. It is of much humbler growth than the Sycamore; and is by no means ornamental; nor is its timber of a good quality, being peculiarly brittle: The texture however is clofe and firm, and it is in good efteem amongft the turners. In the vale of Glocefter, where oak timber is fcarce, Maple is ufed for gate ftuff and other purpofes of hufbandry; and fometimes fcrews for cyder preffes are made of this wood. But the principal value of the Maple, is for *underwood*: it is of quick growth, and affords good fuel.

The method of PROPAGATION is the fame as that of the Sycamore; and, like it, the Maple will grow in almoft any foil and fituation.

3. The ASH-LEAVED MAPLE grows to a large timber tree: its leaves are of a pale green, and well adapted to give variety of tint; but HANBURY fays, this tree is not proper to be planted in expofed fituations, the branches being fubject to be fplit off by the winds. Its ufes are fimilar to thofe of the Sycamore.

It may be propagated from the keys, which are perfected in this country; or by layering; or from cuttings, planted in a moift fituation, in autumn.

4. The NORWAY MAPLE. This alfo grows to a large timber tree. Its leaves are of a fhining green colour, and are as large or larger than thofe of the Sycamore; their edges are acutely and more beautifully indented; they are not fo liable to be eaten by infects in the fummer; and "in the autumn they die to a golden yellow colour, which caufes a delightful effect at that feafon, when the different tints of the decaying vegetable world are difplayed." The flowers are alfo beautiful; they come out early in the fpring, are of a fine yellow colour, and fhew themfelves to advantage before the leaves come out. They are frequently fucceeded by keys, which fometimes arrive at maturity in this climate. There is a *Variety* with ftriped leaves.

The Norway Maple may be PROPAGATED from feed, as the Sycamore; it may alfo be raifed by layers, and cuttings, planted in a moift foil.

5. MONTPELIER MAPLE grows to about twenty feet high, and is a very beautiful tree. The leaves are compofed of three lobes, are of a fhining green, a thickifh fubftance, and retain their verdure later in the year than

moſt of the other ſorts. The flowers come out in the ſpring, but have very little beauty; their blow is ſoon over, and ſometimes they are ſucceeded by ſeeds, which come to perfection in our gardens.

6. CRETAN MAPLE. This grows to about the height of the former. The leaves are downy, compoſed of three lobes, and grow oppoſite to each other on long downy footſtalks. The flowers come out in the ſpring, are inconſiderable to the floriſt, and are very ſeldom ſucceeded by good ſeeds in England.

7. SCARLET-FLOWERING MAPLE. Of this there are two ſorts; called, 1. *Virginian ſcarlet-flowering Maple*; and, 2. *Sir Charles Wager's Maple.* Both of theſe are propagated for the ſake of the flowers, which are of a ſcarlet colour, and come out early in the ſpring. The leaves are compoſed each of five ſharp-pointed lobes, which are ſlightly indented or ſerrated: They are ſmooth, of a pale green on their upper ſurface, glaucous * underneath; and they grow on long, ſimple, taper, reddiſh footſtalks. The flowers come out in cluſters from the ſide of the branches. They appear in April, and the ſeeds ripen in June. The ſort called *Sir Charles Wager*'s produces larger cluſters of flowers than the others; on which account it is in moſt eſteem.

8. SUGAR MAPLE is a large growing tree; will arrive at the height of forty feet; and has broad thin leaves, divided into five principal parts; which are again indented or cut at the edges into ſeveral acute ſegments. Their ſurface is ſmooth, of a light green colour, whitiſh underneath; and they grow on pretty long footſtalks. The flowers come out in the ſpring about the time of the Norway Maple; and they are ſucceeded by long keys, which ſometimes ripen in England. In America, the inhabitants tap this tree in the ſpring, boil the liquor, and the fœces afford a uſeful ſugar. The Sycamore, the Aſh-leaved and the Norway Maples alſo abound with a ſaccharine juice, from which there is no doubt but a uſeful ſugar might be prepared.

9. TARTARIAN MAPLE will grow to upwards of twenty feet high. The leaves are heart-ſhaped, undivided, and their edges are unequally ſerrated. The

* *Glaucous*, of a ſea-green colour.

flowers

flowers come out from the wings of the leaves, in longiſh bunches; they appear early in the ſpring; and ſometimes are ſucceeded by ripe ſeeds in our gardens.

10. MOUNTAIN MAPLE. The *ſtalks* of this ſhrub are ſlender, covered with a whitiſh bark, ſend forth ſeveral red branches, and grow about fifteen feet high. The leaves are three-lobed, pointed, and are unequally and ſharply ſerrated. The flowers come out in longiſh bunches, in the ſpring: They are of a greeniſh yellow colour; and are ſucceeded by ſeeds which (like thoſe of the Norway Maple) generally fall off before they are ripe.

Theſe ſorts are all PROPAGATED, 1. by the ſeeds; but as they do not always ripen in this country, the beſt way will be to procure them from the places where they naturally grow. A cool ſhady part of the ſeminary ſhould be appropriated for the purpoſe; the mould ſhould be made fine; beds ſhould be marked out four feet wide, and in length proportionable to the quantity; and in theſe the ſeeds ſhould be regularly ſown, ſifting over them about half an inch of the fineſt mould. When the plants come up, they muſt be kept clean from weeds, and frequently watered; and this work muſt be duly attended to all ſummer. The ſpring following, the ſtrongeſt may be drawn out, and planted in the nurſery, in rows two feet aſunder, and at the diſtance of a foot from each other in the rows; leaving the others in the ſeminary to gain ſtrength. The ſpring following they alſo muſt receive the ſame culture; and in the nurſery they may remain, with no other trouble than keeping the ground clean from weeds in the ſummer, digging between the rows in the winter, and taking off all ſtrong and irregular ſide ſhoots, till they are planted out. Trees raiſed from ſeeds will grow faſter, and arrive at greater height, than thoſe raiſed from layers; but they will not produce ſuch quantities of flowers; which makes the latter method more eligible for thoſe who want theſe plants for a low ſhrubery. 2. By layers all the ſpecies of this genus are to be propagated; though it is never practiſed for the Common Maple and the Sycamore. The young ſhoots may be at any time laid down in the autumn, winter, or early in the ſpring. By the autumn following, they will have ſtruck root, and become good plants; when the ſtrongeſt may be ſet out in the places where they are to remain; whilſt the weakeſt may be planted in the nurſery, like the ſeedlings, for a year or two, to gain ſtrength. 3. By cuttings alſo theſe trees are to be propagated: But this method is chiefly practiſed on the Aſh leaved and Norway Maples, which more readily take root this way. The cuttings ſhould be the bottom parts of the laſt year's ſhoots: They ſhould be taken off early in October, and plaſited in rows in a moiſt ſhady place. The ſpring and ſummer following they muſt be duly watered as often as dry weather makes it neceſſary, and be kept clean from weeds. By the autumn they will be fit to remove into the nurſery; though if the cuttings are not planted too cloſe, they may remain in their ſituation for a year or two longer, and then be ſet out finally, without the trouble of being previouſly planted in the nurſery. 4. By budding, grafting, and inarching likewiſe Maples are to be propagated: But the other methods being more eligible, theſe are never practiſed, except for the variegated ſorts and the large broad-leaved kind. The latter is to be continued no otherwiſe than by budding it on ſtocks of the common Sycamore; for the ſeeds, though ſo large themſelves, when ſown afford you only the common Sycamore in return.

Seeds of the variegated kinds, however, when ſown will produce variegated plants in return; which renders the propagation of theſe ſorts very expeditious, where plenty of ſeeds may be had. Where theſe are not to be obtained, in order to propagate theſe varieties by budding, let ſome plants of the common Sycamore, one year old, be taken out of the ſeminary, and ſet in the nurſery in rows a yard aſunder, and the plants about a foot and a half diſtance from each other in the rows: Let the ground be kept clean from weeds all ſummer, and be dug, or, as the gardeners call it, *turned in*, in the winter; and the ſummer following the ſtocks will be of a proper ſize to receive the buds, which ſhould be taken from the moſt beautifully ſtriped branches. The beſt time for this work is Auguſt; becauſe if it is done earlier, the buds will ſhoot the ſame ſummer; and when this happens, a hard winter will infallibly kill them. Having, therefore, budded your ſtocks the middle or latter end

of

of Auguſt, with the eyes or buds fronting the north, early in October take off the baſs matting, which before this time will have confined the bark and pinched the bud, but not ſo as to hurt it much. Then cut off the ſtock juſt above the bud, and dig the ground between the rows. The ſummer following, keep the ground clean from weeds; cut off all natural ſide buds from the ſtock as they come out; and by autumn, if the land is good, your buds will have ſhot forth, and formed themſelves into trees five or ſix feet high. They may be then removed into the places where they are deſigned to remain; or a few of them only may be drawn out, leaving the others to be trained up for larger ſtandards, to ſerve for planting out in open places, or ſuch other purpoſes as ſhall be wanting.

The Striped Norway Maple ſhould be budded on ſtocks of its own kind; for on theſe they take beſt, and both kinds are not very liable to run away from their colours. Variegated plants in general muſt be planted in poor, hungry, gravelly, or ſandy ſoils, to feed the diſeaſe which occaſions theſe beautiful ſtripes, and cauſe it to be more powerful. But theſe trees ſhew their ſtripes in greater perfection in a good ſoil: The plant, though in ſickneſs, has the appearance of health; the ſhoots are vigorous and ſtrong; the leaves are large, leſs liable to be hurt by inſects; and the ſtripes appear more perfect, natural, and delightful, than thoſe on ſtunted trees growing on a poor ſoil.

Æ S C U L U S.

LINNEAN Claſs and Order, *Heptandria Monogynia*: Each flower contains ſeven *or* eight males and one female *: There are only two SPECIES:

1. ÆSCULUS *Hippo-caſtanum*: The ESCULUS or HORSE CHESNUT; *a deciduous tree*; native of Aſia.

* The Æſculus is one of the defective genera of Linneus. The *Pavia*, having eight males in each flower, belongs properly to the eighth Claſs.

2. ÆSCULUS *Pavia:* The SCARLET ESCULUS, or SCARLET-FLOWERING HORSE CHESNUT; *a tall deciduous ſhrub*; native of Carolina, the Brazils, and ſeveral parts of the Eaſt.

1. HORSE CHESNUT, or COMMON ESCULUS. This is a large well looking tree; growing to ſeventy or eighty feet high, and throwing out its branches to a conſiderable width; yet forming a cloſe thickſet head; which, if left to nature, takes a moſt beautifully ſtriking parabolic form. Its leaves are large, palmated, and of a dark green colour: they appear very early in the ſpring; their buds ſometimes beginning to ſwell ſo early as Chriſtmas, and anticipate the pleaſures of the coming ſpring. Its flowers are ſingularly beautiful, ſtanding in large ſpikes thick among the leaves. This tree is peculiar in a quick formation of its ſhoots, which are frequently perfected in leſs than three weeks from the time of foliation; "in which time," ſays MILLER, "I have meaſured ſhoots a foot and a half long with their leaves fully expanded." For ſingle trees, the Horſe Cheſnut ſtands amongſt the firſt of the *ornamental* tribe; and in the ſpring of the year, when its flowers are out, we know no tree equal to it in beauty. It is improper however to be planted near gardens or kept walks, as it ſheds its leaves early in autumn, and, being large and numerous, they create a diſagreeable litter. The *uſes* of the Eſculus are few: its timber is of an inferior kind, and its fruit of no great eſtimation: deer are ſaid to affect it much; and MILLER ſays, "in Turkey the nuts of this tree are ground and mixed with the provender for their horſes, eſpecially thoſe which are troubled with coughs or are broken winded, in both which diſorders they are accounted very good." HANBURY tells us, that ſwine will fatten upon them; but does not ſay how they are to be prepared. We have known them offered to hogs raw, alſo boiled, as likewiſe baked in an oven, but without ſucceſs.

The Horſe Cheſnut is PROPAGATED from the nuts: In autumn therefore, when they fall, a ſufficient quantity ſhould be gathered. Theſe ſhould be ſown ſoon afterwards in drills, about two inches aſunder. If the nuts are kept till ſpring, many of them will be faulty; but where the ſeminary ground cannot be got ready before,

fore, and they are kept ſo long, it may be proper to put them in water, to try their goodneſs: The good nuts will ſink, whilſt thoſe which are faulty will ſwim; ſo that by proving them this way you may be ſure of good nuts, and have more promiſing hopes of a crop. In the ſpring the plants will come up; and when they have ſtood one year they may be taken up, their tap roots ſhortened, and afterwards planted in the nurſery. When they are of ſufficient ſize to be planted out finally, they muſt be taken out of the nurſery with care, the great ſide ſhoots and the bruiſed parts of the roots ſhould be taken off, and then planted in large holes level with the ſurface of the ground, at the top of their roots; the fibres being all ſpread and lapped in the fine mould, and the turf alſo worked to the bottom. A ſtake ſhould be placed to keep them ſafe from the winds, and they muſt be fenced from the cattle till they are of a ſufficient ſize to defend themſelves. The beſt ſeaſon for all this work is October. After the trees are planted, neither knife nor hatchet ſhould come near them; but they ſhould be left to Nature to form their beautiful parabolic heads, and aſſume their utmoſt beauty.

The Horſe Cheſnut, like moſt other trees, delights in good fat land; but it will grow exceedingly well on clayey and marley grounds. It prefers a moiſt ſituation.

Miller ſays, "when theſe trees are tranſplanted, their roots ſhould be preſerved as entire as poſſible, for they do not ſucceed well when torn or cut; nor ſhould any of the branches be ſhortened, for there is ſcarce any tree which will not bear amputation better than this; ſo that when any branches are by accident broken, they ſhould be cut off cloſe to the ſtem, that the wound may heal over."

2. The Scarlet Esculus grows to about fifteen or ſixteen feet high; and there is a delicacy in this tree that makes it deſirable. The bark of the young ſhoots is quite ſmooth, and the growing ſhoots in ſummer are of a reddiſh hue. The leaves are palmated, being pretty much like thoſe of the Horſe Cheſnut, only much ſmaller, and the indentures at the edges are deeper and more acute. The lobes of which they are compoſed

posed are spear-shaped; they are five in number, are united at their base, and stand on a long red footstalk. The leaves grow opposite by pairs on the branches, which are spread abroad on every side. The flowers come out from the ends of the branches. The first appearance of the buds is in May; though they will not be in full blow till the middle of June. They are of a bright red colour, and consequently have a pleasing effect among the vast tribe of yellow flowering sorts which shew themselves in bloom at that season. They continue in succession for upwards of six weeks; and sometimes are succeeded by ripe seeds in our gardens.

There are two ways of PROPAGATING this tree; 1. By budding it upon the young plants of the Horse Chesnut. These stocks should be raised as was directed in that article. They should be planted in the nursery way, a foot asunder, and two feet distant in the rows, which should be kept clean of weeds, and must be dug between every winter till the operation is to be performed. After they have stood in the nursery ground about two years, and have made at least one good summer's shoot, the summer following is the time for the operation. Then, having your cuttings ready soon after Midsummer, the evenings and cloudy weather should be made choice of for the work. Whoever has a great number of trees to inoculate, must regard no weather; but keep working on, to get his business over before the season ends; and indeed, a good hand will be always pretty sure of success, be the weather what it will. If the stocks were healthy, the summer following they will make pretty good shoots; and in a year or two after that will flower. This is one method of propagating this tree; and those plants that are propagated this way will grow to a larger size than those raised immediately from seeds. 2. This tree also may be propagated by seeds; which will sometimes ripen with us, and may be obtained out of our own gardens. The manner of raising them this way is as follows: Let a warm border be prepared; and if it is not naturally sandy, let drift sand be mixed with the soil; and in this border let the seeds be sown in the month of March, about half an inch deep. After this, constant weeding must be observed; and when

when the plants are come up, if they could be ſhaded in the heat of the day, it would be much better. Theſe, with now and then a gentle watering in a dry ſeaſon, will be all the precautions they will require the firſt ſummer. The winter following, if the ſituation is not extremely well ſheltered, protection muſt be given them from the hard black froſts, which will otherwiſe often deſtroy them: So that it will be the ſafeſt way to have the bed hooped, to cover them with mats in ſuch weather, if the ſituation is not well defended: if it is, this trouble may be ſaved; for, even when young, they are tolerably hardy. In about two or three years they may be removed into the nurſery, or planted where they are to remain, and they will flower in three or four years after. The uſual nurſery care muſt be taken of them when planted in that way; and the beſt time for planting them there, or where they are to remain, is October; though they will grow exceeding well if removed in any of the winter months; but, if planted late in the ſpring, they will require more watering, as the ground will not be ſo regularly ſettled to the roots, as if they had been planted earlier.

A M O R P H A.

LINNEAN Claſs and Order, *Diadelphia Decandria:* Each flower contains ten males and one female; the males being connected at the baſe in two diviſions: There is only one known SPECIES:

AMO'RPHA *Frutico'ſa:* BASTARD INDIGO; *a deciduous ſhrub*; native of Carolina.

THE AMORPHA has its beauties; but it has alſo ill effects which detract from its value. It is late in the ſpring before the foliage is fully diſplayed. The ends of the branches are generally deſtroyed by the froſt; or, if they recover it, they have the appearance of being dead; whilſt other plants teſtify their effects of the

the reviving months. But notwithſtanding theſe defects, this tree has ſome other good properties that in part make amends for them. The leaves, when out, which will not be before the middle of May, are admired by all: They are of a pleaſant green colour; are very large, beautifully pinnated, the folioles being arranged along the ſtalk by pairs, and terminate by an odd one. The flowers are of a purple colour, and ſhew themſelves in perfection with us the beginning of July. They grow in ſpikes, ſeven or eight inches long, at the ends of the branches, and are of a ſingular ſtructure. In order to make this tree have its beſt effect, it ſhould be planted among others of its own growth, in a well-ſheltered ſituation; by which means the ends will not be ſo liable to be deſtroyed by the winter's froſts; the branches will not ſuffer by the violence of the winds; and as it is ſubject to put out many branches near the root, theſe indelicacies and imperfections will be concealed; whilſt the tree will ſhew itſelf to the utmoſt advantage when in blow, by elevating its purple ſpiked flowers amongſt the others in a pleaſing view.

This tree may be PROPAGATED two ways: firſt, by ſeeds, which muſt be procured from America, where the plant is a native; for they do not ripen with us in England. We generally receive the ſeeds from thence in February: and they ſhould be committed to the ground as ſoon after as poſſible. They will grow in almoſt any ſoil that is tolerably good; though the more ſandy it be, it will be the better. After they are come up, they ſhould have the uſual care of ſeedlings for a year or two, and then be planted, either where they are intended to remain, or elſe in the nurſery, where they will in a year or two make ſtrong plants. This tree may be alſo propagated by layers; and this operation ſhould be performed the latter end of ſummer, whilſt the ſap is in motion; for if it is deferred until winter, the branches are then ſo exceedingly brittle, that it will be with difficulty they are brought down, without breaking, a proper depth into the earth Let the utmoſt care be taken, or many of the young branches that would have made layers will

be

be loſt. In ſummer, then, let the branches be brought down while they are pliable; and by the autumn twelvemonths after they will have taken root, and be fit to remove.

AMYGDALUS.

LINNEAN Claſs and Order, *Icoſandria Monogynia:* Each flower contains about twenty males and one female: There are four SPECIES; three of which are more particularly to our purpoſe.

1. AMY'GDALUS *Commu'nis* The COMMON ALMOND; *a low deciduous tree*; native of Africa.

2. AMY'GDALUS *Na'na:* The DWARF ALMOND; *a deciduous ſhrub*; native of Aſia Minor.

3. AMY'GDALUS *Per'ſica:* The PEACH; *a low deciduous tree*; of what country is uncertain.

1. THE COMMON ALMOND will grow to near twenty feet high; and whether planted ſingly in an open place, or mixed with others in clumps, ſhrubbery quarters, &c. ſhews itſelf one of the fineſt flowering trees in nature. Thoſe who never yet ſaw it, may eaſily conceive what a noble appearance this tree muſt make, when covered all over with a bloom of a delicate red, which will be in March; a time when very few trees are ornamented either with leaves or flowers. No ornamental plantation, therefore, of what ſort or kind ſoever, ſhould be without almond trees. Neither are the beauties of the flowers the only thing deſirable in this tree: The fruit would render it worthy of planting, were there no other motive. It ripens well, and its goodneſs is not unknown to us.

The white flowering Almond, well known in our nurſeries, is a *variety* of this ſpecies, and is cultivated for the ſake of the flowers and the fruit, though the flowers are inferior to the others. Neither is this tree ſo proper to plant ſingly in open places, or near windows, for the ſhow of its flowers; for although they come out early, yet the whole bloom is ſubject to be taken

taken off in one night's nipping weather, which frequently happens at this ſeaſon. Its ſtation, therefore, ſhould be in ſhrubery quarters, in well ſheltered places; and in ſuch it will flower exceedingly well, and ſhew its white bloſſoms to great advantage. When it is deſigned for fruit, it ſhould be ſet againſt a ſouth wall, in a well ſheltered place, otherwiſe there will be little hopes of ſucceſs.

2. The Dwarf Almond. Of this tree there are two ſorts, the ſingle and the double. Both grow to about four or five feet high, and are in the firſt eſteem as flowering ſhrubs. The ſingle ſort has its beauties; but the double kind is matchleſs. In both, the flowers are arranged the whole length of the laſt year's ſhoots; their colour is a delicate red; and they ſhew themſelves early in the ſpring, which ſtill enhances their value.

3. The Peach tree has hitherto been planted againſt walls for the ſake of the fruit; "but, ſays Hanbury, as I hardly ever knew a perſon who was not ſtruck with the beauty of the flowers when in full blow againſt a wall, why ſhould it not have a ſhare in wilderneſs quarters and ſhruberies, amongſt the ſorts of almonds, &c.? It may be kept down, or permitted to grow to the height of the owner's fancy; and the flowers are inferior to none of the other ſorts. Add to this, they frequently, in well ſheltered places, produce fruit which will be exceedingly well flavoured; and thus the owner may enjoy the benefit of a double treat." The above obſervations reſpect the ſingle peach; with regard to the double flowered, it is generally propagated for ornamental plantations, and is univerſally acknowledged to be one of the fineſt flowering trees yet known. Againſt a wall, however, theſe trees are always the faireſt; and if they have this advantage, they are ſucceeded by very good fruit.

All theſe ſorts are propagated by inoculating them into plum ſtocks, in Auguſt. The ſtocks ſhould be firſt planted in the nurſery, when of the ſize of a ſtraw, and the firſt or ſecond ſummer after they will be ready to receive the bud. The uſual method of inoculation muſt be obſerved, and there is no danger of ſucceſs; though it may be proper to obſerve, that the double bloſſomed peach ſhould always be worked into the ſtocks

ſtocks of the muſſel plum. The two ſorts of Dwarf Almond may alſo be propagated by layers, or from the ſuckers, which they ſometimes ſend forth in great plenty.

A N A G Y R I S.

LINNEAN Claſs and Order, *Decandria Monogynia*. Each flower contains ten males and one female. There is only one SPECIES:

ANAGY'RIS *Fœ'tida*: The FETID ANAGYRIS, or Stinking Bean Trefoil; *a deciduous ſhrub*; native of Italy, Sicily, and Spain.

THE ANAGYRIS is a ſhrub of about ten feet growth. The leaves are different in the different varieties: In one ſort they are oval, and moderately broad; in the other, they are oblong and narrow; but all of them are hoary. The flowers are produced from the ſides of the branches, in May, like thoſe of the *Laburnum*. They are numerous, of a bright yellow colour, but ſeldom ſucceeded by good ſeeds in theſe parts.

The beſt method of PROPAGATING theſe plants is, 1ſt, by the ſeeds, which ſhould be procured from the countries where they ripen well. Sow them in a border of good rich earth, in a well ſheltered place, and ſift over them about half an inch of fine mould. March is a very good month for this buſineſs; and when the plants appear, if the weather proves dry, frequently give them water; keep them clean of weeds all ſummer, and at the approach of winter prick round the beds ſome furze buſhes very cloſe: Theſe will break the keen edges of the black winds; for common froſts theſe plants bear moderately well. In the ſpring let them be ſet out in the nurſery ground, a ta foot diſtance from each other. Here let them ſtand a year or two, and they will be of a proper ſize to be finally planted out. 2. Theſe plants may alſo be propagated by layers. For this purpoſe, a few plants ſhould be ſet for ſtools. Let them grow one ſummer, to get good hold of the ground,

ground, and then head them down. The ſummer following they will make ſtrong ſhoots, which in the autumn ſhould be layered. They will readily ſtrike root, and by the autumn following will be good plants. The weakeſt of theſe may be ſet out in the nurſery ground for a ſeaſon or two; but the ſtrongeſt may be immediately planted out.

ANDROMEDA.

LINNEAN Claſs and Order, *Decandria Monogynia*: Each flower contains ten males and one female. There are fifteen SPECIES; three only of which are yet enured to this climate:

1. ANDRO'MEDA *Panicula'ta*. The VIRGINIAN ANDROMEDA; *a deciduous ſhrub*; native of Virginia.

2. ANDRO'MEDA *Calycula'ta*: The CANADIAN ANDROMEDA; *a low deciduous ſhrub*; native of Canada, Siberia, and Ingria.

3. ANDROMEDA *Maria'na*: The MARYLAND ANDROMEDA; *a very low deciduous ſhrub*; native of Maryland and other parts of North America.

1. The VIRGINIAN ANDROMEDA is a branching ſhrub, about four feet high. The leaves are oblong, pointed, plane, and are placed alternately on the branches. The flowers come out in panicles from the ends of the branches: They are of a pale yellow colour, and come out in July, but are rarely ſucceeded by good ſeeds in England.

2. CANADA ANDROMEDA is a low branching ſhrub, hardly a foot and a half high. The leaves are oval, ſpear ſhaped, obtuſe, reclined on their borders, and poſſeſſed of numerous ſmall punctures. The flowers grow in ſhort leafy ſpikes, from the ends of the branches: Their colour is white, they appear in July, and are ſeldom ſucceeded by good ſeeds in this country.

There is a *variety* of this ſpecies, with oval obtuſe leaves, of a thick ſubſtance, and which, in mild ſeaſons, continue on the plants all winter.

3. MARY-

3. MARYLAND ANDROMEDA. This is a ſhrub, about two feet high, ſending forth ſeveral ligneous ſtalks from the root. The leaves are oval, entire, of a pale green colour, and grow alternately on ſhort footſtalks. The flowers come out in ſmall bunches from the points of the ſtalk: They are of a greeniſh colour, come out in June and July, and are ſometimes ſucceeded by five-cornered capſules, full of ſeeds; which, nevertheleſs, ſeldom ripen in England.

PROPAGATION: Theſe plants ſucceed beſt upon boggy and moiſt grounds. You muſt procure the ſeeds from the places where they grow naturally; a year before which a boggy or the moiſteſt part of your garden ſhould be dug, and the roots of all weeds cleared off. As the weeds begin to riſe, ſo conſtantly ſhould the ground be again dug, and ſea or drift ſand ſhould be plentifully mixed with the natural ſoil. By this management till the ſeeds arrive, the ground being made tolerably fine, the ſeeds ſhould be ſown very ſhallow in the moiſt or boggy land; or if the land ſhould be ſo boggy that it cannot be eaſily worked, ſo as to be proper for the reception of the ſeeds, then let a ſufficient quantity of ſoil from a freſh paſture, mixed with drift ſand, be laid over the bog, and let the ſeeds be ſown therein. The bog will in time abſorb this ſoil, but the ſeeds will come up; and this is the moſt effectual method of procuring plants of this kind from ſeeds. The firſt year after they come up they ſhould be ſhaded in very hot weather; and after that they will require little or no care. Another method of increaſing theſe ſhrubs is by layers, or ſuckers; ſo that whoever has not the conveniency of procuring the ſeeds from abroad, ſhould get a plant or two of the ſorts he moſt likes: Theſe he ſhould plant in a boggy ſituation; and in a very little time he will have increaſe enough; for they throw out ſuckers in prodigious plenty, and, if they like the ſituation, to a great diſtance. Theſe ma be taken off, and planted where they are to remain.

ANNONA.

LINNEAN Claſs and Order, *Polyandria Polygynia*: Each flower contains many males and many females: There are nine ſpecies; one only of which is ſufficiently hardy for the open air of this climate.

ANNO'NA *Trilo'ba*: The PAPAW or CUSTARD APPLE; *a tall ſhrub*; native of the Weſt Indies.

The PAPAW grows to about ſixteen or eighteen feet high. The leaves are large, and ſhaped like a ſpear, and they fall off pretty early in the autumn. The flowers, which will ſhew themſelves in the beginning of May, are of a kind of chocolate colour tinged with purple, and grow two or three on a footſtalk. The fruit is large, and never ripens in England; but in the countries where it grows naturally, it is eaten by the meaneſt of the inhabitants. The difference of its ſhape from that of a pear is, that its wideſt part is neareſt the footſtalk; and it contains a number of large ſeeds lying in a row. It is a native of Maryland, Carolina, Virginia, and the Bahama Iſlands; and from thence we have the ſeeds brought, by which numbers of plants are annually raiſed.

The manner of RAISING them is this: Let a bed be prepared in a moiſtiſh part, that is exceedingly well ſheltered, and naturally ſandy, or inclined thereto. If the ſoil is oppoſite to this, let a fourth part of drift ſand be mixed with the mould; and having obtained the ſeeds from abroad, ſow them in this bed about half an inch deep, letting the ſeeds be at ſome diſtance from each other. It is probable they will come up in the ſpring, though they ſometimes remain till the ſecond, nay the third ſpring before they make their appearance. When this happens, the beds muſt be weeded all the time, and the mould at the ſurface gently looſened, if it ſhould be inclined to cruſt over. After the plants are come up in the ſpring, no other than the uſual care of ſeedlings need be taken, until the autumn, when the beds muſt be hooped over, to be covered with mats at the

the approach of any froſt; and the gardener muſt conſtantly obſerve the weather, whether the air hath the leaſt tendency to it, that he may cover the bed over; for one night's hard froſt, while they are ſo very young, would deſtroy them all. With this careful eye he muſt conſtantly watch over theſe plants all winter. He muſt double his covering as the froſt increaſes, and muſt always uncover them again in mild and open weather. The ſecond winter the ſame care muſt be obſerved, though ſo ſtrict an eye will not be neceſſary; for although they will be ſubject to be deſtroyed by hard froſts, yet if a gentle froſt ſhould catch them unawares to the gardener in the night, there will not be much danger of their ſuffering; for they will be got tolerably ſtrong by the ſecond ſummer's ſhoot: They will, nevertheleſs, be too tender to ſtand the brunt of a winter's froſt for a year or two after that; and conſequently muſt have a proportional ſhare of this attention every year during theſe months. By this time the plants will have grown to be tolerably ſtrong, and may be taken up and planted where they are to remain; though their ſituation ſhould be well defended; for a ſevere froſt in an expoſed place would ſtill overpower them; though, after they have grown to be of larger ſize, they are hardy enough.

If a perſon has the conveniency of a greenhouſe, or ſome ſuch room, he may ſow his ſeeds in boxes or pots filled with maiden earth, from a rich paſture, mixed with drift ſand. Theſe boxes or pots ſhould be afterwards plunged into the natural mould, in a ſhady part of the garden; and the autumn after the plants are come up, they may be removed into the greenhouſe, where they will be naturally protected from the injuries of weather. This protection may be afforded them every winter, till they are ſtrong enough to defend themſelves, when they may be turned out of the boxes or pots, mould and all, into the places where they are deſigned to remain.

ARALIA.

LINNEAN Clafs and Order, *Pentandria Pentagynia:* Each flower contains five males and five females: There are five SPECIES; only one of which is adapted to our purpofe:

ARA'LIA *Spino'fa:* The ANGE'LICA TREE, or PRICKLY ANGELICA: *a deciduous fhrub*; native of Virginia.

The PRICKLY ANGE'LICA: The height to which this tree will grow, if the foil and fituation wholly agree with it, is about twelve feet; and the ftem, which is of a dark brown colour, is defended by fharp fpines, which fall off; nay, the very leaves, which are branching, and compofed of many wings, and are of a pleafant green colour, have thefe defenders, which are both crooked and ftrong, and ftand as guards to them till the leaves fall off in the autumn. The flowers are produced in large umbels from the ends of the branches: They are of a greenifh yellow colour; and their general characters indicate their ftructure. They make their appearance the end of July or beginning of Auguft; but are not fucceeded by ripe feeds in our gardens.

PROPAGATION: This tree will what gardeners call *fpawn*; *i. e.* after digging among the roots young plants will arife, the broken roots fending forth frefh ftems; nay, if the roots are planted in a warm border, and fhaded in hot weather, they will grow; but if they are planted in pots, and affifted by a moderate warmth of dung, or tanners bark, they will be pretty fure of fuccefs; fo that the propagation of this tree is very eafy. But the general method of propagating it, and by which the beft plants may be had, is from feeds, which muft be procured from America, for they do not ripen in England; and, after having obtained them, they muft be managed in the following manner: The time that we generally receive them is in the fpring; fo that againft their coming we muft be furnifhed with a fuf-

ficient

ficient number of large pots. These, when the seeds are come, must be filled with fine mould, which, if taken from a rich border, will do very well. The seeds must be sown in these pots as soon as possible after their arrival, hardly half an inch deep, and then the pots should be plunged in a warm place their whole depth in the soil. Care must be taken to break the mould in the pots, and water them as often as it has a tendency to crust over; and if they are shaded in hot weather, the plants will frequently come up the first summer. But as this does not often happen, if the young plants do not appear by Midsummer, the pots should be taken and plunged in a shady place; nay, if they should, there will be still more occasion for this being done; for they will flourish after that better in the shade; and the design of plunging them in a warm place at first was only with a view of setting the powers of vegetation at work, that, having natural heat, artificial shade also may be given them, and water likewise, the three grand necessaries for the purpose. The pots, whether the plants are come up in them or not, should be removed into shelter in October, either into a green house, some room, or under a hotbed frame; and in the spring, when all danger of frost is over, they should be plunged into the natural ground their own depth in a shady place. Those that were already come up will have shot strong by the autumn following; and if none of them have appeared, they will come up this spring; and whether they are young seedlings, or small plants of a former summer's growth, they must be constantly kept clean of weeds, and duly watered in the time of drought; and this care must be observed until the autumn. In October they must be again removed into shelter, either into a greenhouse, &c. as before, or fixed in a warm place, and hooped, that they may be covered with mats in frosty weather. In the latter end of March following, they should be planted in the nursery way, to gain strength before they are finally planted out. The ground for this purpose, besides the natural shelter, should have a reed hedge, or something of the like nature, the more effectually to prevent the piercing winds from destroying the young plants. In this snug place the plants may be set in rows: in each of which rows furze bushes should be stuck the whole length; and all these together will ensure their safety. But here one caution is to be observed; not to stick the furze so thick, but that the plants may enjoy the free air in mild weather, and not to take them away too early in the spring, lest, being kept warm the whole winter, and being deprived of their protection, a cutting frost should happen, as it sometimes does even in April, and destroy them. Weeding and watering in dry weather must be their summer's care. They may be stuck again with furze bushes in the winter; though it will not be necessary to do it in so close a manner; and with this care, still diminishing in proportion the number of furze bushes, they may continue for three or four years, when they may be planted out into the warmest parts of the plantation. With this management these plants will be inured to bear our winters, in well sheltered places.

The spines which grow on the branches and the leaves admonish us, for our own safety, not to plant this tree too near the sides of frequented walks; and the consideration of the nature of the tree, which is rather tender at the best, directs us (if we have a mind to retain the sort) to plant it in a warm and well sheltered situation; where the piercing frosts, come from what point they will, will lose their edge; for without this, they will be too tender to stand the test of a severe winter; though it has often happened, that after the main stem of the plant has been destroyed, it has shot out again from the root, and the plant by that means been both encreased and preserved.

A R B U T U S.

LINNEAN Class and Order, *Decandria Monogynia*: Each flower contains ten males and one female. There are ten SPECIES; two of which are proper for our purpose.

1. ARBU'TUS *Une'do*: The ARBUTUS, or the COMMON

MON STRAWBERRY TREE; *an evergreen tree or ſhrub*; native of Ireland and many parts of Europe.

2. ARBU'TUS *Andra'chne:* The ANDRACHNE; or the ORIENTAL ARBUTUS; or the ORIENTAL STRAWBERRY TREE; *an evergreen tree or ſhrub*; native of the Eaſt.

1. The ARBUTUS. Of this Species there are four *Varieties:* namely,

THE OBLONG-FRUITED,	THE RED-FLOWERED, AND
THE ROUND-FRUITED,	THE DOUBLE-BLOSSOMED.

One deſcription is nearly common to them all: And their inconſiderable variation is almoſt ſufficiently ſhewn in their reſpective appellations.

The oblong-fruited Arbutus will grow to be a middling-ſized tree in ſome countries; for we read of the large uſes its wood has been applied to; ſuch as, *Arbuteæ crates*, &c. Arbutean harrows, &c. With us it is rather a tall ſhrub, and may be kept down to any ſize. The main ſtems are covered with a light brown bark, rough, and falling. The younger branches are of a kind of purple colour, whilſt the laſt year's ſhoots are of a fine red, and a little hairy. The leaves grow alternately on the branches, and are of an oblong oval figure. They ſtand on ſhort footſtalks, and the oldeſt leaves make a contraſt with the younger by having their footſtalk and mid rib of a fine ſcarlet colour. They are ſmooth, and beautifully ſerrated. Their upper ſurface (as in moſt trees) is of a ſtronger green than their under; and the young twigs are garniſhed with them in plenty. Theſe are beauties in common to moſt trees, in ſome degree or other; but every thing elſe almoſt of this tree that preſents itſelf to conſideration is ſingular: The time of its flowering will be in November and December; when it is rather ſingular to ſee a tree in the open ground in full blow; and the fruit ripens by that time twelvemonth after. The manner and nature of the fruit, which look like very large red ſtrawberries, give it alſo a ſingular and delightful look; and this is heightened as they appear all over the tree among the flowers; for that is the time of its being ripe, when the flowers for the ſucceeding crop are fully out. The flowers themſelves make no great figure; they are of a kind of whitiſh yellow colour; and are ſucceeded by the abovementioned Strawberry fruit, which will require a revolution of twelve months, before they perfectly arrive at their maturity and colour. The flowers of the first sort are larger than those of the second; and the fruit is oval, and much larger than our Common Scarlet Strawberry.

The round-fruited sort has its pitcher shaped flowers, which are succeeded by round scarlet fruit, as wide as they are long; and this is all the difference between these sorts.

The Strawberry tree with red flowers differs in no respect from the common sort, only the flowers are red, and these constitute a variety from the other sorts of flowers; but the contrast is not so great between their fruit and them, as of the other sorts, their colour approaching too near to a sameness.

The Double-blossomed Strawberry tree differs in no respect, only that the flowers are double; but this difference is so inconsiderable, that it will not be seen without looking into the flower; and even then the doubleness will appear so trifling as scarcely to merit notice; so that a plant or two, to have it said that the collection is not without it, will be sufficient. Neither ought any more to be admitted; for they will not produce the same plenty of fruit, which constitutes the greatest beauty of these trees, as the single sorts.

The method of PROPAGATING the *Varieties* of the Arbutus is by layers and cuttings: the *Species* itself may be raised from seed. By layers they will all grow: The operation must be performed on the youngest twigs; and in some soils they will strike root pretty freely, whilst in others they can hardly be made to grow at all: But before they have lain two summers, you may scarcely venture to look for any. When the roots are struck, the layers should be carefully taken off in the spring, and planted in separate pots; and after well watering them, they should be plunged up to the rims in a hotbed, and this will set them forward; for without this assistance, many of the layers will be lost; since they are difficult plants to make grow. After the hotbed has forced the seeds into a state of vegetation, the pots may be taken out, and plunged up to the rims in some natural mould, to keep them cool and moist; and here they may stand for two or three years, or longer, if the

the pots are large enough, without ever removing or ſheltering in winter; for they are hardy enough to reſiſt our ſevereſt cold. When they are to be finally ſet out, all the mould may be turned out of the pots hanging to the roots; and having proper holes made ready, they may be planted in them, and the plant will be ignorant of its new ſituation.

Theſe plants may be encreaſed by cuttings, which muſt be planted in pots, and have the benefit of a good bark bed; in which being conſtantly ſhaded and duly watered, many of them will grow. As the plants raiſed this way will be rather tender by being forced in the bark bed, it will be neceſſary to remove them into the greenhouſe, or to place them under a hotbed frame during the firſt winter: and after that, the pots may be ſet up to the rims in the ground, and, like the layers, the plants may be turned out at a convenient time into the places where they are to remain.

Next we proceed to the beſt way of raiſing the *Common Arbutus*; and that is from ſeeds. Let theſe be taken from the oblong or round fruited ſort. The ſeeds, which will be ripe ſome time in November or the beginning of December, for they will not be ripe at the ſame time in all places, muſt be then gathered; and as they ſhould not be ſowed until the ſpring, it will be proper to put them into a pot or jar, mixing with them a quantity of drift ſand; and this will preſerve them ſound and good. The beginning of March is the beſt time for ſowing the ſeeds; and the beſt ſoil for them is maiden earth, taken from a rich paſture at leaſt a year before, with the ſward; and this, by conſtant turning, being well rotted and mixed, will be ready to receive them. Having filled a different quantity of pots with this fine mould, let the ſeeds be ſown, and but juſt covered, ſcarcely a quarter of an inch deep. A dry day ſhould be choſen for the buſineſs; and no watering by the hand ſhould be given them, as it will endanger the ſetting the mould hard in the pots. Leave them abroad until ſome rain falls, which at that time may be hourly expected; and after that, having a hotbed ready, plunge the pots therein. In leſs than ſix weeks you may

may expect your plants to appear; when much air ſhould be afforded them, and frequent waterings, in ſmall quantities, gently ſprinkled over them. After this, they may be hardened to the air by degrees, and the pots ſet up to the rims in the natural mould, in a ſhady place. In October they ſhould be removed into the green-houſe, or ſome ſhelter, in froſty weather; though they ſhould always be ſet abroad in mild open weather. In the ſpring they may be ſhook out, and planted in ſeparate pots; and they ſhould have the advantage alſo of a hotbed to ſet them a-growing their future management may be the ſame as was directed for the layers. When theſe trees are to be planted out, very little regard need be paid to the ſoil or ſituation; for they will grow almoſt anywhere, and reſiſt our ſevereſt northern blaſts. One thing, however, the gardener muſt conſtantly obſerve, in order to continue his trees in their beauty; viz. as often as a heavy ſnow falls, ſo conſtantly ſhould he go and ſhake the boughs; for it will lodge amongſt the leaves and branches in ſuch great quantity, as to weigh down and ſplit the largeſt branches; the deformity of which afterwards may be eaſily conceived. Beſides, many years muſt expire before the tree will, if ever it ſhould, grow to its former beauty; to preſerve this, therefore, makes the narrowly watching theſe trees in ſnowy weather highly neceſſary.

2. The ANDRACHNE will grow to a larger ſize than the Arbutus. The leaves are ſmooth, and nearly of the ſame figure as the preceding ſort; though they are larger, and have their edges undivided. The flowers grow like the other ſorts; are of the ſame colour; and they are ſucceeded by large, oval, ſcarlet fruit. It is called the Oriental Strawberry Tree, becauſe this ſort grows plentifully in many parts of the Eaſt, and is uſeful to the inhabitants for many purpoſes in life.

The *Andrachne* may be PROPAGATED in the ſame manner as the *Arbutus*.

ARTEMISIA,

LINNEAN Claſs and Order, *Syngeneſia Polygamia Superflua*: Hermaphrodite florets containing five males and one female, and female florets containing one piſtil, in the ſame diſcous flower: There are twenty-five SPECIES, which are principally herbaceous; one only being intitled to a place amongſt the tribe of ornamentals.

ARTEMI'SIA *Arboreſcens*: The TREE WORMWOOD: *a non-deciduous ſhrub*: native of Italy and the Eaſt.

The TREE WORMWOOD riſes with an upright ſtalk to the height of about ſix feet. The leaves are its chief excellence; and of theſe there are two or three ſorts: One ſort is very much divided, or cut into ſeveral narrow ſegments; thoſe of the other are broader. They are very hoary; and as they continue on the branches all winter, they have a ſingular and an agreeable effect among the evergreens at that ſeaſon. The flowers are ſmall, and have very little beauty; they are collected into roundiſh heads, and we never perceived them to be followed by good ſeeds.

This plant is eaſily PROPAGATED by cuttings. Plant them in May, June, July, or Auguſt, in a ſhady place, and they will readily grow, eſpecially if they are watered a few times at the firſt planting. In the autumn theſe cuttings, which will then have become good plants, ſhould be each ſet in a ſeparate ſmall pot, and placed under a hotbed frame, or in the greenhouſe, to be preſerved all winter. In the ſpring they may be turned out into the places where they are deſigned to remain, which muſt be naturally warm and well ſheltered, or they will be liable to be deſtroyed by the ſeverity of the following winter. In ſuch a ſituation they will live for many years; though it may be adviſeable to keep a plant or two in the greenhouſe, to keep up the ſtock, if a more than common hard winter ſhould put a period to thoſe that are planted abroad.

ATRIPLEX.

LINNEAN Clafs and Order, *Polygamia Monoecia*: Hermaphrodite flowers containing five males and one female, and female flowers containing one piftil, on the fame plant. There are twelve SPECIES; two only of which are to our purpofe.

1. A'TRIPLEX *Ha'limus*: BROAD-LEAVED SEA PURSLAIN TREE: *a non-deciduous fhrub*; grows naturally upon the fea-coaft of Spain and Portugal; as alfo in Virginia.

2. A'TRIPLEX *Portulaco'ides*: NARROW-LEAVED SEA PURSLAIN TREE; *a non-deciduous fhrub*; native of our own fea coaft, and of the North of Europe.

1. The BROAD-LEAVED PURSLAIN TREE generally grows to about five or fix feet; and will fend forth its branches fo as to fpread around, and form a large broad head. The young branches are covered with a fmooth white bark; that of the older is of a light gray colour, which will be peeling lengthways, and falling, efpecially in the fpring. The branches are exceedingly brittle, and their infide is green to the very pith, of which there is very little. The leaves are foft, white, and filvery, and nearly of the fhape of the Greek letter *Delta*. They have their edges entire; and look well at all times, efpecially in winter, when they caufe as great a variety as poffible among thofe trees that retain their leaves at that time. This fhrub feldom flowers in our gardens; and when that happens, it is poffeffed of no beauty to recommend it to the florift.

2. The NARROW-LEAVED PURSLAIN TREE commonly grows to about four feet high. The branches are numerous and grey; and they naturally fpread abroad in a bufhy manner. The leaves are filvery; though not fo white as the other fort; but they are narrower, which occafions its being fo diftinguifhed; and of an oval figure; and by them the fhrub receives no fmall ornament. The flowers have little beauty.

Thefe

Theſe ſhrubs are PROPAGATED by cuttings; which will grow, if planted at any time of the year; though the beſt way is to take the cuttings in March, of the ſtrongeſt former ſummer's ſhoots, to cut them into lengths about a foot each, and to plant them a third part deep in the mould. Theſe will all readily take root, and be good plants by the autumn following. In ſummer, ſlips and cuttings may be planted; but then it will be adviſeable to plant them pretty cloſe together in beds, and afterwards to hoop the beds, and ſhade them from the heat at that time. They will ſoon take root; and after that will require no further trouble: But until that is effected, they ſhould be watered and ſhaded in the hot weather, and the mats ſhould be conſtantly taken off in the evening, and alſo in rainy, moiſt, or cloudy weather; and by this means plenty of plants may be raiſed. If it happened to be a dripping day when they were firſt planted, much trouble in ſhading and watering will be ſaved, as they may be nearly upon ſtriking root before the weather clears up. Theſe ſhrubs ſhould be always raiſed at a diſtance from farm yards, barns, &c. where there are ſparrows; for theſe birds are ſo exceedingly fond of the leaves, that when once they find them out, they will never leave nor forſake them until they have entirely ſtripped the plants; and though the ſhrub will ſhoot out afreſh, yet they will as conſtantly repair to their repaſt, and will thus continue to prey upon them until they have entirely deſtroyed them. I am obliged (continues HANBURY) to give this precaution, becauſe all my plants of theſe ſorts are thus conſtantly eat up by the ſparrows in my gardens at Church-Langton, as often as I plant them; ſo that I am obliged to keep them at Gumley, and in my other diſtant nurſeries, where they remain free from ſuch devourers.

Theſe plants require a warm ſheltered ſituation, being ſubject to be cut by the early froſts.

AZALEA.

LINNEAN Clafs and Order, *Pentandria Monogynia*. Each flower contains five males and one female. There are fix SPECIES; two of which are proper for the fhrubbery.

AZALE'A *Nudiflo'ra*: The RED AMERICAN UPRIGHT HONEYSUCKLE: or the RED AZALE'A; *a deciduous fhrub*; native of Virginia.

AZALE'A *Vifcofa*: The WHITE AMERICAN UPRIGHT HONEYSUCKLE: or The WHITE AZALEA; *a low deciduous fhrub*; native of Virginia.

1. The Red AZALEA has feveral ftems arifing from the fame root, which will grow to feven or eight feet high. The leaves are of an oval figure, fmooth, entire, and placed alternately on the branches. The flowers are produced in clufters from the fides of the branches, on long naked footftalks: Their colour is red, and they are agreeably fcented; each compofed of a long naked tube, cut at the top into five fpreading fegments. They will be in blow in July; but they feldom ripen their feeds in our gardens. There is a variety of this, with yellow flowers.

2. The WHITE AZALEA. From the root of this arife feveral flender brown ftems, to three or four feet high. The leaves are fpear fhaped, narrow at their bafe, have a rough border, and grow in clufters. The flowers terminate the branches in clufters, coming out between the leaves. They are finely fcented, and each of them has a tube of near an inch long, divided at the top into five fegments, two of which are reflexed. Their colour is white, with a bad yellow on their outfide; they will be in blow in July, but are never fucceeded by feeds in our gardens.

Thefe forts are PROPAGATED, 1. By layering the young fhoots; and for this purpofe, a flit muft be made on each, as is practifed for carnations: The autumn is the beft feafon for the work. When the layers have ftruck good root, they may be removed into the nurfery, and

and planted in lines at a ſmall diſtance from each other; where after having ſtood a year or two at the moſt, they will be proper plants to be planted out. 2. Theſe ſorts alſo propagate themſelves very faſt; for as they throw up many ſtems from the ſame roots after they have ſtood a few years, ſome of theſe may eaſily be taken off, with ſome root at each, and either planted in the nurſery ground, or the places where they are to remain.

B E R B E R I S.

LINNEAN Claſs and Order, *Hexandria Monogynia*: Each flower contains ſix males and one female: There are three SPECIES; two of which are here treated of:

1. BE'RBERIS *Vulgáris*: The COMMON BERBERY: a well known *deciduous ſhrub*, common in our hedges.

2. BER'BERIS *Cre'tica*: The CRETAN, or BOX-LEAVED BERBERY: *a low deciduous ſhrub*; native of Crete.

1. The COMMON BERBERY. This ſhrub is diſtinguiſhed by the acidity of its leaves, the ſharpneſs of its ſpines, the yellowneſs of its inner bark, and the ſcarlet colour of its berries, which add a beauty to our hedges in winter, and afford a favourite pickle and garniſh for our tables. HANBURY enumerates other uſes of the Berbery, particularly in medicine, and recommends the cultivation of it in the warmeſt manner. There is however an evil attendant on the Berbery buſh which ought to confine it within the pale of our gardens and ſhruberiés; we mean its poiſonous effect upon corn; more particularly upon wheat. This is a circumſtance which has been long known to the common farmers in different parts of the Kingdom, eſpecially in Norfolk, where the farmers are more obſervant and much more enlightened than thoſe uſeful members of ſociety in general are. The idea, nevertheleſs, has been treated by theoretical writers on Huſbandry

Huſbandry as chimerical and ſuperſtitious; and has been brought forward as one of thoſe vulgar errors of farmers which ought to induce gentlemen and men of genius to reſcue ſo uſeful a ſcience as that of Agriculture out of the hands of ignorance. Being however always ready to hear the opinion of *profeſſional* men, and having been aſſured by many ſenſible farmers of the truth of this matter, we had a few years ago a Berbery buſh planted, in the month of February, in the centre of a large piece of wheat. No obvious effect took place until the corn began to change its colour before harveſt, when a long blackening ſtripe became ſo conſpicuous amongſt the growing whiteneſs of the wheat, that it might have been diſtinguiſhed at a mile's diſtance. It reſembled the tail of a comet; the buſh repreſenting the comet itſelf; and what rendered the experiment ſtriking, whilſt on one ſide the effect did not reach more than three or four feet, on the oppoſite ſide it was obvious to the diſtance of ten or twelve yards; notwithſtanding the top of the ſhrub planted was not much larger than a man's head. At harveſt, the ears which grew in the immediate neighbourhood of the buſh, ſtood erect, the grains ſhriveled and empty;——as the diſtance from the Berbery increaſed the effect leſſened, vaniſhing imperceptibly: whilſt the grain of the reſt of the field was of a good quality. We do not mean in this place to comment upon the fact, or to attempt to account for ſo ſingular an effect by the help of *reaſoning* only; having in our intentions a ſuite of *experiments* in order to endeavour to come at the cauſe. Our motive for mentioning the fact at preſent is to induce others to make ſimilar experiments, as well as to ſtimulate gentlemen to extirpate from their eſtates ſo pernicious a plant; more particularly from the hedges and borders of arable fields.

There are three *Varieties* of the Common Berbery:

The Berbery with white fruit.

The Berbery with black fruit.

The Berbery without ſtones: which laſt is the ſort principally cultivated for the berries.

2. The BOX-LEAVED BERBERY grows to a yard or four feet high, and is poſſeſſed of many ſharp ſpines at the joints. The leaves are like thoſe of the box tree between

between which the flowers come out, on ſlender foot-ſtalks. But as this ſort never produces any fruit in England, and being alſo liable to be killed by hard froſts, it is ſeldom propagated in our gardens.

The PROPAGATION of the Berbery is as follows. 1. When a quantity of the common Berbery is wanted, the beſt way is to raiſe it from the ſeeds, which ſhould be ſown, ſoon after they are ripe, in a bed made in any part of the garden. Theſe will frequently remain till the ſecond ſpring before the plants come all up; till which time the beds ſhould be weeded as often as the weeds appear; for if they are neglected ſo as to get ſtrong, by pulling them up many of the ſeeds will alſo be drawn out of the bed by their roots. After the plants have grown one year in the ſeed bed, they ſhould be planted out in the nurſery, where they may remain for about two years, when they will be fit to plant out finally. This is the moſt expeditious method of raiſing a large quantity of theſe trees when wanted. 2. Another method of propagating the Berbery is by layers; a method by which all the ſorts may be encreaſed; and in the performance of which, no other art or trouble need be uſed, than laying the branches down in the ground, without either ſlit or twiſt. If this be done any time in the winter, by the autumn following they will have taken good root; the ſtrongeſt of which layers will be then fit to plant out; whilſt thoſe that are weaker may be planted in the nurſery ground, to gain ſtreng h. 3. The cuttings alſo of theſe trees will grow: for if they be planted in October, in a moiſtiſh good earth, they will moſt of them ſtrike root; ſo that the propagation of this tree by any of theſe ways is very eaſy. Whoever is deſirous of the Box-leaved Berbery muſt afford it a warm dry ſoil, in a well ſhel-tered place. The Common Berbery alſo diſlikes a wet ſituation.

BETULA.

LINNEAN Claſs and Order, *Monoecia Tetrandria*: Male flowers containing four ſtamens, and female flowers containing two piſtils, diſpoſed in ſeparate

 cylindrical

cylindrical catkins upon the ſame plant. There are five SPECIES:

1. BE'TULA *A'lba:* The COMMON WHITE BIRCH: a well known *deciduous tree*; native of this country, as alſo of moſt of the colder parts of Europe.

2. BE'TULA *Ni'gra:* The VIRGINIAN BLACK BIRCH: *a deciduous tree*; native of Virginia, as alſo of Canada.

3. BE'TULA *Le'nta:* The CANADA BIRCH: *a deciduous tree*; native of Canada, and of Virginia.

4 BE'TULA *Na'na:* The DWARF BIRCH: *a deciduous ſhrub*; native of Lapland, Ruſſia, Sweden, and Scotland.

5. BE'TULA *A'lnus:* The ALDER: a well known aquatic *deciduous tree*; common in our ſwamps and low grounds; it is alſo common in moſt parts of Europe, and in America.

1. The COMMON WHITE BIRCH. This tree is ſo common, and its uſes ſo well known throughout the kingdom, that any deſcription of it ſeems unneceſſary. It is in general of a humble growth; however, in a ſoil and ſituation it affects, it will riſe to a great height, and ſwell to a conſiderable ſize. There is a ſpruceneſs in its general appearance in ſummer; and in winter its bark ſometimes exhibits, in its variegations of red and white, no inelegant object. Were it not for its being ſo commonly ſeen upon poor ſoils, and in bleak inhoſpitable ſituations, as well as for the mean and degrading purpoſes to which it is univerſally put, the Birch would have ſome claim to being admitted to a place among the *ornamentals*. Its *uſes* are chiefly for brooms, fuel, and charcoal: if it be ſuffered to grow to a proper ſize, it will make tolerable gates, hurdles, rails, &c.: it is alſo uſed by the patten-makers. HANBURY ſays, it is alſo applicable to larger uſes; and is highly proper for the fellies of broad wheel waggons, it being inlocked ſo as not to be cleaved. "I have been informed (ſays he) by an old experienced wheelwright, that old Birch-trees cannot be cleft, as the grains run croſſways, and that he prefers it for ſeveral uſes in his way to moſt wood; and as I have ſeen ſeveral of theſe trees more than two feet ſquare, the timber of the Birch may perhaps be of more value than it has hitherto been eſteemed."

esteemed." Its use in making wine is well known. But although we enumerate the uses to which the Birch is applicable when it is already in possession, we do not mean to recommend in general terms the planting of Birch; except in bleak and barren situations where no other tree will thrive; and except as a skreen and guardian to nurse-up and defend from chilling blasts plants of greater value.

A striking and elegant *variety* of this species of tree is the *Weeping Birch*: not uncommon in the midland counties of England; but most frequent in the Highlands of cotland; where it often takes a picturable form; being, even in polished scenery, singularly *ornamental*

It is observable, that in the bleaker districts of the Highlands, the Birch is in a manner necessary to the habitableness of the country. The timber of houses are of Birch; implements of husbandry are of Birch; and Birch is the favourite fuel; its flame being clear, and its smoke less offensive to the eyes, than that of most, or all other woods.

This plant seems to accommodate itself to every situation. It not only flourishes on dry barren mountains, but thrives well, in low wet situations.

The PROPAGATION of the Birch is easy: it may be raised either from seeds, or by layering; and it will flourish in almost any soil or situation.

2. The BLACK VIRGINIA BIRCH will grow to upwards of sixty feet in height. The branches are spotted, and more sparingly set in the trees than the common sorts. The leaves are broader, grow on long footstalks, and add a dignity to the appearance of the tree; and as it is naturally of upright and swift growth, and arrives at so great a magnitude, HANBURY thinks it ought to have a share among our forest trees, and to be planted for standards in open places, as well as to be joined with other trees of its own growth in plantations more immediately designed for relaxation and pleasure.

There are several *varieties* of this species, differing in the colour, size of the leaves, and shoots; all of which have names given them by nurserymen, who propagate the different sorts for sale; such as, 1. *The*

 Broad-

Broad-leaved Virginian Birch; 2. *The Poplar-leaved Birch*; 3. *The Paper Birch*; 4. *The Brown Birch*, *&c.*

3. CANADA BIRCH. This grows to a timber tree of ſixty or more feet in height. The leaves are heart-ſhaped, oblong, ſmooth, of a thin conſiſtence, pointed, and very ſharply ſerrated. They differ in colour; and the *varieties* of this ſpecies go by the names of, 1. *Duſky Canada Birch*; 2. *White Paper Birch*; 3. *Poplar-leaved Canada Birch*; 4. *Low-growing Canada Birch*, &c. The bark of this ſpecies is very light, tough, and durable; and the inhabitants of America uſe it for canoes.

4. DWARF BIRCH. This is a low branching ſhrub, about two feet high. The leaves are round, and their edges are ſerrated. It hardly ever produces either male or female flowers, and is chiefly coveted when a general collection of plants is making.

The method of PROPAGATING all the foreign ſorts is, 1. From ſeeds. We receive the ſeeds from America, where they are natives; and if we ſow them in beds of fine mould, covering them over about a quarter of an inch deep, they will readily grow. During the time they are in the ſeminary, they muſt be conſtantly weeded, watered in dry weather, and when they are one or two years old, according to their ſtrength, they ſhould be planted in the nurſery, in rows, in the uſual manner. Weeding muſt always be obſerved in ſummer, and digging between the rows in winter; and when the plants are about a yard or four feet high, they will be of a good ſize to be planted out for the ſhrubery quarters. A part, therefore, may be then taken up for ſuch purpoſes; whilſt the remainder may be left to grow for ſtandards, to anſwer ſuch other purpoſes as may be wanted. 2. Theſe trees may alſo be propagated by layers; and this is the way to continue the peculiarities in the varieties of the different ſorts. A ſufficient number of plants ſhould be procured for this purpoſe, and ſet on a ſpot of double-dug ground, three yards diſtance from each other. The year following, if they have made no young ſhoots, they ſhould be headed to within half a foot of the ground, to form the ſtools which will then ſhoot vigorouſly the ſummer following; and in the autumn the young

young ſhoots ſhould be plaſhed near the ſtools, and the tender twigs layered near their ends. They will then ſtrike root, and become good plants by the autumn following; whilſt freſh twigs will have ſprung up from the ſtools, to be ready for the ſame operation. The layers,therefore,ſhould be taken up,and the operation performed afreſh. If the plants deſigned for ſtools have made good ſhoots the firſt year, they need not be headed down, but plaſhed near the ground, and all the young twigs layered Thus may an immediate crop be raiſed this way; whilſt young ſhoots will ſpring out in great plenty below the plaſhed part, in order for layering the ſucceeding year. This work, therefore, may be repeated every autumn or winter; when ſome of the ſtrongeſt layers may be planted out, if they are immediately wanted; whilſt the others may be removed into the nurſery, to grow to be ſtronger plants, before they are removed to their deſtined habitations. 3. Cuttings alſo, if ſet in a moiſt ſhady border the beginning of October, will frequently grow: But as this is not a ſure method, and as theſe trees are ſo eaſily propagated by layers, it hardly deſerves to be put in practice.

5. The Alder. This well known aquatic will grow to a large timber tree. The Alder, like the Birch, ſuffers, as an *ornamental*, from an aſſociation of ideas; we not only ſee it very common, but we ſee it in low, dreary, dirty ſituations: nevertheleſs, if the Alder be ſuffered to form its own head in an open advantageous ſituation, it is by no means an unſightly tree: in Stowe Gardens, in what is called the old part, there are ſome very fine ones; and in coming round from the houſe by the road leading to Buckingham, there is one which is truly ornamental. Hacked and diſfigured in the manner in which Alders in general are, they have but little effect in doing away the unſightlineſs of a ſwamp; but if they were ſuffered to riſe in groups and ſinglets, open enough to have room to form their full tops, and cloſe enough to hide ſufficiently the unſeemlineſs of the ſurface, even a moor or a moraſs ſeen from a diſtance might be rendered an agreeable object. Many *uſes* of Alders have been enumerated by authors: they were, indeed, more numerous than they are at preſent. Leaden pipes have ſuperſeded them as pump trees and water pipes, and logwood has rendered their

bark of little value. They are however ſtill uſeful as piles, and make tolerable boards; they are alſo convenient as poles, and make good charcoal: great quantities are cut up for patten wood, and for the wooden heels and ſoles of ſhoes. It is alſo uſed for wooden veſſels by the diſh turners. But upon the whole the conſumption is too inconſiderable to make them an object of the planter's notice, except in particular ſituations. For ſecuring the banks of rivers we know of nothing better than the Alder; its roots are ſtronger and more interwoven with each other than thoſe of the Salix tribe: alſo in low ſwampy ſituations, where the ground cannot be drained but at too great an expence, the Alder may be planted with propriety and advantage: but wherever the ſoil is or can be made paſturable, the Alder ſhould by no means be permitted to gain a footing. Its ſuckers and ſeedlings poiſon the herbage; and it is a fact well known to the obſervant huſbandman, that the roots of the Alder have a peculiar property of rendering the ſoil they grow in more moiſt and rotten than it would be if not occupied by this aqueous plant. Plantations of Alders ſhould therefore be confined to ſwampy, low, unpaſturable places; except when they are made for the purpoſes of ornament; and in this caſe the native ſpecies ought to give place to its more ornamental *varieties*, of which HANBURY makes five;- namely. 1. The Long-leaved Alder. 2. The White Alder. 3. The Black Alder. 4. The Hoary-leaved Alder. 5. The Dwarf Alder.

The PROPAGATION of the Alder, like that of the other aquatic natives, is very eaſy: it may be raiſed either from ſuckers, from cuttings, or by layering; and no doubt from ſeed, though this mode of propagation is ſeldom practiſed in this country. EVELYN mentions a peculiar method of raiſing this tree from cuttings or truncheons, which he calls the *Jerſey manner*: he ſays, "I received it from a moſt ingenious gentleman of that country: it is, to take truncheons of two or three feet long at the beginning of the winter, and to bind them in fagots, and place the ends of them in water till towards the ſpring, by which ſeaſon they will have contracted a ſwelling ſpire or knur about that part, which being ſet does (like the Gennet-moil Apple-

Tree)

Tree) never fail of growing and ſtriking root." MILLER recommends truncheons of three feet long, two feet of which to be thruſt into the ground. HANBURY ſays, that truncheons are uncertain, and ſtrongly recommends layering; which, for preſerving the varieties, at leaſt, is the beſt method.

BIGNONIA.

LINNEAN Claſs and Order, *Didynamia Angioſpermia:* Each flower contains four males and one female; two of the males being ſhorter than the other two; and the ſeeds being incloſed in a pod. There are eighteen SPECIES; five of which are enured to this climate.

1. BIGNONIA *Cata'lpa:* The CATALPA; *a deciduous tree*; native of Carolina.

2. BIGNONIA *Sempervi'rens:* The EVERGREEN BIGNONIA, or the VIRGINIA JASMINE, or the VIRGINIA CLIMBER; *an evergreen climber*; native of Virginia.

3. BIGNO'NIA *U'nguis:* The CLAW BIGNONIA, or the QUADRIFOLIATE BIGNONIA; *a deciduous climber*; native of Barbadoes, and other Weſt-India Iſlands.

4. BIGNO'NIA *Capreola'ta:* The TENDRIL BIGNONIA, or the CAPREOLATE BIGNONIA; *a deciduous climber*; native of North America.

5. BIGNO'NIA *Ra'dicans:* The SCARLET TRUMPET FLOWER; *a deciduous climber*; native of Carolina, Virginia, and Canada.

1. The CATALPA will grow to the height of thirty or forty feet; and as the ſtem is upright, and the leaves fine and large, it ſhould be planted as a ſtandard in the midſt of fine openings, that it may without moleſtation ſend forth its lateral branches, and ſhew itſelf to every advantage in view. Theſe opens, nevertheleſs, ſhould be ſuch as are well ſheltered, otherwiſe the ends of the branches will be deſtroyed by the ſeverity of the winter's froſt, which will cauſe an unſightly appearance; and

 the

the leaves, being very large, make ſuch a reſiſtance to the ſummer's high winds, as to occaſion whole branches to be ſplit off by that powerful element. The bark of the Catalpa is brown and ſmooth, and the leaves are cordated. They are about five or ſix inches in breadth, and as many in length. They ſtand by threes at the joints, are of a blueiſh caſt, and are late in the ſpring before they come out. The flowers are tubulous; their colour is white, having purple ſpots, and yellowiſh ſtripes on the inſide. They will be in full blow in Auguſt; but are not ſucceeded by good ſeeds in England.

Whoever has the conveniency of a bark bed may PROPAGATE this tree in plenty, 1. By cuttings, which being planted in pots, and plunged into the beds in the ſpring, will ſoon ſtrike root, and may afterwards be ſo hardened to the open air, that they may be ſet abroad in the ſhade before the end of ſummer: in the beginning of October, they ſhould be removed into a greenhouſe, or under ſome ſhelter, to be protected from the winter's froſt. In the ſpring, after the bad weather is paſt, they may be turned out of the pots, and planted in the nurſery way, in a well ſheltered place; and if the ſoil be rich, and rather inclined to be moiſt, it will be the better. Here they may ſtand for four or five years, the rows being dug in winter, and weeded in ſummer, when they will be of a proper ſize to be planted out to ſtand. Theſe cuttings will often grow in a rich, ſhady, moiſt border; ſo that whoever can have plenty of them, ſhould plant them pretty thick in ſuch a place, and he may be tolerably ſure, by this way, of raiſing many plants. 2. From ſeed, which muſt be procured from America, and ſhould be ſown in a fine warm border of light rich mould, or elſe in pots or boxes; the ſeedling plants requiring more than a common care.

2. The EVERGREEN BIGNONIA has almoſt every perfection to recommend it as a climber; for though the plants are ſmall, yet if they are trained up to a wall, or have buſhes or trees on which to climb, they will mount to a great height, by their twining ſtalks, and over-top hedges, and even trees and will form at a diſtance a grand figure from the ſway they will bear.

The

The leaves of Bignonia are ſingle and of a lanceolate figure. They grow from the joints, are of a fine ſtrong green colour, and very ornamental: but the flowers conſtitute the greateſt value of this plant, on account of the fine odour Nature has beſtowed on them; which is to ſo great a degree as to perfume the circumambient air to a conſiderable diſtance. Theſe flowers are of a yellow colour, and leſs beautiful than ſome of the other ſorts, which is ſufficiently recompenſed by their extraordinary fragrance. They grow in an erect manner, from the wings of the leaves at each joint, and their figure nearly reſembles that of a trumpet. The pods that ſucceed theſe flowers are ſmall.

There is a *variety* of this ſpecies, which over-tops whatever plants are near it, to a great height. The leaves are of a lanceolate figure, and grow from the joints. often four oppoſite. They are of a fine green; but their flowers are produced rather thinly, and ſtand each on its own footſtalk; and are not poſſeſſed of the heightened fragrance of the other.

3. The CLAW BIGNONIA is another noble climber. It riſes by the help of claw like tendrils, the branches being very ſlender and weak; and by theſe it will over-top buſhes, trees, &c. twenty or thirty feet high. The branches, however, ſhew their natural tendency to aſpire, for they wind about every thing that is near them; ſo that, together with the aſſiſtance Nature has given them of tendrils, it is no wonder they arrive at ſo great a height. Theſe branches, or rather ſtalks, have a ſmooth ſurface, are often of a reddiſh colour, particularly next the ſun, and are very tough. The tendrils grow from the joints; they are bowed, and are divided into three parts. The leaves grow in pairs at the joints, and are four in number at each. Theſe are of an oblong figure, have their edges entire, and are very ornamental to the plant; for they are of an elegant green colour: their under ſurface is much paler than their upper, and their footſtalks, midrib. and veins, alter to a fine purple. The flowers are monopetalous and bell-ſhaped. The tube is very large, and the rim is divided and ſpreads open. They grow from the wings of the leaves in Auguſt, two uſually at each joint;

joint; and they are succeeded in the countries where they grow naturally by long pods.

4. The TENDRIL BIGNONIA is another fine climber, which rises by the assistance of tendrils or claspers. The leaves grow at the joints opposite by pairs, though those which appear at the bottom frequently come out singly. They are of an oblong figure, and continue on the plant all winter. The flowers are produced in August, from the wings of the leaves: they are of the same nature, and of the shape nearly of the former; are large, of a yellow colour, and succeeded by short pods.

5. The SCARLET TRUMPET FLOWER will arrive to a prodigious height, if it has either buildings or trees to climb up by; for it strikes root from the joints into whatever is near it, and thus will get up to the tops of buildings, trees, &c. be they ever so high. This species has pinnated leaves, which grow opposite by pairs at the joints. These leaves are composed of about four pair of folioles, which end with an odd one. They are of a good green colour, have their edges deeply cut, and drawn out into a long point. The flowers are produced in August, at the ends of the branches, in bunches: they are large, and, like the other, are composed of one tube; but they are shaped more like a trumpet than any of the sorts. They are of a fine red colour, and make a grand show. This is the sort chiefly known by the name of the Scarlet Trumpet Flower.

There is another sort called, the *Smaller Trumpet Flower*. It differs from the last only in that the leaves and flowers are smaller, and some fancy their colour to be a finer red; the colour of the former, in some situations, often approaching to that of an orange colour. These two sorts are more hardy than any of the others, and consequently more proper to be set against old walls, &c. in exposed situations: they will all, however, bear our climate very well; though it would be adviseable to set the tender sorts in well sheltered places, as they will otherwise be in danger of suffering by severe frosts, especially while young, if there be nothing to break them off

The

The PROPAGATION of the Climbers. 1. If the ſhoots are laid upon the ground, and covered with a little mould, they will immediately ſtrike root, and become good plants for ſetting out where they are wanted. 2. They will all grow by cuttings The bottom part of the ſtrongeſt young ſhoots is the beſt; and by this method plenty may be ſoon raiſed. 3. They are to be raiſed by ſeeds; but this is a tedious method, eſpecially of the pinnated-leaved ſorts; for it will be many years before the plants raiſed from ſeeds will blow,

BUPLEURUM.

LINNEAN Claſs and Order, *Pentandria Digynia:* Each flower contains five males and one female. There are ſixteen SPECIES; but they are principally herbaceous: There is only one fit for open grounds in this country.

BUPLEU'RUM *Frut'co'ſum:* The ETHIOPIAN HARTWORT, or the SHRUBBY BUPLEURUM; *an evergreen ſhrub*; growing naturally amongſt the rocks on the coaſt of the South of France, and alſo in ſome parts of Italy.

The ETHIOPIAN HARTWORT is of low growth; it ſeldom riſes more than eight feet high; and will produce plenty of flowers before it gets to the height of one yard. The bark of the oldeſt ſtems is of a brown, that on the younger ſhoots of a reddiſh, colour; but this is not conſtant, for ſometimes it will be grayiſh, at others of a purpliſh blue. The leaves are of a fine pale green colour, and placed alternately on the branches. They are of an oblong, oval figure, and have their edges entire. They are ſmooth, and being of a delicate pale green, are very ornamental to the ſhrub. The flowers are produced from the ends of the branches, in longiſh umbels. They make no great figure (having but a bad yellow colour); appear in July and Auguſt; and are ſucceeded by ſeeds, which will often,

often, though not always, ripen with us; and by which, when they do, plenty of plants may be raised.

The method of PROPAGATING this shrub is either from seeds sown in pots of rich light loam in March; or from cuttings, in the following manner: The latter end of July is the time; and if the weather be moist or rainy, so much the better; if not, some beds must be well dug, and made moist by watering. The cuttings should be planted in the evening, and the beds must be hooped, to be covered with mats in the heat of the day. On their being first planted, no sun should come near them; but after they have been set a fortnight, they may have the morning sun until nine o'clock, and afterwards shading; observing always to uncover them in the evening, as also in moist, cloudy, or rainy weather. Many of these cuttings will grow; and in winter it will be proper to protect them from the frost with mats in the like manner: After that they will require no farther trouble until they are planted out.

This evergreen is scarcely hardy enough to struggle with our severest weather; whenever therefore it is introduced into plantations, it should always have the advantage of a dry soil and a well sheltered situation.

B U X U S.

LINNEAN Class and Order, *Monoecia Tetrandria*: Male flowers containing flour stamens, and female flowers containing three pistils upon the same plant. LINNEUS makes only one SPECIES of *Bu'xus* (BU'XUS *semper-vi'rens*): of this however there are several *varieties*; some of which in their present state have every appearance of distinct species*:

* MILLER says, "The two sorts of Tree Box have been frequently raised from seeds, and constantly produced plants of the same kind with those the seeds were taken from, and the Dwarf Box will never rise to any considerable height with any culture." (Art. BUXUS.)

1. The

1. The BROAD-LEAVED TREE BOX: A tall *ever-green shrub*; native of the southern parts of Europe, and, *it is said*, of this island.

2. The NARROW LEAVED TREE BOX.

3. The Gold-striped Box.

4. The silver striped Box.

5. The Gold-edged Box.

6. The Curled leaved Striped Box.

7. The DWARF BOX.

1. The BROAD-LEAVED TREE BOX This we will consider as the TRUE BUXUS, and the rest as *varieties*. The Box Tree will grow to the height of fifteen or twenty feet. The leaves are smooth and shining, and the branches of a yellowish hue. There is a swelling softness and a peculiar delicacy and richness in the general appearance of the Box, which, in winter more especially, affords the eye a delicious repast As an *ornamental* it stands first among the evergreens; and its *uses* are very many. Indeed, we know of no shrub or tree whatever, the Oak, the Ash, the Elm, and the Beech excepted, so deserving of the planter's notice as the Box. It will flourish upon barren soils and in bleak situations. The only extensive *plantations* of Box in this kingdom are those upon Box Hill; and the soil there is a poor thin-skinned chalky loam, and the situation high, unsheltered, and bleak in the extreme; yet the plants thrive with great luxuriance. The Box however is by no means partial to poor land and an open country; it thrives in every soil and in the closest situations, being remarkably patient of the shade and drip of other trees: we have seen it in a neglected grove, growing under a perfect canopy of foliage with the same healthfulness and luxuriancy as if it had stood in the open air. his naturally points out a situation and use proper for the Box, which does not seem to have been thought of: we mean that of UNDERWOOD TO THE OAK Thus employed, what an admirable cover to game; and how friendly to the sportsman! what a delightful passage in cultivated nature; and how profitable to the planter! Box wood is now (1785) worth 16s per cwt.

2. The NARROW-LEAVED TREE BOX. Of this beautiful plant there are some *sub-varieties*, that differ in the

the ſize of their leaves; but it is the ſmalleſt leaved ſort that is here meant; and as this ſort is not very common, it is valued on that account. It is rather of a lower growth than the former ſort, and its branches are more ſlender and numerous. It forms itſelf naturally into a regular head, and the whole ſhrub aſſumes an air of delicacy. The leaves grow oppoſite by pairs, as in the other ſort; but are produced in great plenty. They are very ſmall and narrow; and their ſurface is not ſo ſhining as the Broad-leaved Box. As the branches and leaves are the only ornament theſe trees afford, nothing farther need be added to the deſcription of this ſort.

3. 4. *The two ſorts with ſtriped leaves* are the Common Tree Box variegated; though they have a different appearance in their manner of growth, as well as in their ſtriped leaves. They will grow, indeed, to be as tall; but the branches will be naturally more ſlender and weaker, and many of them will often hang downwards, which gives the tree a much different appearance from the plain Tree Box, whoſe branches are naturally ſtraight and upright. The leaves of theſe ſorts being beautifully ſtriped, makes them coveted by thoſe who are fond of variegated trees.

5. *The Gold-edged Box* is ſtill the Tree Box, in the ſame natural upright growth. The branches of this are not ſo weak as thoſe of the former ſorts, but are upright and ſtrong. Their bark is rather yellower than the green ſort: in other reſpects there is no difference, except that the leaves are tipped or edged with yellow; which is thought by many to be very ornamental to the ſhrub.

6. *The Curl-leaved Striped Box* is ſo called on account of its leaves being a little waved. This, together with the Narrow-leaved, is the ſcarceſt of all the ſorts; and is indeed, like that, a very elegant ſhrub. It is certainly a variety of the Common Tree Box; but it ſeems rather of lower growth. Its leaves are waved; and they are variegated in ſuch a manner as to cauſe the ſhrub to have what HANBURY calls a luſcious look. It makes a variety from all the other ſorts, and is truly beautiful and pleaſing.

7. The

7. The DWARF BOX is a plant ſo well known as an edging to borders, and through thickets of ſhrubs, that it needs no deſcription It may be planted as an evergreen ſhrub among the lower ſorts.

The method of PROPAGATING the Box is perfectly eaſy: it may be raiſed from cuttings, or from the ſeed, or by layering. 1. For planting the cuttings, HANBURY ſays, the month of Auguſt is the beſt time, if any rain falls. If none ſhould happen, then the work muſt be deferred till it does. Indeed the cuttings may be planted with ſucceſs any time in the winter, even till the middle of April; but it is moſt prudent, if the ground is ready, to have this work done as ſoon as the firſt autumnal rains fall. Theſe cuttings ought to be of one and two years wood, ſhould be about a foot long, rather more than the half of which muſt be planted in the ground. A ſlip of the laſt year's wood, ſtripped from an older branch, is an excellent ſet, of which there will be little fear of its growing. The cuttings for the firſt raiſing of theſe trees ſhould be at about four inches diſtance in the beds; and, after they are planted, will need no trouble except watering in dry weather, and keeping clean from weeds, till about the third year after planting; for in all that time they will not be got too big for the *ſeed beds.* The ſeaſon for tranſplanting theſe trees from the *ſeed beds* to the nurſery is any time from Auguſt to April, though if they are to be tranſplanted early in the autumn, or late in the ſpring, moiſt weather ſhould be made choice of for this purpoſe. The diſtance theſe plants ſhould be placed at in the nurſery muſt be a foot aſunder, and two feet in the rows; and here they may ſtand till they are planted out. 2. The Box Tree may be alſo propagated from ſeeds; and trees raiſed this way will often grow to a larger ſize. In order to raiſe this tree from ſeeds, let them be gathered when they are quite ripe, and juſt ready to burſt out of their cells, and ſoon after ſow them in a border of light ſandy earth, about half an inch deep. In the ſpring the plants will appear; though it ſometimes happens that they lie in the beds one whole ſeaſon before they come up, eſpecially if they happen to have been kept long before they were ſowed after being gathered. If they ſhould not appear in in the ſpring, the beds muſt remain undiſturbed till the next, only keeping them free from weeds, and now and then giving them a gentle watering in dry weather. After they have made their appearance, they ſhould ſtand two or three years in the ſeed bed, the firſt of which will require attendance by watering in dry weather. When they are ſtrong enough to plant out, they may be ſet in rows in the nurſery, as was directed for the cuttings. 3. The Box propagates itſelf by *layering*; for whether it be borne down by the weight of its own foliage, or be broken down by a fall of ſnow lodging upon its leaves and branches, it no ſooner comes into contact with the ground than it ſends forth fibres, and the branch layered (whether by nature, by accident, or by art) preſently forms to itſelf a detached root, which being ſevered from the main tree, a ſeparate plant is produced.

CALLICARPA.

LINNEAN Claſs and Order, *Tetrandria Monogynia*: Each flower contains four males and one female. There are two SPECIES; one of which is enured to our climate.

CALLICA'RPA *America'a*: The CALLICARPA; *a low deciduous ſhrub*; native of many parts of America, but particularly of Virginia and Carolina.

The CALLICARPA. The leaves are roundiſh, acute, pointed, and are near three inches in length. They are of a hoary caſt, being, like the youngeſt ſhoots, covered with a kind of woolly matter. They ſtand oppoſite by pairs on moderate footſtalks, and their edges are made delicate by beautiful ſmall ſerratures. The flowers are produced in whirls round the twigs, at the ſetting-on of the leaves, and are of a reddiſh-purple colour. Each flower ſeparately is ſmall and inconſiderable; though the whole number of which the whirls are compoſed form, together with the leaves and nature of the growth of the tree, a ſingular and pleaſing

pleaſing aſpect. Their appearance is uſually in July, and they are ſucceeded by ſucculent berries, which are at firſt red, and afterwards of a deep purple when ripe.

It is PROPAGATED, 1. By cuttings. When by cuttings, they ſhould be planted, in the ſpring, in a moiſt ſandy border. As the hot weather comes on, they ſhould be conſtantly ſhaded, and watered if the bed is not naturally very moiſt; and by this means many of the cuttings will ſtrike root, and become good plants. 2. By layers, which is a certain method, theſe plants may alſo be increaſed. If a few plants are obtained for this purpoſe, they ſhould be planted in a warm well ſheltered ſituation; and if the ſoil be naturally ſandy, it will be the better. The autumn after theſe ſtools have ſhot forth young wood, theſe young ſhoots ſhould be laid in the ground, and by the autumn following they will be fit to take off, either for the nurſery, or where they are to remain. 3. By ſeeds, which ſhould be ſown in a warm border of ſandy earth, a quarter of an inch deep, and ſhould be carefully ſhaded and the ſeedlings ſheltered; theſe plants being tender when young, though afterwards they are ſufficiently hardy.

CALYCANTHUS.

LINNEAN Claſs and Order, *Icoſandria Polygynia*: Each flower contains twenty males and numerous females. There are two SPECIES; namely, CALYCANTHUS *Præcox*; not enured, we believe, to this climate; and

CALYCANTHUS *Floridus*: The FLOWERING CALYCANTHUS, or CAROLINA ALLSPICE TREE; *a deciduous aromatic ſhrub*; native of Carolina.

The FLOWERING CALYCANTHUS is a ſhrub which ſeldom grows, at leaſt with us, to more than five feet high. It divides into many branches irregularly near the ground. They are of a brown colour, and being bruiſed emit a moſt agreeable odour. The leaves that

 garniſh

garnish this delightful aromatic are of an oval figure, pointed: They are near four inches long, and are at least two and a half broad, and are placed opposite by pairs on the branches. At the end of these stand the flowers, of a kind of chocolate purple colour, and which are possessed of the opposite qualities of the bark on the branches. They stand single on their short footstalks, come out in May and June, and are succeeded by ripe seeds in England.

The PROPAGATION of this shrub is not very difficult; though more than a common care must be taken, after small plants are obtained, to preserve them till they are of a size to be ventured abroad. The last year s shoots of this tree, if laid in the ground, the bark especially being a little bruised, will strike root within the compass of twelve months, particularly if the layers are shaded, and now and then watered in the summer's drought. In the spring they should be taken off, and planted in pots; and if these are afforded a small degree of heat in a bed, they will strike so much the sooner and stronger. After they have been in this bed a month or six weeks, they should be taken out In the heat of the summer they should be placed in the shade; and if the pots are plunged into the natural ground, it will be so much the better. At the approach of the succeeding winter's bad weather, the pots should be removed into the greenhouse, or some shelter, and in the spring may resume their old stations: and this should be repeated till they are of a proper size and strength to be planted out to stand. If the pots in which they were first planted were small, they may be shifted into larger a spring or two after; and, when they have got to be pretty strong plants, they may be turned out, mould and all, into the places where they are to remain. By this care of potting them, and housing them during the severe weather in winter, the young crop will be preserved; otherwise, if they were planted immediately abroad, the first hard frost the ensuing winter would destroy them all: Tanners' bark about their roots will be the most proper security; as they are at best, when full grown, but tender plants, and must have the warmest situation and the driest soil.

CAR-

CARPINUS.

LINNEAN Claſs and Order, *Monoecia Polyandria:* Male flowers containing many ſtamens, and female flowers containing two piſtils diſpoſed in ſeparate catkins, upon the ſame plant: There are only two SPECIES:

1. CA'RPINUS *Be'tulus:* The COMMON HORNBEAM: *a deciduous tree*; native of Europe and America.

2. CA'RPINUS *O'ſtrya:* The HOP HORNBEAM; *a low deciduous tree*; native of Italy and of Virginia.

1. The COMMON HORNBEAM. This tree, it is ſaid, will grow ſo high as ſixty or ſeventy feet: we ſeldom ſee it, however, arrive at ſo great a height. Its leaves are of a darkiſh green, and about the ſize of thoſe of the Beech, but more pointed and deeply ſerrated. Its branches are long, flexible, and crooked; yet in their general appearance very much reſemble thoſe of the Beech: indeed there is ſo great a likeneſs between thoſe two trees, eſpecially in the ſhrubby underwood ſtate, that it would be difficult to diſtinguiſh them at the firſt glance, were it not for that gloſſy varniſh with which the leaves of the Beech are ſtrongly marked. In the days of EVELYN, when topiary work was the Gardener's idol, the Hornbeam might be conſidered as deſerving of thoſe *endearing* expreſſions which that enthuſiaſtic writer has been pleaſed to laviſh upon it: nevertheleſs, as an *ornamental* in *modern* gardening it ſtands low; and its *preſent uſes* are few. As an underwood it affords ſtakes and edders, fuel and charcoal. Its timber ranks with that of Beech and the Sycamore. The only ſuperior excellency of the Hornbeam lies in its fitneſs for SKREEN-FENCES for ſheltering gardens, nurſeries, and young plantations from the ſeverities of the winter ſeaſon. It may be trained to almoſt any height, and by keeping it trimmed on the ſides it becomes thick of branchlets, and conſequently thick of leaves; which being by their nature retained upon the

plant after they wither, a Hornbeam hedge occaſions a degree of ſhelter nearly equal to that given by a brick wall. Indeed, being leſs reflective than that expenſive ſcreen, it affords a more uniform temperature of air to the plants which ſtand near it. In this point of view, too, the Hornbeam is uſeful to be planted promiſcuouſly, or in alternate rows, amongſt more tender plants in expoſed ſituations, in the ſame manner as the Birch; to which it has more than one preference: namely, it is warmer in winter.—And, HANBURY ſays, the Hornbeam is peculiarly grateful to hares and rabbits; conſequently it may prevent their injuring its more valuable neighbours: yet, like EVELYN, he ſeems to be of opinion that is is diſaffected by deer. If this be really the caſe, the Hornbeam may upon many occaſions be introduced into deer parks with ſingular propriety.

The Common Hornbeam may be PROPAGATED either by layering (at almoſt any time of the year) or from ſeeds, in the following manner: In the autumn the ſeeds will be ripe; when, having gathered a ſufficient quantity for the purpoſe, let them be ſpread upon a mat a few days to dry. After this, they ſhould be ſown in the ſeminary ground, in beds four feet wide, with an alley of about two feet, and from one to two inches deep. In this bed they muſt remain till the ſecond ſpring before they make their appearance; and all the ſummer they lie concealed, the weeds ſhould conſtantly be plucked up as ſoon as they peep; for if they are neglected they will get ſo ſtrong, and the fibres of their roots will be ſo far ſtruck down among the ſeeds, as to endanger the drawing many ſeeds out with them, on weeding the ground. After the young plants appear, they ſhould conſtantly be kept clear of weeds during the next ſummer; and if they were to be now and then gently refreſhed with water in dry weather, it would prove ſerviceable to them. In the ſpring following they may be taken out of theſe beds, and planted in the nurſery, in which ſituation they may remain till they are of a ſufficient ſize to plant out for ſtandards.

Of the Common Hornbeam there are three *Varieties*: The *Eaſtern Hornbeam*, *Flowering Hornbeam*, *American Hornbeam*.

The

The *Eastern Hornbeam* arrives to the least height of all the sorts: about ten feet is the farthest of its growth, and it looks pretty enough with trees of the same growth. The leaves are by no means so large as the common sort; and as the branches are always closer in proportion to the smallness of the leaves, where a low hedge is wanted of the deciduous kind, this would not be an improper tree for the purpose, either to be kept sheered, or suffered to grow in its natural state. The bark of this sort is more spotted than that of the Common.

The *Flowering Hornbeam* is the most free shooter of any of the sorts; and will arrive to be the highest, the Common Hornbeam only excepted. It will grow to be thirty or forty feet high. The branches of this tree are less spotted with grayish spots than any of the other sorts. The leaves are very rough, of a dark green colour, and are longer than the common sort. The property which the Common Hornbeam is possessed of, of retaining its leaves all winter, does not belong to this sort, the leaves of which constantly fall off in the autumn with other deciduous trees.

American Hornbeam is a more elegant tree than any of the former sorts. The branches are slender, covered with a brownish speckled bark, and are more sparingly sent forth than from any of the others. The leaves are oblong, pointed, and of a palish green, and are not nearly so rough as the Common Hornbeam, though the flowers and fruit are produced in the same manner.

2. HOP HORNBEAM is of taller growth than the Eastern kind. It will arrive to the height of twenty feet, or more. The leaves are nearly the size of the common sort, and some people admire this tree on account of the singular appearance it makes with its seeds, before they begin to fall. There is a *Variety* of this tree, which grows to thirty feet high, shoots freely, has long rough leaves like those of the elm, and longish yellow coloured flowers, called the *Virginian Flowering Hop Hornbeam*.

These different sorts of Hornbeam are to be PROPAGATED by layers; for which purpose a few plants for stools must be procured. The stools of the Eastern Hornbeam should be planted a yard, and the other sorts

 a yard

a yard and a half or two yards asunder. After these plants have made some young shoots, they should be layered in the autumn, and by that time twelvemonth they will have struck root; at which time, or any time in the winter, or early in the spring, they should be taken off, and planted in the nursery way, observing always to brush up the stool, that it may afford fine young shoots for fresh layering by the autumn following. The distance the plants should be allowed in the nursery need be no more than one foot, in rows that are two feet asunder; and here they may stand, with the usual nursery care of weeding and digging the rows in winter, until they are to be finally planted out; though the Virginian Hornbeam will frequently send forth two shoots, which will seem to strive for mastery in the lead. When this is observed, the weakest should always be taken away, otherwise the tree will grow forked.

CEANOTHUS.

LINNEAN Class and Order, *Pentandria Monogynia*: Each flower contains five males and one female. There are three SPECIES; one of which will bear the open air of this climate.

CEANOTHUS *Americana*: The NEW-JERSEY TEA, or the REDTWIG; *a low deciduous shrub*; native of North America.

The NEW-JERSEY TEA. The height to which it will grow in our country is about a yard. The stem, which is of a pale brown colour, sends out branches from the bottom. These are thin, flexible, and of a reddish colour, which may have occasioned this tree to go by the name of *Redtwig*. The leaves which ornament these branches stand on reddish pedicles, about half an inch in length. They are oval, serrated, pointed, about two inches and a half long, are proportionably broad, and have three nerves running lengthways. From the footstalk to the point they are of of a light green colour, grow irregularly on the branches, and not oppoſite by pairs, as has been aſſerted. They are late in the ſpring before they ſhoot. The flowers grow at the ends of the twigs in cluſters: They are of a white colour, and when in blow give the ſhrub a moſt beautiful appearance. Indeed, it ſeems to be almoſt covered with them, as there is uſually a cluſter at the end of nearly every twig; and the leaves which appear among them ſerve as ornaments only, like myrtle in a diſtant noſegay: nature however has denied them ſmell. This tree will be in blow in July; and the flowers are ſucceeded by ſmall browniſh fruit, in which the ſeeds will ſometimes ripen in England

The PROPAGATION of this plant is by layering; or from ſeeds ſown in pots of compoſt conſiſting of two parts virgin earth well tempered, and one part ſand, about a quarter of an inch deep; being equally careful to defend the young ſeedlings from an extremity of cold in winter as from the parching drought of the ſummer months. The beſt time of layering them is in the ſummer, juſt before they begin to flower: At that time lay the tender twigs of the ſpring ſhoots in the earth, and nip off the end which would produce the flowers. By the autumn twelvemonth ſome of them will be rooted. At the ſtools, however, the plants ſhould remain until the ſpring, when they ſhould be taken off, and the beſt rooted and the ſtrongeſt may be planted in the nurſery way, or in a dry ſoil and well ſheltered place, where they are to remain; while the bad rooted ones and the weakeſt ſhould be planted in pots; and if theſe are plunged into a moderate warmth of dung, it will promote their growth, and make them good plants before autumn. In the winter they ſhould be guarded againſt the froſts; and in the ſpring they may be planted out where they are to remain.

CELASTRUS.

LINNEAN Clafs and Order, *Pentandria Monogynia*. Each flower contains five males and one female. There are eleven SPECIES; two of which are to our purpofe.

1. CELA'STRUS *Bulla'tus:* The STAFF TREE; *an uncertain deciduous fhrub*; native of Virginia.

2. CELA'STRUS *Sca'ndens:* The CLIMBING STAFF TREE, or BASTARD EVONYMUS; *a climber*; native of Canada.

1. The STAFF TREE is a fhrub of about four feet in growth, rifing from the ground with feveral ftalks, which divide into many branches, and are covered with a brownifh bark. The leaves are of a fine green colour, and grow alternately on the branches. They are of an oval figure, and have their edges undivided. The flowers are produced in July, at the ends of the branches, in loofe fpikes. They are of a white colour, and in their native countries are fucceeded by very ornamental fcarlet fruit; but with us this feldom happens.

It is eafily PROPAGATED from feeds fown, about an inch deep, in beds of good frefh mould made fine. They feldom come up until the fecond, and fometimes not before the third fpring. This fpecies is alfo PROPAGATED by layers; and, to be concife, the work muft be performed on the young wood, in the autumn, by a flit at the joint. Thefe layers may be expected to ftrike root by the autumn following; when they may be taken up and planted in the nurfery ground. This fhrub muft have a well fheltered fituation, otherwife the leaves are apt to fall off at the approach of frofty weather. And MILLER fays, that, growing naturally in moift places, it will not thrive well in a dry foil.

2. The CLIMBING STAFF TREE. The ftalks are woody, twining, and will rife by the help of neighbouring trees or bufhes to the height of twelve feet. The leaves are oblong, ferrated, of a pleafant green colour, pale, and veined underneath, and grow alternately on the

the

the branches. The flowers are produced in ſmall bunches, from the ſides of the branches, near the ends. They are of a greeniſh colour, appear in June; and are ſucceeded by roundiſh, red, three-cornered capſules, containing ripe ſeeds, in the autumn.

The plant is exceeding hardy, and makes a beautiful appearance among other trees in the autumn, by their beautiful red berries, which much reſemble thoſe of the Spindle-tree, and will be produced in vaſt profuſion on the tops of other trees, to the height of which theſe plants by their twiſting property aſpire. They ſhould not be planted near weak or tender trees, to climb on; for they embrace the ſtalks ſo cloſely as to bring on death to any but the hardieſt trees and ſhrubs.

It is PROPAGATED, 1. By laying down the young ſhoots in the ſpring. By the autumn they will have ſtruck root, and may then be taken off and ſet in the places where they are deſigned to remain. 2. They are alſo propagated by ſeeds. Theſe ſhould be ſown ſoon after they are ripe, otherwiſe they will be two, and ſometimes three years before they come up. When they make their appearance, nothing more need be done than keeping them clear from weeds all ſummer and the winter following; and in the ſpring the ſtrongeſt plants may be drawn out, and ſet in the nurſery for a year and then removed to the places where they are deſigned to remain; whilſt the weakeſt, being left in the ſeed-bed one year more, may undergo the ſame diſcipline.

CELTIS.

LINNEAN Claſs and Order, *Polygamia Monoecia*. Hermaphrodite flowers, containing five ſtamens and two piſtils, and male flowers containing five ſtamens. There are three SPECIES.

1. CE'LTIS *Auſtra'lis:* The SOUTHERN CELTIS, or the BLACK-FRUITED NETTLE TREE, or LOTE TREE; *a deciduous tree*; native of Africa and the South of Europe.

2. CE'LTIS

2. CE'LTIS *Occidenta'lis:* The WESTERN CELTIS, or the PURPLE-FRUITED or OCCIDENTAL NETTLE TREE; *a deciduous tree*; native of Virginia.

3. CE'LTIS *Orienta'lis:* or the EASTERN CELTIS; or the YELLOW-FRUITED or ORIENTAL NETTLE TREE; *a deciduous shrub*; native of Armenia.

1. The SOUTHERN ELTIS.

2. The WESTERN CELTIS.

"These two species grow with large, fair, straight stems; their branches are numerous and diffuse; their bark is of a darkish gray colour; their leaves are of a pleasant green, three or four inches long, deeply serrated, end in a narrow point, nearly resemble the leaves of the common stinging nettle, and continue on the trees till late in the autumn: So that one may easily conceive what an agreeable variety these trees would make. Add to this, their shade is admirable. The leaves are late in the spring before they shew themselves; but they make amends for this, by retaining their verdure till near the close of autumn, and then do not resemble most deciduous trees, whose leaves shew their approaching fall by the change of their colour; but continue to exhibit themselves of a pleasant green, even to the last.

HANBURY speaks highly of the Celtis as a timber-tree: he says, "The wood of the Lote Tree is extremely durable. In Italy they make their flutes, pipes, and other wind instruments of it. With us the coach-makers use it for the frames of their vehicles." MILLER mentions also the wood of the *Occidentalis* being used by the coachmakers.

The two species of Tree Celtis are PROPAGATED from seeds, which ripen in England, if they have a favourable autumn; but the foreign seeds are the most certain of producing a crop. These seeds should be sown, soon after they are ripe, either in boxes, or in a fine warm border of rich earth, a quarter of an inch deep; and in the following spring many of the young plants will appear, though a great part often lie till the second spring before they shew their heads. If the seeds in the beds shoot early in the spring, they should be hooped, and protected by mats from the frosts; which would nip them in the bud. When all danger from frosts is over, the mats should be laid aside till the

parching

parching beams of the ſun get powerful; when, in the day time, they may be laid over the hoops again, to ſcreen the plants from injury. The mats ſhould be conſtantly taken off every night, and the young plants ſhould never be covered either in rainy or cloudy weather. During the whole ſummer, theſe ſeedlings ſhould be frequently watered in dry weather, and the beds kept clean of weeds, &c. In the autumn, they muſt be protected from the froſts, which often come early in that ſeaſon, and would not fail to deſtroy their tops. The like care ſhould be continued all winter, to defend them from the ſame enemies. In this ſeminary they may remain being kept clean of weeds and watered in dry weather, till the end of June, when they ſhould be taken out of their beds, and planted in others at ſix inches diſtance. And here let no one (continues HANBURY) be ſtartled at my recommending the month of June for this work; for I have found by repeated experience, that the plants will be then almoſt certain of growing, and will continue their ſhoots till the autumn; whereas I have ever perceived, that many of thoſe planted in March have frequently periſhed, and that thoſe which did grow made hardly any ſhoot that year, and ſhewed the early figure of a ſtunted tree. In June, therefore, let the ground be well dug, and prepared for this work; and let the mould be rich and good: But the operation of removing muſt be deferred till rain comes; and if the ſeaſon ſhould be dry, this work may be poſtponed till the middle of July. After a ſhower, therefore, or a night's rain, let the plants be taken out of their beds, and pricked out at ſix inches diſtance from each other After this, the beds in which they are planted ſhould be hooped, and covered with mats when the ſun ſhines; but theſe muſt always be taken away at night, as well as in rainy or cloudy weather. With this management, they will have ſhot to a good height by the autumn, and have acquired ſo much hardineſs and ſtrength as to need no farther care than to be kept clear of weeds for two or three years; when they may be planted out in places where they are to remain, or ſet in the nurſery, to be trained up for large ſtandards.

The beſt ſeaſon for planting out theſe ſtandard trees is

is the latter end of October, or beginning of November and in performing that operation, the usual rules must be observed, with care.

The soil for the Lote tree should be light, and in good heart; and the situation ought to be well defended, the young shoots being very liable to be destroyed by the winter's frosts.

3. The EASTERN CELTIS. The height to which this species will grow is no more than about twelve feet; and the branches are many, smooth, and of a greenish colour. The leaves are smaller than those of the other sorts, though they are of a thicker texture, and of a lighter green. The flowers come out from the wings of the leaves, on slender footstalks: They are yellowish, appear early in the spring, and are succeeded by large yellow fruit.

The CULTURE of this species is the same, and the plants may be raised in the same manner as the other two sorts; only let this all along have a peculiarly dry soil, and a well sheltered situation, otherwise it will not bear the cold of our winters.

CEPHALANTHUS.

LINNEAN Class and Order, *Tetrandria Monogynia* Each flower contains four males and one female. There is only one SPECIES.

CEPHALA'NTHUS *Occidenta'lis* · The CEPHALANTHUS, or BUTTON WOOD; *a deciduous shrub*; native of North America.

The CEPHALANTHUS grows to about five or six feet high. It is not a very bushy plant, as the branches are always placed thinly in proportion to the size of the leaves, which will grow more than three inches long, and one and a half broad, if the trees are planted in a soil they like. The leaves stand opposite by pairs on the twigs, and also sometimes by threes, and are of a light green colour: Their upper surface is smooth; they have a strong nerve running from the footstalk to the

the point, and ſeveral others from that on each ſide to the borders: Theſe, as well as the footſtalks, in the autumn die to a reddiſh colour. The flowers, which are aggregate flowers, properly ſo called, are produced at the ends of the branches, in globular heads, in July. The florets which compoſe theſe heads are funnel ſhaped, of a yellow colour, and faſtened to an axis which is in the middle.

The PROPAGATION of the Cephalanthus is from ſeeds, which we receive from America. Theſe ſhould be ſown as ſoon as they arrive, and there will be a chance of their coming up the firſt ſpring; though they often lie till the ſpring after before they make their appearance. They may be ſown in good garden mould of almoſt any ſoil, if ſomewhat moiſt the better, and ſhould be covered about a quarter of an inch deep. This ſhrub is alſo propagated by layers. If the young ſhoots are laid in the autumn, they will have ſtruck good root by the autumn following, and may be then taken up, and ſet in the places where they are deſigned to remain. Cuttings of this tree, alſo, planted in the autumn in a rich, light, moiſt ſoil will grow: and by that means alſo plenty of theſe plants may be ſoon obtained.

C E R C I S.

LINNEAN Claſs and Order, *Decandria Monogynia:* Each flower contains ten males and one female. There are only two SPECIES.

1. CE'RCIS *Siliqua'ſtrum:* The COMMON JUDAS TREE, or the ITALIAN CERCIS; *a tall deciduous flowering ſhrub;* native of Italy and other parts of the South of Europe.

2. CE'RCIS *Canadénſis:* The CANADIAN JUDAS TREE, or the REDBUD, or the CANADIAN CERCIS; *a deciduous flowering ſhrub;* native of Canada, Virginia, and other parts of America.

1. The COMMON JUDAS TREES differ in the height of

of their growth in different places: In ſome they will arrive to be fine trees, of near twenty feet high; whilſt in others they will not riſe to more than ten or twelve feet, ſending forth young branches irregularly from the very bottom. The ſtem of this tree is of a dark grayiſh colour, and the branches, which are few and irregular, have a purpliſh caſt. The leaves are ſmooth, heart-ſhaped, and roundiſh, of a pleaſant green on their upper ſurface, hoary underneath, and grow alternately on long footſtalks. The flowers are of a fine purple: They come out early in the ſpring, in cluſters, from the ſide of the branches, growing upon ſhort footſtalks; and in ſome ſituations they are ſucceeded by long flat pods, containing the ſeeds, which, in very favourable ſeaſons, ripen in England. Some people are fond of eating theſe flowers in ſallads, on which account alone in ſome parts this tree is propagated. The *varieties* of this ſpecies are, 1. The Fleſh-coloured; 2. The White-flowered; and, 3. The Broad-podded Judas tree.

2. The CANADIAN JUDAS TREE will grow to the ſize of the firſt ſort in ſome places. The branches are alſo irregular. The leaves are cordated, downy, and placed alternately. The flowers uſually are of a paliſh red colour, and ſhew themſelves likewiſe in the ſpring, before the leaves are grown to their ſize. Theſe too are often eaten in ſallads, and afford an excellent pickle. There is a variety of this with deep red, and another with purple flowers. The pleaſure which theſe trees will afford in a plantation may be eaſily conceived, not only as they exhibit their flowers in cluſters, in different colours, early in the ſpring, before the leaves are grown to ſuch a ſize as to hide them; but from the difference of the upper and lower ſurface of the leaves; the one being of a fine green, the other of a hoary caſt; ſo that on the ſame tree, even in his reſpect, is ſhewn variety; an improvement whereof is made by the waving winds, which will preſent them alternately to view.

As theſe ſpecies will not take root by layers, they muſt be PROPAGATED by ſeeds, which may be had from abroad. They are generally brought us ſound and good, and may be ſown in the months of February or March. Making any particular compoſt for their reception is unneceſſary; common garden mould, of almoſt

almoſt every ſort, will do very well: And this being well dug, and cleared of all roots, weeds, &c. lines may be drawn for the beds. The mould being fine, part of it ſhould be taken out, and ſifted over the ſeeds, after they are ſown, about half an inch thick. Part of the ſeeds will come up in the ſpring, and the others will remain until the ſpring following; ſo that whoever is deſirous of drawing the ſeedlings of a year old to plant out, muſt not deſtroy the bed, but draw them carefully out, and after that there will be a ſucceeding crop. However, be this as it will, the ſeeds being come up, they muſt be weeded, and encouraged by watering in the dry ſeaſon; and they will require no farther care during the firſt ſummer. In the winter alſo they may be left to themſelves, for they are very hardy; though not ſo much but that the ends of the branches will be killed by the froſt, nay, ſometimes to the very bottom of the young plant, where it will ſhoot out again afreſh in the ſpring. Whoever, therefore, is deſirous of ſecuring his ſeedling plants from this evil, ſhould have his beds hooped, in order to throw mats over them during the hard froſts. Toward the latter end of March, or beginning of April, the plants having been in the ſeed bed one or two years, they ſhould be taken out, and planted in the nurſery: The diſtance of one foot aſunder, and two feet in the rows, ſhould be given them. Hoeing the weeds down in the ſummer muſt alſo be allowed, as well as digging between the rows in the winter. Here they may ſtand until they are to be removed finally; but they muſt be gone over in the winter with the knife, and ſuch irregular branches taken off as are produced near the root; by which management the tree may be trained up to a regular ſtem. Such, continues HANBURY, is the culture of the ſpecies of Cercis; ſorts that are not to be omitted where there are any pretenſions to a collection Beſides, the wood itſelf is of great value, for it poliſhes exceedingly well, and is admirably veined with black and green.

CHIO-

CHIONANTHUS.

Linnean Clafs and Order, *Diandria Monogynia*: Each flower contains two males and one female. There are two Species: Chiona'nthus *Zeylo'nica*; and

Chiona'nthus *Virgi'nica*: The Snow-Drop Tree, or the Fringe Tree, or the Virginian Chionanthus; *a tall deciduous fhrub*; native of Virginia and other parts of North America.

The Snow-Drop Tree. This fhrub will grow to the height of about fifteen feet, and, until late years, was very rarely to be met with in our gardens. The ftem of it is rough, and of a dark brown colour. The leaves are large, fhaped like a laurel, broad and roundifh, of a fine deep green on their upper furface, but rather hoary. The flowers come out in bunches, in May, from every part of the tree: They are of a pure white; and, in the places where it grows naturally, this muft be a moft delightful plant; for at that feafon it exhibits its white flowers in bunches all over it, fo as to refemble a tree covered with fnow. The few trees we have feldom flower; and even when they do, the flowers are few, and make no great figure. Whoever is defirous of raifing this fhrub muft plant it in a moift part of the garden, which is well defended with other trees; for there he will have a chance of feeing the flowers (which are fucceeded by black berries, of a moderate fize) in more plenty, and in greater perfection.

The culture of this tree is not very eafy; for if we attempt to propagate it by layers, thefe are with difficulty made to ftrike root; and if we obtain good feeds from abroad, great care and management muft be ufed, to make them to be ftrong plants, fit to be fet out to ftand. By layers and feeds, however, this tree may be encreafed; and, 1. When layers is the method adopted, let the plants defigned for ftools be fet in a very moift place, where the foil is rich and good. After thefe ftools have thrown out young fhoots, they

fhould

ſhould be layered in the autumn. If there be many twigs of the ſummer's growth to be layered, different methods may be uſed on the different twigs; for no one particular method can be depended on, and yet they will grow by almoſt all. One time the layering has been performed by a ſmall ſlit at the joint; another twig has had a gentle twiſt, ſo as to juſt break the bark; a third has been wired. The ſlit-layers, after three or four years, have only ſwelled to a knob, without any fibres; while the twiſted parts have ſhot out fibres, and become good plants. At other times, the twiſted part, after waiting the ſame number of years, has ſtill remained in the ground as a branch without any root; whilſt the ſlit twig, in the mean time, has become a good plant. The like uncertainty has been found to attend the other manner of layering. To propagate the Snow-drop tree this way, every method ſhould be uſed; and then there will be a greater chance of having ſome plants; but, at the beſt, you muſt not expect them with good roots, until they have lain in the ground about three years; for it is very rarely that they are to be obtained ſooner. The layers ſhould be taken from the ſtools the latter end of March, and planted in pots. Theſe ſhould be plunged into a hotbed; and, after they have ſtruck root, ſhould be uſed to the open air. In May they may be taken out, and plunged in the natural ſoil, in a moiſt ſhady place. When the froſt comes on, they ſhould be removed into the greenhouſe, or ſet under a hotbed frame for protection; and in the ſpring they may be turned out of the pots, with the mould, into the places where they are to remain, which ought to be naturally moiſt and well ſheltered. 2. From ſeeds: they muſt be ſown in large pots, about half an inch deep, in a ſtrong ſandy loam, plunging the pots into a moiſt ſhady place in ſummer, and in winter removing them into the greenhouſe or under hotbed frames.

MILLER ſays, "This ſhrub delights in a moiſt, ſoft, loamy ſoil; and is ſubject to decay in dry ſoils and hot ſeaſons."

CISTUS.

LINNEAN Clafs and Order, *Polyandria Monogynia:* Each flower contains numerous males and one female. There are no lefs than forty-three SPECIES of this genus of plants, moft of which are herbaceous, or herbaceous-fhrubby; of the thirteen arborefcent fpecies, twelve are naturalized to this climate.

1. CISTUS *Populifólius:* The POPLAR-LEAVED CISTUS, or ROCK ROSE; *an evergreen fhrub*; native of Spain and Portugal.

2. CI'STUS *Laurifólius:* The BAY-LEAVED CISTUS, or ROCK ROSE; *an evergreen fhrub*; native of Spain.

3. CI'STUS *Ladaníferus:* The LADANUM CISTUS, or ROCK ROSE; *an evergreen fhrub*; native of Spain, Italy, Crete, and the South of France.

4. CI'STUS *Incánus:* The HOARY CISTUS, or HOARY-LEAVED ROCK ROSE; *a low evergreen fhrub*; native of Spain and the South of France.

5. CI'STUS *Monfpeliénfis:* The MONTPELIER CISTUS, or GUM CISTUS of MONTPELIER; *an evergreen fhrub*; native of the South of France.

6. CI'STUS *A'lbidus:* The WHITE CISTUS, or OBLONG WHITE-LEAVED ROCK ROSE; *an evergreen fhrub*; native of Spain, Portugal, and France.

7. CI'STUS *Salvifólius:* The SAGE-LEAVED CISTUS, or ROCK ROSE; *a low evergreen fhrub*; native of France, Italy, and Sicily.

8. CI'STUS *Críſpus:* The CURLED CISTUS, or WAVED-LEAVED ROCK ROSE; *an evergreen fhrub*; native of Lufitania.

9. CI'STUS *Halimifólius:* The HALIMUS-LEAVED CISTUS, or SEA PURSLAIN-LEAVED ROCK ROSE; *a low evergreen fhrub*; grows common near the fea fhore in Spain and Portugal.

10. CI'STUS *Villófus:* The SHAGGY-LEAVED CISTUS, or SPANISH ROUND-LEAVED ROCK ROSE; *a low evergreen fhrub*; native of Italy and Spain.

11. CI'STUS *Créticus:* The CRETAN CISTUS, or ROCK

ROCK ROSE; *a low evergreen ſhrub*; native of Crete and Syria.

12. CI'STUS *Libano'tis*: The FRANKINCENSE CISTUS, or NARROW-LEAVED SPANISH ROCK ROSE; *a low evergreen ſhrub*; native of Spain.

1. The POPLAR-LEAVED CISTUS is a ſhrub of about ſix feet in height, though it begins its bloom when lower than two feet. The branches have no regular way of growth, and are covered with a brown bark, which will be lighter or darker according to the different ſoils. The leaves are cordated, ſmooth, pointed, have footſtalks, and a little reſemblance to thoſe of the Black Poplar. Old Botaniſts have diſtinguiſhed two ſpecies of this ſort, which they called the Major and the Minor, the one being of larger growth than the other; but modern improvements ſhew theſe to be varieties only. The flowers are white, and produced about Midſummer, in plenty, at the ends and ſides of the branches. They are of ſhort continuance; but there will be a ſucceſſion kept up for near ſix weeks, during which time the ſhrub will have great beauty.

2. BAY-LEAVED CISTUS is an irregular branching ſhrub, of about the ſame height with the former. The leaves are oval, pointed, and in the Midſummer months are very clammy. Their upper ſurface is of a ſtrong green, but their under is white, and they grow on footſtalks which join together at their baſe. The flowers are produced from the ends and ſides of the branches, about Midſummer. They are white, and ſtand on naked footſtalks; and being large, and produced in plenty at that time, make a good figure. This ſpecies is rather tender, and requires a warm, dry ſoil, and a well ſheltered ſituation.

3. The LADANUM CISTUS is ſo called, becauſe the Ladanum of the ſhops is collected from this ſhrub. There are many varieties of it, differing in the colour of the flowers, or in ſome reſpect or other; and the tree, with its varieties, will grow to be ſix or more feet high; though it produces its flowers and exhibits great beauty when very low. It riſes with a woody ſtem; and though it produces its branches in no regular manner, yet it has the appearance of a well faſhioned ſhrub. The leaves are of a lanceolate figure. Their upper

ſurface is ſmooth, and of a fine green colour, but their under is whitiſh and veined. They are ſcented; and have footſtalks that join together at their baſe. The flowers are very large and delicate, and are produced all over the ſhrub in plenty. They exhibit themſelves about the uſual time: Many of them are of a pure white, with a deep purple ſpot at the bottom of each petal; whilſt others again from theſe afford a variety, being of a purple colour, or having their edges of a reddiſh tinge. The beauty of this tree, when in blow, is often over, in very hot weather, by eleven o'clock in the morning; but that is renewed every day; and for about ſix weeks ſucceſſively a morning's walk will be rendered delightful by the renewed bounties which they beſtow.

4. The HOARY CISTUS is a ſhrub of about four feet high, and forms itſelf into a buſhy head. There are four or five varieties of this ſort, that have been looked upon by ſome authors as diſtinct ſpecies; but experience now teaches us better. The leaves of all are hoary; but they differ often in ſhape, ſize, or figure; and this has occaſioned their being named accordingly, and to be diſtinguiſhed by the names of Common Hoary-leaved Ciſtus; the Long-leaved Hoary Male Ciſtus: the Rounder-leaved Male Ciſtus; the Large Hoary-leaved Male Ciſtus, &c. When theſe different ſorts can be procured, they make the plantations more agreeable. The leaves of theſe ſorts of Ciſtus ſit cloſe to the branches, are hairy, and rough on both ſides. Their figure will be different on the ſame plant, and be produced in different manners: thoſe on the tops of the branches are ſpear-ſhaped, and grow ſingly; but the lower ones are oval, and joined together at their baſe. All of them are hoary, though ſome of the ſorts are whiter than others; and theſe leaves make a good contraſt with the ſtronger greens during the winter months. Theſe ſhrubs produce their flowers earlier than the other ſorts; they often ſhew ſome in May. They are of a purple colour, which, in different ſorts, will be ſtronger or lighter. They fall away in the evening; but are conſtantly renewed, for a month or longer, by a ſucceſſion every morning.

5. The GUM CISTUS OF MONTPELIER is commonly of about four feet growth, though, like the others, it is very beautiful when no higher than one or two feet. The branches proceed from the bottom of the plant, in plenty; they are hairy, tough, and ſlender. Their leaves are lanceolated, exſude a very fragrant matter, are hairy on both ſides, have three veins running lengthways, are of a dark green colour, and ſit cloſe to the branches. The flowers are produced in their greateſt plenty about Midſummer, and ſometimes earlier, on long footſtalks, at the ends of the branches. They are white, and the ſucceſſion of the blow will be continued often longer than ſix weeks.

6. WHITE CISTUS will grow to be five or ſix feet high; and the younger branches, which will grow in an upright manner, are tough, and covered with a woolly ſubſtance. The leaves are oblong, very white, downy, trinervous, and ſit cloſe, ſurrounding the ſtalk at the baſe. The flowers are produced from the ends of the branches, at the beginning of June. They are large, of a fine purple colour, and look very beautiful.

7. The SAGE-LEAVED CISTUS is a much lower ſhrub, and the branches are many, ſpreading, and ſlender. The leaves reſemble thoſe of ſome of the ſorts of ſage plants. They are oval, on both ſides hairy, and have very ſhort footſtalks. The flowers are produced in June, from the wings of the leaves. They are white, and ſtand on naked footſtalks; and though they are ſmaller than ſome of the other ſorts, yet being produced all over the ſhrub, they make a fine ſhow.

8. The CURLED CISTUS is of about four or five feet growth. The branches are very many, and ſpreading. The leaves are ſpear-ſhaped, waved, hairy, naturally bend backwards, and grow oppoſite by pairs on the branches. The flowers are produced from the wings of the leaves in June. Their colour is white. The ſucceſſion will be kept up for a month, or longer.

9. The SEA PURSLAIN-LEAVED CISTUS is a ſhrub of about four feet growth, and ſends forth many branches in an upright pretty manner. The younger branches are downy, and the leaves have ſome little reſemblance to the Sea Purſlain; though there are varieties of this ſpecies with broader and narrower

leaves; ſome that approach to an oval, and others that are ſharp-pointed They grow oppoſite by pairs, and make a good variety by their white and hoary look. The flowers are produced in June and July on very long, naked footſtalks, which ſupport others alſo with ſhorter footſtalks. They are of a fine yellow colour, and make a good figure when in blow. This is the moſt tender of all the ſorts, and is generally treated as a greenhouſe plant; but if the ſoil be naturally dry and warm, and the ſituation well ſheltered, it will do very well abroad in our tolerably open winters. It may be adviſeable, however, to ſecure a plant or two in the greenhouſe, that, in caſe a very ſevere winter ſhould happen to kill thoſe abroad, a freſh ſtock may be raiſed from the thus preſerved plants.

10. Spanish Round leaved Cistus. This is a branching ſhrub, of about a yard or four feet high. The leaves are oval, round, hairy, and placed on footſtalks on the branches. The flowers come out in plenty from the tops and ſides of the branches, in July. Their colour is purple; and though they are very fugacious, yet there will be a ſucceſſion of them for a long time.

11. Cretan Cistus. This is a branching ſhrub, of about the ſame height with the former. The leaves are ſpatulated, oval, enervous, rough, and grow on footſtalks on the branches. The flowers are red; and they make their appearance about the ſame time with the former.

12. Spanish Narrow-leaved Cistus. This riſes with a ſhrubby, naked, purple coloured ſtalk, to about four feet high The leaves are narrow, light, reflexed on their ſides, and grow oppoſite to each other without any footſtalks. The flowers grow in ſmall umbels, and come out from the ends and ſides of the branches, on long ſlender footſtalks. Their colour is white; and their appearance is about the ſame time with the former.

All the ſorts of Ciſtus are propagated by ſeeds and cuttings. 1. Seeds is the beſt way, as by them the moſt handſome plants are produced, though they will not always afford ſo great a plenty of flowers as the plants raiſed from cuttings. When they are to be raiſed

raiſed by ſeeds, a moderate hotbed ſhould be in readineſs for their reception by the beginning of March; and they ſhould be ſown in drills a quarter of an inch deep. A dry day ſhould be made choice of for the purpoſe, and pegs ſhould be ſtuck to ſhew the extremity of the drills. The drills may be made two inches aſunder; and the bed being neated up, no other covering will be neceſſary than an old mat, to guard the plants, when coming up, from the ſpring froſts which may happen; for if the ſeeds are good, you may expect many plants to appear in leſs than a month; at which time they ſhould be covered in the night, but be always kept uncovered in open and fine weather. As the dry weather comes on, they muſt be watered moderately every other morning, and the weeds conſtantly cleared off; and as the ſummer heat increaſes, the mats uſed to guard them from the froſt in the night, muſt change their office: They muſt never come near them in the night, but only protect them from the ſcorching heat in the middle of the day. By the latter end of Auguſt many of the plants will be four or five inches high; when they may be thinned, and thoſe drawn out either pricked in the nurſery ground, in beds at ſmall diſtances, in well ſheltered places, or planted in pots, to be ſecured in the winter, and turned out at leiſure. Of all the ſorts, the Bay-leaved and the Sea Purſlain-leaved ſpecies, with all their varieties, require this treatment. The reſt are all very hardy. Thoſe that are pricked out in rows in the nurſery will immediately ſtrike root: and, as well as thoſe left in the old hotbed, if they are in well ſheltered places, will do without any protection. If the place is not well defended, either by trees or hedges, it will be proper to prick ſome furze buſhes all around, to break the keen edge of the ſevere froſts. Thoſe left in the old bed ſhould be planted out in the ſpring in the nurſery ground; and in a ſpring or two after this, they ſhould all be planted out where they are to remain; for none of theſe plants ſucceed ſo well if removed when grown old and woody. 2. Theſe plants are eaſily raiſed by cuttings; and plants raiſed this way are often the beſt flowerers, though their manner of growth is not always ſo upright and beautiful. Auguſt is the month for this work; and if a dripping day happens in that month, it muſt be made choice of; if not, a bed of fine mould muſt be prepared, and the cuttings ſhould be planted a few inches aſunder; and after that, ſhould be watered to ſettle the mould to them. The beds ſhould be hooped; and the next day, as the heat of the ſun comes on, they ſhould be covered with mats: This covering ſhould be repeated, obſerving always to uncover them in the evenings, and alſo in moiſt and cloudy weather. Theſe cuttings will take root in a very little time; and their after management may be the ſame as the ſeedlings.

CLEMATIS.

LINNEAN Claſs and Order, *Polyandria Polygynia*. Each flower contains many males and many females. There are thirteen SPECIES, ten climbing, and three erect: Eight of the former have been introduced into this country.

1. CLE'MATIS *Vitice'lla:* The VIRGIN'S BOWER; *a deciduous climber*; native of Italy and Spain.

2. CLE'MATIS *Vio'rna:* The VIRGINIA CLIMBER, or the PURPLE CLIMBER; *a deciduous climber*; native of Virginia and Carolina.

3. CLE'MATIS *Cri'spa:* The CAROLINA CLIMBER, or the CURLED PURPLE CLIMBER; *a deciduous climber*; native of the Eaſt.

4. CLE'MATIS *Orienta'lis:* The ORIENTAL CLIMBER; *a deciduous climber*; native of the Eaſt.

5. CLE'MATIS *Vita'lba:* The TRAVELLER'S JOY, or OLD MAN'S BEARD, or BINDWITH; *a deciduous climber*; growing naturally in the hedges of England, and moſt of the northern parts of Europe; alſo in Virginia and in Jamaica.

6. CLE'MATIS *Cirrho'ſa:* The EVERGREEN CLEMATIS, or EVERGREEN SPANISH CLIMBER; *an evergreen climber*; native of Spain and Portugal.

7. CLE'MATIS *Fla'mmula:* The CREEPING CLIMBER; *a deciduous climber*; native of the South of Europe.

8. CLE-

8. Cle'matis *Virginia'na:* The Sweet-scented Clematis, or the Sweet-scented American Climber; *a deciduous climber*; native of North America.

1. Virgin's Bower. Of this species of Clematis there are the following *Varieties:*

Double Purple Virgin's Bower.
Single Purple Virgin's Bower.
Single Blue Virgin's Bower.
Single Red Virgin's Bower.

Double Purple Virgin's Bower. This sort stands first on the list, not only because it is an admirable climber, but also is possessed of a large double flower. It will grow to the height of twenty or thirty feet, if supported; and is very proper to cover arbours, as well as walls, hedges, &c. The branches are of a dark brown or dusky colour, angular and channelled. The younger branches are of a fine green colour, and nearly square: They are very numerous, and grow from the joints of the older; and thus they multiply in that manner from the bottom to the top of the plant. The leaves also grow from the joints: They are both compound and decompound *. The folioles, of which each is composed, are of an oval figure, and their edges are entire; and in summer, when the plant is in full leaf, if set alone to form an arbour, after it is said to be grown strong, the branches and large leaves will be produced in such plenty, as not only effectually to procure shade, but even to keep off a moderate shower; so excellently is this plant adapted to this purpose; and more particularly so, as it will grow, when it has properly taken to the ground, fifteen or sixteen feet in one year. The flowers are double, and of a purple colour: They blow in July and August, and are succeeded by no seeds, the multiplicity of the petals entirely destroying the organs of generation.

The *Single Purple Virgin's Bower* is rather a stronger shooter than the Double, and will climb to rather a still greater height. The Double is only a sub-variety of this, which ought not to be neglected; for this ex-

* Doubly compound.

hibits

hibits a fair flower, compofed of four large petals, in the center of which are feated the numerous ftamina.

The *Single Blue Virgin's Bower* produces its fhoots, leaves, and flowers, in the fame manner as the other; and makes a variety only in that the flowers are of a blue colour.

The *Single Red* is of much lower growth, and feems of a more delicate and tender nature; not but it is hardy enough to endure any weather; but its fhoots are weak, and fhort in proportion. They are angular, and channelled in the manner of the other; but they are of a reddifh colour. The leaves are fmaller than the other forts, and the flowers alfo are fmaller, though they make a fine variety, by their colour being red. Thefe all flower at the fame time; but are fucceeded by no ornamental feeds.

2. VIRGINIA CLIMBER. The branches are flender and numerous; and the leaves, as in the Virgin's Bower, are both compound and decompound. The folioles grow by threes, and thefe are often multiplied to form a decompound leaf of nine in number. They are nearly cordated, of a good green, and fome of them are trifid. The flowers are produced in July and Auguft, from the wings of the leaves. They are a kind of blue colour; and the petals (which are four in number) of which each is compofed, are of a thick coriaceous fubftance. This fort will fometimes ripen its feeds in England.

3. CAROLINA CLIMBER. This is by fome called the Curled flowering Climber; and indeed by that name it is chiefly diftinguifhed in our gardens. It is one of the lower kind of climbers; feldom arifing, by the affiftance of its clafpers, to more than fix feet. The ftalks are very weak and flender. The leaves afford great variety, being fometimes trifoliate and fometimes fingle. The folioles alfo differ much; for fome of them are found whole and entire, whilft others again are divided into three lobes. Thefe leaves are of a dark green colour, and are produced oppofite, from the joints of the ftalks. The flowers are produced in July and Auguft, on fhort footftalks, below which a pair or more of oblong pointed leaves often grow. Thefe flowers are compofed of four thick, coriaceous,

purple,

purple, curled petals. This ſpecies will for the moſt part produce ripe ſeeds in our gardens.

4. ORIENTAL CLIMBER is no great rambler; for notwithſtanding its ſlender ſtalks are well furniſhed with claſpers, it is ſeldom found to climb higher than about ten feet. The leaves of this ſort are compound. The folioles are cut angularly, and the lobes are ſhaped like a wedge. They are of a good green colour, and are very ornamental to the plant. The flowers are produced from the wings of the leaves early; for it will often be in blow in April. They are of a kind of yellowiſh green colour, and the petals naturally turn backwards. Theſe flowers differing in colour from the above ſorts, and coming earlier in the ſpring, make it more deſirable, as it teſtifies how many months in the ſummer are ornamented with the blow of ſome one or other ſpecies of Clematis. The ſeeds of this ſort alſo will often ripen with us.

5. TRAVELLER'S JOY is a noble climber, and well known in many parts of England; the hedges where it abounds being frequently covered with it: But its greateſt ſingularity is in winter; at which time it more peculiarly invites the traveller's attention. The branches of this ſpecies are very thick and tough, ſufficient to make withs for faggots; and for this purpoſe it is always uſed in the woods where it can be got. Theſe are ſo numerous, and produce ſide branches in ſuch plenty, which divide alſo into others, that they will overtop hedges, or almoſt any thing they can lay hold of to climb by. Beſides the claſpers with which it is furniſhed, the very leaves have a tendency to twine round plants. Theſe leaves are pinnated; and a variety is occaſioned by them; for the folioles of ſome ſorts are indented at their edges, whilſt others are found with their edges entire. They are of a blueiſh green, and moderately large. The flowers are produced in June, July, and Auguſt, all over the plant, in cluſters. They are ſucceeded by flat ſeeds, each of which, when ripe, is poſſeſſed of a white hairy plume, and growing in cluſters will exhibit themſelves in winter all over the tops of buſhes, hedges, &c. which at that time will look beautiful and ſingular. This is the *Viorna* of old Botaniſts; and is called Traveller's Joy from its thus orna-

ornamenting hedges, bushes, &c. to the entertainment of the traveller.

6. Evergreen Clematis. This is but a low climber, seldom growing higher than six or eight feet. The branches are very numerous, weak, and slender; but it rises by claspers, which naturally lay hold on any thing near them. The footstalks of the leaves, also, will twine round twigs, &c. so that they become claspers, and ensure the hold of the plant. Nay, if there be no hedge or plant near, by which they may hold and rise, they will twine among themselves; and as the branches are produced in great plenty, they will be so mixed one amongst another, as to form a low thicket, which makes this plant well adapted to produce variety in evergreen shrubery quarters, where, if planted singly, at a distance from other trees, it will naturally form itself into a thick bush. These leaves are sometimes cut into three lobes, sometimes into two, and many of them are undivided. The lobes when most perfect are nearly lanceolate, have their edges indented, and are of as fine a shining green as can be conceived. The flowers are produced in the midst of winter, from the sides of the branches: They are of a greenish colour, though inclined to a white; but the petals being pretty large, and blooming at that unusual season, makes this plant highly valuable.

7. Creeping Clematis, or *Flammula*, will mount by the assistance of other plants to a good height, sometimes near twenty feet. The stalks are slender and numerous; and the leaves are in this respect singular; for the lower ones are pinnated, and their edges are jagged; but the upper ones grow single. They are of a lanceolate figure, and their edges are entire. The flowers of this species are exhibited in June, July, and August. They are white, and extremely elegant.

8 Sweet-scented Clematis. · This sort will rise, by the assistance of neighbouring bushes and trees, to a great height. The branches are many, spread themselves all around, and lay hold of every thing that is near them. The leaves are ternate. The folioles are heart-shaped, angular, and nearly cut into three lobes. The flowers are white, and, being possessed of a most agreeable fragrance, render this climber highly proper

for

for arbours, and to be ſtationed near ſeats and places of reſort.

Theſe are all the hardy climbing ſpecies of this genus yet known. The *varieties* of the firſt kind are notable, and afford as much diverſity in a garden as if they were diſtinct ſpecies. The other ſorts alſo admit of varieties; but the difference is very inconſiderable, and makes little variety, as they nearly agree with ſome or other of the above ſorts.

The PROPAGATION of all theſe ſorts is by layers; and this is beſt done in ſummer on the young ſhoots as they grow. As ſoon, therefore, as they have ſhot about a yard or four feet in length, let the ground be well dug about each ſtool, and made fine, and a gentle hollow made about a foot from the ſtool. In this hollow let the young ſhoots be preſſed, and covered with mould, leaving their ends out to continue growing. In a very little time they will be a yard or more in length; when a ſecond hollow may be made, at a diſtance from the other, and the ſhoots preſſed down and covered with mould as before, the ends being ſtill left out to grow. On ſome of the long ſhooting ſorts this may be repeated again, and even again; and theſe ſhoots, thus layered, will ſtrike root. Many of the ſorts will have good root by the autumn; and others muſt be waited for until the autumn following. This ſummer method of layering is highly neceſſary; becauſe ſome of the ſorts, particularly the Virgin's Bower, if layered in winter in the common way, will be often two whole years, nay ſometimes three, before they will ſtrike root. Any time from autumn to ſpring the layers may be taken up; and from one ſtool ſome ſcores are often obtained. Thoſe with good roots may be ſet out to remain; and every bit that has a fibre ſhould be cut off below that fibre, and ſhould be headed to one eye or joint above the part that had been out of the ground; and thus all the layers being collected together ſhould be planted in the nurſery at ſmall diſtances, and in a year or two they alſo will be good plants for uſe.

The TRAVELLER'S JOY may be layered at any time, for the roots will eaſily ſtrike; nay, they will grow by cuttings.

The

The EVERGREEN SPANISH CLIMBER requires no art or trouble to increase it; for it will increase itself if the ground is left undisturbed a year or two, and will throw out plenty of suckers, which will have roots, and be good plants.

CLETHRA.

LINNEAN Class and Order, *Decandria Monogynia*: Each flower contains ten males and one female. There is only one known SPECIES:

CLE'THRA *Alnifo'lia:* The ALDER-LEAVED CLETHRA, or the AMERICAN ALDER; *a deciduous aquatic shrub*; growing naturally in Virginia, Carolina, and Pennsylvania, like our Alder, by the sides of rivers and watery places.

The CLETHRA is a shrub, with us, about four or five feet high, though in its native soil it is sometimes found so high as eight or ten feet. The branches it sends forth are not numerous, and these are garnished with leaves, which are spear-shaped and serrated. They are about three inches long, an inch and a half broad, and have short footstalks. The Clethra usually flowers in July. The flowers are produced at the ends of the branches, in long spikes: They are white, and possessed of a strong scent. This plant, at present, is not very common in our gardens.

The CULTURE of this shrub is by layers, seeds, and suckers. 1. The plants designed to be increased by layers should be set in the moistest part of the garden, and managed like those of the CHIONANTHUS. 2. By seeds,—which also should be sown and managed the same as CHIONANTHUS. 3. These shrubs will very often send out suckers, by which they may likewise be propagated. These may be taken off in the autumn, if they have good roots, and planted out in the nursery way: if they have not, they should be let alone till March; then taken up, and planted in pots of good loamy

loamy ſoil, and afterwards plunged into a moderate warmth of dung; which will promote their growth. The autumn following they will be fit to be planted out to ſtand.

C N E O R U M.

LINNEAN Claſs and Order, *Triandria Monogynia:* Each flower contains three males and one female. There is only one SPECIES:

CNEO'RUM *Trico'ccon:* The CNEORUM, or WIDOW-WAIL; *a low evergreen ſhrub*; native of dry gravelly places in Spain, Italy, and France.

The CNEORUM, or WIDOW-WAIL, is a ſhrub of about a yard in growth, and is an excellent one for the front of evergreen quarters, where the loweſt ſhrubs are to be placed. The wood of this tree is very hard, and the older branches are covered with a brown bark. The ſtem naturally divides into many branches; and the bark on the youngeſt is ſmooth, and of a pale green colour. The leaves are ſmooth, of a fine dark green colour, and conſtitute the greateſt beauty of this ſhrub. They are of an oblong figure, and very long in proportion to the breadth: They will be two inches or more long, and about half an inch in breadth. Their under ſurface is of rather a paler green than their upper, and their baſe joins to the young branches without any footſtalk. The flowers are yellow, and make no great ſhow. A healthy plant may be expected to be in blow moſt part of the ſummer. They grow from the wings of the leaves, towards the ends of the branches; and are ſucceeded by the ſeeds, which grow together by threes; which will be of a dark brown or black when they are ripe.

CNEORUM may be PROPAGATED by ſeeds or by cuttings. 1. By ſeeds. Theſe ſhould be gathered in October, and be thoſe which have grown from the firſt flowers of the ſhrub that ſummer, and which will be then black, or nearly ſo, if ripe. They ſhould be ſown in in a bed of common garden mould made fine, about half an inch deep. One may expect to see the plants come up in the spring; though it often happens that the greatest part of them remain until the second spring before they appear. 2. These plants may be increased by cuttings; but they never make such beautiful shrubs; neither is the method worth practising if seeds can be obtained. The cuttings may be planted in spring; then it will be necessary to set them in pots, and give them the assistance of a hotbed; and this will set them a-growing. The beginning of August is a very good time for planting these cuttings or slips. They should be planted in beds of good fine mould; and these should be hooped, and matted from nine o'clock in the morning until near sun-set. Then they should be uncovered, and remain so in all cloudy and rainy weather. Most of these cuttings will grow; and there they may remain without removing until they are set out for good.

When these shrubs are to be planted out, the most dry and gravelly spots must be chosen for them; and in these places they will bid defiance to our severest weather; though in such a soil they will not grow so high as in a moist fat soil, by a foot or more, which is considerable in a shrub of such a natural low growth; but it is necessary for them to be planted in a dry or gravelly soil, because there they will be secure from injury by frosts.

COLUTEA.

LINNEAN Class and Order, *Diadelphia Decandria* Each flower contains ten males and one female, the males standing in two divisions. There are three SPECIES; one of which is herbaceous, and another a shrub, too delicate for the open air of this climate; the third has long been an ornament to the English garden.

COLUTE'A *Arboréscens*: The BLADDER SENNA; a well known *deciduous shrub*; native of the South of Europe, particularly about Mount Vesuvius.

The

The BLADDER SENNA ſports in the following *varieties*; all of which are beautiful in their kind, and afford delight both by their flowers and leaves; viz. 1. The Common Bladder Senna. 2. The Oriental Bladder Senna. 3. Pocock's Bladder Senna. 4. The Red-podded Bladder Senna.

The *Common Bladder Senna* is the talleſt grower of all the ſorts. It will arrive to the height of about ten or twelve feet. The branches are of a whitiſh colour, which diſtinguiſh it in the winter, and the leaves in the ſummer have a pleaſing effect. They are pinnated; the folioles are oval, and indented at the top; they conſiſt of ſometimes four, ſometimes five pair, placed oppoſite, and are terminated by an odd one. The flowers are of the butterfly kind: They are produced in June, July, and Auguſt, in cluſters; are numerous, of a yellow colour, and the footſtalk that ſupports them is long and ſlender. The flowers are ſucceeded by large inflated pods, like bladders, which catch the attention of thoſe who have never before ſeen them. This tree has variety enough of itſelf to make it eſteemed; but it ſhould always be planted among other trees of the ſame growth, to break the force of the ſtrong winds; not but that it is hardy enough to reſiſt our ſevereſt winters, but the branches will eaſily ſplit, which will make it unſightly, unleſs they are ſheltered in ſome degree by other trees. This ſort will ripen its ſeeds in the autumn.

The *Oriental Colutea* will grow to the height of about five or ſix feet. The branches of this tree alſo are grayiſh, and the leaves pinnated, as well as terminated by an odd one, and the lobes are obverſely cordated and ſmall. The flowers are reddiſh, ſpotted with yellow, and grow from the ſides of the branches on footſtalks, each of which is formed ſometimes with two, ſometimes with three flowers. This tree is extremely hardy; and as it does not grow to the ſize of the common ſort, nor in ſo luxuriant a manner, the branches will not be ſo liable to be ſplit off by the winds; and therefore the precaution neceſſary for that, in this ſort may be the leſs obſerved.

Pocock's Bladder Senna is another variety, of lower growth than the common ſort. The leaves are pinnated. and the folioles ſtand oppoſite by pairs in both the kinds. They are indented in the ſame manner at the top; neither can I perceive any other difference between this and the Common Bladder Senna, only that the one is larger than the other, and the flowers come out earlier in the year.

The *Red-podded Bladder Senna* is alſo a variety, which will happen in common to all the ſorts, more or leſs, when raiſed from ſeeds.

Theſe trees are all very eaſily PROPAGATED. 1. By ſeeds. Any time in the ſpring will do for the work, though the month of March is the beſt ſeaſon; and no other compoſt will be required than garden mould of almoſt any ſort, dug and raked fine. If the ſeeds are ſown about half an inch deep, they will come up like corn in a month or two after. Keep the beds weeded until the ſpring following; and then plant them out in the nurſery way, obſerving always to ſhorten the tap-root which they often have. In a year or two they will be good and proper plants for the ſhrubery. 2. Theſe trees may alſo be propagated by layers; and that is the method generally practiſed with Pocock's ſort, to continue it in its low growth.

CORNUS.

LINNEAN Claſs and Order, *Tetrandria Monogynia*: Each flower contains four males and one female. There are eight SPECIES; three of which are adapted to ornamental gardening.

1. CO'RNUS *Ma'ſcula*: The CORNELIAN CHERRY; *a tall deciduous ſhrub*; growing naturally in the hedges of Auſtria.

2. CO'RNUS *Sangui'nea*: The COMMON DOGWOOD, or BLOODY TWIG; *a deciduous ſhrub*; common in our hedges, and is natural to moſt parts of Europe, Aſia, and America.

3. CO'R-

3. CORNUS *Flo'rida*: The VIRGINIAN DOGWOOD, or FLOWERING CORNUS; *a deciduous shrub*; native of Virginia.

1. The CORNELIAN CHERRY will riſe to twenty feet high. Its principal merit as an *ornamental* lies in its flowering early in the ſpring, and in exhibiting its beautiful ſcarlet berries in autumn Its *uſes* are held out as numerous. Its fruit was formerly in good eſteem; and its wood is ſaid to be uſeful for wheel-work, pins, hedges, &c. It is arranged by EVELYN and HANBURY among Foreſt trees.

2. The COMMON DOGWOOD is well known all over England, as it grows naturally in moſt parts of the kingdom; a few of theſe trees are nevertheleſs admiſſible into the ſhrubery, if they are not already too common in its neighbourhood; for the young twigs are red, eſpecially in winter, which look well at that ſeaſon, as do alſo its flowers in the ſummer, and its leaves in the autumn. The redneſs of theſe young ſhoots has occaſioned this ſort to go by the name *Bloody Twig*. The leaves are about two inches long, and an inch and a half broad: theſe have large nerves, which terminate in a point, and they often die in the autumn to a reddiſh colour. The flowers are white, produced in umbels at the ends of the branches, and are ſucceeded by black berries, like thoſe of the Buckthorn, but have in each only one ſtone. The wood, it is ſaid, makes the beſt kind of charcoal in the world for gunpowder. It is brittle, exceedingly white, and when growing is covered with a dark brown bark, the twigs being red.

3. VIRGINIAN DOGWOOD will grow rather higher than our Common Dogwood. The twigs are of a beautiful red. The leaves are obverſely cordated. The flowers are produced in large bunches ſomewhat like thoſe of the Elder: Their colour is white; they come out in May and June, and the berries ripen in autumn. Mixed among evergreens, the Dogwoods have a beautiful effect in the winter months.

From theſe ſpecies, the following beautiful *Varieties* figure in our nurſeries; viz. Female Virginian Dogwood; American Blue-berried Dogwood; White-berried Dogwood of Pennſylvania; and Swamp Dogwood.

Female Virginian Dogwood, during the winter months, exhibits its branches of ſo beautiful a red colour, as to diſtinguiſh itſelf to all at that ſeaſon. It grows to eight or ten feet high; the leaves are ſomewhat ſpear-ſhaped, acute, nervous, and in the autumn die to a fine red. The flowers come out in umbels, at the ends of the branches: They appear in May and June, and the berries ripen in the autumn.

The *American Blue-berried Dogwood* arrives at the height of about eight or nine feet. The twigs of this tree alſo are of a delightful red. The leaves are large, oval, and hoary on their under-ſide. The flowers are white, come out in umbels from the extremity of the branches, and are ſucceeded by large, oval, blue berries, which make a fine appearance in the autumn.

White berried Dogwood arrives at the ſame ſize with the others. The young ſhoots, like thoſe of the former, are of a beautiful red colour during the winter. Like them, alſo, it produces its white flowers in large umbels in May; but they are ſucceeded by white berries in the autumn.

Swamp Dogwood grows naturally in moiſt places, almoſt all over America; and it will grow with us in almoſt any ſoil or ſituation. The leaves of this are of a much whiter colour than any of the other ſorts; though the flowers and fruit are produced in the ſame manner.

One method of PROPAGATION is common to all theſe ſorts of Cornus; though this may be effected three ways; by ſeeds, layers, and cuttings. 1. The ſeeds of the common ſort ſhould be ſown in the autumn, ſoon after they are ripe; and theſe will come up in the ſpring. The ſeeds of the American ſorts we generally receive in the ſpring: Theſe ſhould be ſown directly; but they will not come up till the ſpring following; nor would thoſe of our common ſort, if they were kept until the ſpring before they were ſown. No particular art is required for theſe ſeeds. They will grow in common garden mould of almoſt any ſort, though the richer it is the better. This muſt be made fine, cleared of all roots, weeds, &c. and the ſeeds ſhould be ſown about

about half an inch deep. The ſpring after the plants come up, they ſhould be planted in the nurſery, at a ſmall diſtance from each other, where they may ſtand for two or three years, and then be planted out to ſtand. 2. Theſe trees may be eaſily propagated by layers; for after having obtained ſome plants for the purpoſe, if the ſhoot that were made the preceding ſummer be only laid in the ground in the autumn, they will have good roots by the autumn following. Theſe may be taken off, and planted in the nurſery for a year or two, as the ſeedlings; and the ſtools being cleared of all ſtraggling branches, and refreſhed with a knife, they will make ſtrong ſhoots for a ſecond operation by the autumn next enſuing. 3. By cuttings likewiſe theſe ſorts may be propagated. This work ſhould be done in October; and the cuttings for the purpoſe ſhould be the ſtrongeſt part of the laſt year's ſhoot, that had ſhot vigorouſly from a healthy ſoil. If theſe are cut into lengths of about a foot long, and planted in a moiſtiſh ſoil, three parts deep, they will grow, and make good ſhoots the ſummer following; and theſe will require no removing before they are planted out finally.

CORIARIA.

LINNEAN Claſs and Order, *Dioecia Decandria:* Male flowers containing ten ſtamina, and female flowers containing five piſtils upon diſtinct plants: There are two SPECIES; one of which will bear the open air of this climate.

CORIA'RIA *Myrtifo'lia:* The MYRTLE-LEAVED SUMACH, or TANNER'S SUMACH; *a deciduous ſhrub;* grows naturally about Montpelier in France, where it is ſaid to be uſed by the tanners in tanning of leather.

The MYRTLE-LEAVED SUMACH is a ſhrub of lowiſh growth, ſeldom arriving to more than four or five feet high. The bark is of a grayiſh colour, and ſpotted. The wood is very brittle, and very full of light pith.

The young ſhoots are produced in great plenty from the bottom to the top: They are ſquare, and come out three or four together, from one ſide of the ſtem, whilſt the other ſide is often furniſhed with an equal number. The leaves reſemble ſome of the ſorts of Myrtle, which gave occaſion for its being called the Myrtle-leaved Sumach: They are oblong, pointed, of a bright green, and ſtand oppoſite by pairs on the twigs. The flowers grow in ſpikes, at the ends and ſides of the branches, and have little beauty to recommend them. The tree is planted, however, as a flowering ſhrub, amongſt others of its own growth; but the place in which it is ſet ſhould be well ſheltered; for notwithſtanding this is a very hardy ſhrub, yet the ends of the branches are often killed in the winter, which makes the plant unſightly in the ſpring.

The PROPAGATION of the Coriaria is very eaſy. No other art need be uſed, than, after having obtained a few plants, to plant them in a lightiſh ſoil of any ſort. Here they will propagate themſelves in great plenty; for they will (what gardeners call) *ſpawn*; *i. e.* their creeping roots will ſend forth many young plants, at more than three yards diſtance from the real plant. The ſtrongeſt of theſe may be taken up, and planted where they are to remain, whilſt the weaker may be ſet in the nurſery way, to gain ſtrength, before they are ſet out for good. In this eaſy manner may plenty of theſe ſhrubs be obtained; and every winter after they are taken up, if the mould about the mother plant be raked ſmooth, and weeded in ſummer, ſhe will afford you a freſh crop by the autumn following, which may be taken off and planted as before.

CORONILLA.

LINNEAN Claſs and Order, *Diadelphia Decandria*: Each flower contains ten males and one female, the males being divided at the baſe into two ſets. There are eleven SPECIES; two of them herbaceous, the reſt of a ligneous

ligneous nature; but only one of them has been introduced into our ſhruberies.

CORONI'LLA *E'merus*: The SCORPION SENNA, or JOINTED-PODDED COLUTEA; *a deciduous ſhrub*; native of the South of Europe.

The SCORPION SENNA ſends out numerous irregular branches from the root and on all ſides; the oldeſt and moſt woody of which are of a grayiſh colour, whilſt the youngeſt are ſmooth, and of a dark brown. The leaves are pinnated, and conſtitute a great beauty in this ſhrub, being of a pleaſant green, and are compoſed of three pair of folioles, which are terminated by an odd one; theſe ſtand oppoſite on the midrib, and each has an indenture at the top. Theſe leaves, by a proper fermentation, will afford a dye nearly like that of indigo. However, beautiful as the leaves are, it is the flowers which conſtitute the beauty of theſe ſhrubs; and, indeed, of all the ſhrubby tribe, there is none more ſtriking or pleaſing than this when in full blow. This uſually happens in May; when it will be covered all over with bloom, the ſhrub itſelf appearing as one large flower divided into many looſe ſpikes; for the flowers come out all along the ſides of the branches by the leaves, on long footſtalks, each ſupporting two or three flowers, which are butterfly-ſhaped, of a yellowiſh colour, and large in proportion to the ſize of the ſhrub. They are ſucceeded by longiſh pods, in which the ſeeds are contained. This ſhrub often flowers again in the autumn.

There is a *Variety* of lower growth, called *Dwarf Scorpion Senna.*

This beautiful ſhrub is very readily PROPAGATED, either by ſeeds, layers, or cuttings; any of which may be eaſily made to grow. 1. By ſeeds. Theſe ſhould be ſown, in the ſpring, in beds of common garden mould made fine, and cleared of the roots of all weeds, &c. They ſhould be covered about half an inch deep; and, if a very dry ſpring does not enſue, they will be up in about a month or ſix weeks. If this ſhould happen, the beds muſt be now and then watered, and ſhaded from the heat of the ſun, which ſometimes is very intenſe and parching, even at the beginning of May. They may ſtand in the ſeed bed two years before they are taken up; all which time they will want no other care than weeding; and if they have watering the firſt ſummer, ſhould it prove a dry one, they will grow the faſter. After this, they may be taken out of the ſeed bed, planted in the nurſery way, and in about two or three years will be good plants to join in the ſhrubery. 2. By layers. This buſineſs may be performed any time in the winter; but as the ſhrub ſends forth numerous branches, many of them ſhould be taken off, and only ſuch a number left, as that they may be laid into the ground without crowding one another. The branches ſhould be of the laſt year's ſhoot; and the operation ſhould be performed by a gentle twiſt, ſo as juſt to break the bark; for, ſays HANBURY, without this I have found them in the autumn juſt as they were when layered; and with this, they have always ſtruck root, ſo as to be fit to take off the winter following. Theſe layers ſhould be planted out in the nurſery; and after having ſtood about two years, they alſo will be grown to be good plants. 3. By cuttings. The cuttings ſhould be the ſtrongeſt of the laſt year's ſhoots. They ſhould be planted cloſe, in October, in a ſhady border of good fine mould. If the ſpring and ſummer prove dry, watering muſt be afforded them every other day; and by this means many plants may be raiſed. If the cuttings are planted cloſe, and moſt of them grow, they ſhould be thinned, by taking up ſo many as may leave the others at a foot or more aſunder; and theſe plants alſo, thus taken up, ſhould be ſet out in the nurſery ground a foot aſunder, in rows at a foot and a half diſtance; where they may ſtand until they are finally taken up. It diſlikes a very moiſt ſituation.

CORYLUS.

LINNEAN Claſs and Order, *Monoecia Polyandria*; Male and female flowers upon the ſame plant: The males, containing ten ſtamina each, are collected in cylindrical catkins; the females, containing two piſtils each,

each, issue from the point of the leaf bud. There are two SPECIES.

1. CORYLUS *Avella'na:* The HAZEL; a well known *tall deciduous shrub*; very common in this country, and in most parts of Europe.

2. CO'RYLUS *Colu'rna:* The BYZANTINE NUT, or DWARF NUT TREE; *a low deciduous shrub*; growing naturally near Constantinople.

The HAZEL will grow to twenty feet high and upwards. A particular description of it here would be superfluous. LINNEUS considers the various kinds of FILBERTS as *Varieties* of the common Hazel, improved by culture. MILLER was of a different opinion: he says, "I have several times propagated both from the nuts, but never have found them vary from the other, though they have altered in the size and colour of their fruit from the sorts which were sown;" he therefore divides them into two distinct species: But HANBURY on the other hand says, that they "are varieties only of the same species; for I have planted the nuts of all the sorts, and sorts of all kinds have been produced from them." (Page 111.) As an *Ornamental*, the Hazel is of an inferior class; nevertheless, in recluse quarters, the Filbert may be introduced with propriety: the idea of utility associated with that real ornament which is undoubtedly given by the various tints of the leaves of the different kinds and colours of Filberts, may probably afford more real satisfaction, especially to the owner, than the transient glare of a useless exotic. Be this as it may, the Hazel in point of *use* stands high; as an underwood it has no superior: indeed, the Oak and Ash excepted, the husbandman knows not so useful a wood as the Hazel. For stakes, edders, and withs, it is in use every where. In Surry, Kent, and other southern counties, where numerous flocks of sheep are kept, the Hazel alone supplies the farmer with folding hurdles; and in Yorkshire and other parts of the North of England, from whence great quantities of butter are sent to the London market, the hoops or firkin rods are gathered almost wholly from this useful shrub.

2. The BYZANTINE NUT. This is distinguished from the other species chiefly by the stipulæ, which are very narrow and acute, whereas those of the common nut

nut are oval and obtuse. It differs also in the size of its growth, the true Byzantine Nut tree seldom growing higher than four or five feet; and hence the name *Dwarf Nut tree* has been used for this plant. In other respects, it is like our common nut tree; it flowers at the same time, the fruit is produced in clusters, and it ripens accordingly.

The method of PROPAGATING the Hazel kind is from seeds, by layering, or from the suckers, which it spontaneously sends up in great plenty. The *Nuts* should be sown about two inches deep, in February; until which time they should be kept in a cool, moist place to prevent the kernels from becoming dry and shrivelled, yet sufficiently airy to prevent their growing mouldy. The *Varieties* are best preserved by layering; for which purpose a few plants should be procured of the most valuable kinds, and planted for stools. They will grow on almost any soil; and the young twigs being laid in the ground in the autumn, will have struck root by the autumn following. These should be taken off, and planted in the nursery, a foot asunder, and two feet distant in the rows; and if there be any young shoots made the intermediate summer, they also may be laid down, or the plant headed within half a foot of the ground, to send forth young shoots for a second operation the autumn following. By this means the sorts may be propagated, and kept distinct; for the seeds sown of any of them will not in general come to good; though it is observable, that from the best nuts there will be the best chance of having good nuts again; and "I have (says HANBURY) sometimes known some few trees, raised from seeds, which have produced nuts better than those they were raised from. This may, perhaps, induce a gardener desirous of obtaining a great variety to try this method, when he may extirpate the worst sorts, and, if any should be worthy of it, may propagate the others in the manner directed." The Hazel, like the Birch, accommodates itself to every situation.

CRA-

CRATÆGUS.

LINNEAN Clafs and Order, *Icofandria Digynia:* Each flower contains about twenty males and two females: There are ten SPECIES; eight of which add confiderable beauty to the modern garden.

1. CRATÆ'GUS *Oxyaca'ntha:* The HAWTHORN, or WHITE THORN; a well known *deciduous tree or fhrub*; common with us, and growing naturally all over Europe.

2. CRATÆ'GUS *Aza'rolus:* The AZAROLE; *a tall deciduous fhrub*; native of Italy and the South of France.

3. CRATÆ'GUS *A'ria:* The WHITE LEAF; or the WHITE BEAM, or the ARIA, or the ARIA THEOPHRASTI; *a deciduous tree or fhrub*; grows naturally upon the hills of Kent and Surry, particularly near Box Hill; and in moft of the cold parts of Europe.

4. CRATÆ'GUS *Tormina'lis:* The WILD SERVICE, or the MAPLE-LEAVED SERVICE TREE; *a deciduous tree*; native of England, Germany, Switzerland, and Burgundy.

5. CRATÆ'GUS *Cocci'nea:* The VIRGINIA AZAROLE; *a tall deciduous fhrub*; native of Virginia and Canada.

6. CRATÆ'GUS *Crus Ga'lli:* The COCKSPUR HAWTHORN; *a tall deciduous fhrub*; native of Virginia.

7. CRATÆ'GUS *Tomento'fa:* The GOOSEBERRY-LEAVED VIRGINIA HAWTHORN; *a deciduous fhrub*; native of Virginia.

8. CRATÆ'GUS *Vi'ridis:* The GREEN-LEAVED VIRGINIA HAWTHORN; *a deciduous fhrub*; native of Virginia.

1. The HAWTHORN, in the ftate in which we are ufed to obferve it, is nothing better than a tall, uncouth, irregular fhrub; but trained up as a ftandard, it fwells to a large timber fize, with a tall ftem and a full fpreading head; though we believe it feldom rifes to a great height; perhaps not often fo high as thirty feet. We have meafured the ftem of a youthful thriving Hawthorn eight

eight feet high, and five feet and a half in circumference, with a head proportionable. Mr. Marſham * mentions one near Bethel Church, in the neighbourhood of Norwich, which, at four feet high, girted, in the year 1755, nine feet one inch and a quarter, one of its arms extending more than ſeven yards. The Standard Hawthorn, whether we view its flowers in the ſpring, its foliage in the ſummer, or its fruit in the autumn and winter, is one of the moſt *ornamental* plants, ſtanding ſingly, that can be ſcattered over a park or lawn. Its *uſes* will be explained when we come to treat of HEDGES.

In order to PROPAGATE a quantity of *Quick*, one method is generally practiſed; namely, firſt burying the haws, and taking them up to ſow the October following; though, ſays HANBURY, there is another way more preferable; namely, to prepare the beds, and ſow the haws ſoon after they are gathered. Whoever purſues the former method, having gathered what quantity of haws will anſwer his purpoſe, ſhould in ſome by-corner of the kitchen garden or nurſery dig a hole or pit capacious enough to receive them; ſome of the earth which came out of the hole, after the haws are put in it, ſhould be laid upon them; and, being thus carefully covered down, they may remain there till October. Then, having ground well dug, and cleared of the roots of all troubleſome weeds, and the mould being fit for working, the beds ſhould be made for the haws. Four feet is a very good width for theſe beds, as they may be eaſily reached over to be weeded; and if the alleys between be each one foot and a half wide, they will be of a good ſize. The beds being marked out with a line, ſufficient mould muſt be raked out to cover the haws an inch and a half deep. This being done, and the bottom of the beds being made level and even, the haws ſhould be ſown, and afterwards gently tapped down with the back of the ſpade; and then the fine mould, which had been raked out of the beds, muſt be thrown over them, covering them an inch and a half deep. In the ſpring the plants will come up, and in the ſummer following

* Of Norfolk, in a Letter publiſhed in the Firſt Volume of the Papers of the Bath Agriculture Society.

ſhould

ſhould be kept clean from weeds; though it does ſometimes happen, that few of them will appear till the ſecond ſpring after ſowing. Sometimes the young plants are planted out from the ſeed beds at one, two, or three years old; but the beſt plants are obtained by tranſplanting them into freſh mould the firſt or ſecond year, letting them remain in the nurſery two or three years longer. The practice of the London Nurſerymen is this: The ſtrongeſt of the ſeed bed plants having been drawn at two or three years old for ſale, they clear the beds entirely by drawing the remaining weak underling plants, and tranſplanting them into freſh beds in this manner (which they call *bedding* them): The ground having been trenched, and the tips of the plants as well as the lower fibres of their roots having been taken off with a ſharp knife, they ſtrain a line along one ſide of the bed; and, by chopping with a ſpade by the ſide of the line, leave a cleft or drill, of a depth proportioned to the length of the plants to be laid in; and, drawing the looſe mould ſomewhat towards them, leave the ſide of the drill next to the line with a ſmooth poliſhed face. Againſt this face the plants are ſet up, leaning towards the line, about three inches aſunder, leaving their heads about an inch above the mould, and placing their roots at ſuch a depth as to bury their ſtems from two to three inches deeper than they ſtood in the ſeed bed. The looſe mould being returned and preſſed gently to the roots with the foot, the line is removed, and another row planted in the ſame manner, about a foot from the firſt.

The Common Hawthorn ſports in the following *Varieties:*

The Large Scarlet Hawthorn.
The Yellow Hawthorn.
The White Hawthorn.
The Maple-leaved Hawthorn.
The Double-bloſſomed Hawthorn.
The Glaſtonbury Thorn.

The Large Scarlet Hawthorn is no more than a beautiful variety of the Common Haw. It is exceedingly large, oblong, perfectly ſmooth, and of a bright ſcarlet; and, from the additional ſplendor it acquires by the berries,

berries, it is propagated to cause variety in plantations for observation and pleasure.

Yellow Haw is a most exquisite plant. The buds, at their first coming out in the spring, are of a fine yellow, and the fruit is of the colour of gold. The tree is a great bearer, and retains its fruit all winter, causing a delightful effect in plantations of any kind. It was originally brought from Virginia, is greatly admired, and no collection of hardy trees should be without it.

White Haw is but a paltry tree, compared with the former. It hardly ever grows to the height of the Common Hawthorn, is an indifferent bearer, and the fruit is small, and a very bad white.

Maple-leaved Hawthorn will grow to be near twenty feet high, and has very few thorns. The leaves are larger than the Common Hawthorn, resemble those of the Maple, and are of a whitish green colour. The flowers are produced in large bunches, in June, and are succeeded by remarkable fruit, of a shining red, which looks beautiful in the winter.

Double-blossom'd Hawthorn produces a full flower, and is one of the sweetest ornaments in the spring. Nature seems to have peculiarly designed this sort for the pleasure garden; for though it be the Common Hawthorn only, with the flowers doubled, yet it may be kept down to what size the owner pleases; so that it is not only suitable for wilderness quarters, shruberies, and the like, but is also useful for small gardens, where a tree or two only are admitted. These beautiful double flowers come out in large bunches in May, and the tree is so good a bearer, that it will often appear covered with them. Their colour, at their first appearance, is a delicate white: They afterwards die to a faint red colour, and are frequently succeeded by small imperfect fruit.

Glastonbury Thorn differs in no respect from the Common Hawthorn, only that it sometimes flowers in the winter. It is said to have originally been the staff of Joseph of Arimathea, that noble counsellor who buried Christ. He, according to the tradition of the abbey of Glastonbury, attended by eleven companions, came over into Britain, and founded, in honour of the Blessed

Virgin,

Virgin, the firſt Chriſtian Church in this iſle. As a proof of his miſſion, he is ſaid to have ſtuck his ſtaff into the ground, which immediately ſhot forth and bloomed. This tree is ſaid to have bloſſomed on Chriſtmas day ever ſince, and is univerſally diſtinguiſhed by the name of the Glaſtonbury Thorn. HANBURY ſays, I have many plants that were originally propagated from this thorn; and they often flower in the winter, but there is no exact time of their flowering; for in fine ſeaſons they will ſometimes be in blow before Chriſtmas, ſometimes they afford their bloſſoms in February, and ſometimes it ſo happens that they will be out on Chriſtmas day.

2. AZAROLE. The Azarole Thorn will grow to be fifteen or ſixteen feet high. The leaves are large, nearly trifid, ſerrated, and obtuſe. The flowers are large, come out in May, and, in the different varieties, are ſucceeded by fruit of different ſize, ſhape, and reliſh.

The principal *Varieties* of this ſpecies are, The *Azarole with ſtrong thorns*; the *Azarole with no thorns*; the *Jagged-leaved Azarole*; the *Oriental Medlar*.

3. The WHITE LEAF. The *Aria Theophraſti*, called the White leaf tree, will grow to be more than twenty feet high *. This tree is engaging at all times of the year, and catches the attention, even in the winter: for then we ſee it ſtand, though naked of leaves, with a fine ſtraight ſtem, with ſmooth branches, ſpotted with white, at the end of which are the buds, ſwelled for the next year's ſhoot, giving the tree a bold and fine appearance. In the ſpring the leaves come out of courſe, and look delightfully, having their upper ſurface green, and the lower white. Their figure is oval; they are unequally ſerrated, about three inches long, and half as wide. Several ſtrong nerves run from the midrib to the border, and they are placed alternately on the branches, which appear as if powdered with the fineſt meal. The flowers are produced at the ends of the branches, in May; they are white, grow in large bunches, having meally footſtalks, and are ſucceeded by red berries, which will be ripe in autumn.

* At Blair of Athol, a ſeat of the Duke of Athol, in the Highlands of Perthſhire, this Tree grows to a timber ſize.

4. The

4. The Wild Service. The Maple leaved Service is a large growing tree It will arrive to near fifty feet, and is worth propagating for the ſake of the timber, which is very white and hard. This tree grows naturally in ſeveral woods in England; and it is the fruit of this ſpecies that is tied in bunches, and expoſed for ſale in the autumn: It is gathered in the woods, and by ſome perſons is much liked. The leaves in ſome degree reſemble thoſe of the Maple tree in ſhape; their upper ſurface is a fine green, their under hoary; and they grow alternately on the branches. The flowers come out in May, exhibiting themſelves in large cluſters at the ends of the branches: They are white, and are ſucceeded by the aforeſaid eatable fruit, which, when ripe, is of a brown colour, and about the ſize of a large haw.

5. Virginia Azarole. This ſpecies will grow to be near twenty feet high. The ſtem is robuſt, and covered with a light coloured bark. The branches are produced without order, are of a dark brown colour, and poſſeſſed of a few long ſharp thorns. The leaves are ſpear-ſhaped, oval, ſmooth, and ſerrated; of a thickiſh conſiſtence, and often remain on the tree the greateſt part of the winter. Each ſeparate flower is large; but as few of them grow together, the umbels they form are rather ſmall. They come out in May, and are ſucceeded by large dark red coloured fruit, which ripens late in the autumn.

The *Varieties* of this ſpecies are, The *Pear-leaved Thorn*; the *Plum-leaved Thorn with very long ſtrong ſpines and large fruit*; the *Plum-leaved Thorn with ſhort ſpines and ſmall fruit*.

6. Cockspur Hawthorn. The Virginia Cockſpur Thorn will grow to about twenty feet high. It riſes with an upright ſtem, irregularly ſending forth branches, which are ſmooth, and of a browniſh colour, ſpotted thinly with ſmall white ſpots. It is armed with thorns, that reſemble the ſpurs of cocks, which gained it the appellation of Cockſpur Thorn. In winter, the leaf buds appear large, turgid, and have a bold and pleaſant look among others of different appearances. In ſummer, this tree is very delightful. The leaves are oval, angular, ſerrated ſmooth, and bend backwards. They

are

are about four inches long, and three and a half broad; have five or six pair of ſtrong nerves running from the midrib to the border; and die to a browniſh red colour in the autumn. The flowers are produced in very large umbels, making a noble ſhow, in May; and are ſucceeded by large fruit, of a bright red colour, which have a good effect in the winter. It will bear a very moiſt ſituation.

The principal *Varieties* of this ſpecies are, The *Cockſpur Hawthorn with many thorns*; the *Cockſpur Hawthorn with no thorn*; the *Cockſpur with eatable fruit*. The latter was ſent me, ſays HANBURY, from America with that name, and I have raiſed ſome trees from the ſeed; but they have not yet produced any fruit, ſo that I cannot pretend to ſay how far it may be deſirable; though I have been informed it is reliſhed in America by ſome of the inhabitants there.

7. GOOSEBERRY-LEAVED VIRGINIA HAWTHORN. This ſpecies grows to about ſeven or eight feet high. The branches are ſlender, and cloſely ſet with ſharp thorns. The leaves are cuneiform, oval, ſerrated, and hairy underneath. The flowers are ſmall, and of a white colour: They are produced from the ſides of the branches, about the end of May; and are ſucceeded by yellow fruit, which ripens late in autumn.

There is a *Variety* of this, called the *Carolina Hawthorn*, which has longer and whiter leaves, larger flowers and fruit, and no thorns.

8. GREEN-LEAVED VIRGINIA HAWTHORN. The ſtem and branches of this ſpecies are altogether deſtitute of thorns. The leaves are lanceolate, oval, nearly trilobate, ſerrated, ſmooth, and green on both ſides. The flowers are white, moderately large, come out the end of May, and are ſucceeded by a roundiſh fruit, which will be ripe late in the autumn.

The reſpective ſpecies are all PROPAGATED by ſowing of the ſeeds; and the varieties are continued by budding them upon ſtocks of the White Thorn. This latter method is generally practiſed for all the ſorts; though, when good ſeeds can be procured, the largeſt and moſt beautiful plants are raiſed that way. 1. In order to raiſe them from ſeeds, let theſe be ſown ſoon after they are ripe, in beds of freſh, light, rich earth.

Let alleys be left between the beds, for the conveniency of weeding, and let the seeds be covered over with fine mould, about an inch deep. The summer following, the beds must be kept clean from weeds, and probably some few plants will appear: But this is not common in any of the sorts; for they generally lie till the second spring after sowing before they come up. At the time they make their appearance they must be watered, if the weather proves dry; and this should be occasionally repeated all summer. They should also be constantly kept clean from weeds; and in the autumn the strongest may be drawn out, and set in the nursery ground, a foot asunder, in rows that are two feet distant from each other; while the weakest may remain until another year. During the time they are in the nursery, the ground between the rows should be dug every winter, and the weeds constantly hoed down in the summer; and this is all the trouble they will require until they are planted out for good, which may be in two, three, or more years, at the pleasure of the owner, or according to the purposes for which they are wanted.

2. These trees are easily propagated by budding also; they will all readily take on one another; but the usual stocks are those of the Common awthorn In order to have these the best for the purpose, the haws should be got from the largest trees, such as have the fewest thorns and largest leaves. After they are come up, and have stood one year in the seed bed, the strongest should be planted out in the nursery, a foot asunder, and two feet distant in the rows; and the second summer after, many of them will be fit for working. The end of July is the best time for this business; and cloudy weather, night and morning, are always preferable to the heat of the day. Having worked all the different sorts into these stocks, they may be let alone until the latter end of September, when the bass matting should be taken off. In the winter the ground between the rows should be dug, and in the spring the stock should be headed about half a foot above the bud. he young shoots the stocks will always attempt to put out, should be as constantly rubbed off, for these would in proportion starve the bud, and stop its progress. With this care, several of the sorts have been known to shoot six

fix feet by the autumn; and as they will be liable to be blown out of their fockets by the high winds which often happen in the fummer, they fhould be flightly tied to the top of the ftock that is left on for the purpofe, and this will help to preferve them.

CUPRESSUS.

LINNEAN Clafs and Order, *Monoecia Monadelphia:* Male flowers containing four ftamens connected at the bafe, and female flowers containing many piftils; the males being difpofed in oval catkins; and the females, collected in roundifh cones upon the fame plant. There are five SPECIES (one of them lately difcovered in Japan):

1. CUPRE'SSUS *Sempervi'rens:* The COMMON CYPRESS; *an evergreen tree*; native of Italy, Spain, Portugal and Crete.

2. CUPRE'SSUS *Thyo'ides:* The AMERICAN CYPRESS; or the ARBOR VITÆ-LIKE CYPRESS, or the SMALL BLUE-BERRIED CYPRESS; *an evergreen tree or fhrub*; native of Maryland and Canada.

3. CUPRE'SSUS *Juniperoi'des:* The AFRICAN CYPRESS, or the JUNIPER-LIKE CYPRESS, or the CAPE CYPRESS; *a deciduous tree or fhrub*; native of the Cape of Good Hope.

4. CUPRESSUS *Di'fticha:* The DECIDUOUS CYPRESS; *a deciduous tree*; native of North America:

1. The COMMON CYPRESS. There are two ftriking *Varieties* of this plant (MILLER makes them two diftinct SPECIES); namely,

The Upright or Female Cyprefs; and

The Spreading or Male Cyprefs.

There is alfo a third *Variety* (which the fame profeffional writer confiders likewife as a diftinct fpecies); namely,

The Small-fruited Cyprefs.

The *Upright Cyprefs* is a moft elegant plant, and, notwithftanding it has of late years been fomewhat unfashionable, it certainly merits a place amongst *ornamental* evergreens. Its conical, or rather somewhat obeliscal, form makes an agreeable variety with fuller headed plants. It aspires to a considerable height, though we believe it seldom swells to a large girt. However, EVELYN and HANBURY speak of this kind of Cypress as a timber tree; but both of them seem to give preference to

The *Spreading Cypress*. This grows with a fuller and less regular head than the upright sort. MILLER tells us, that in the Levant this is the common timber; and recommends the planting of it in England very strongly; especially upon hot, sandy, or gravelly soils.

The *Small-fruited Cypress* is still more spreading than the other, and produces its boughs in an irregular manner. If it is not crowded by other trees, and is left to nature, it will be feathered from the top to the bottom. It will grow to about the height of the Common Cypress, and is a sort that looks well if planted singly on grass plats, &c. as well as when assisting to form clumps, or larger quarters of evergreens.

2. AMERICAN CYPRESS. This is the lowest grower of all the sorts with us; though in America, where it grows naturally, it arrives to timber, which serves for many excellent purposes. The tallest of these trees seldom rise much higher than fifteen feet; and as this tree is increased by cuttings, those plants raised this way seldom rise higher than about nine or ten feet. The branches stand two ways, and are pretty numerous; and the tree naturally forms itself into a regular head. The leaves of this sort are imbricated, like the *Arbor Vitæ*, though small, and are of a browner kind of green than the Common Cypress. The fruit is very small, and of a blue colour, and will be produced in great plenty all over the plant. They are of the size of the juniper berry, and much resemble it; though they are cones, and like the other species of this genus, but much smaller. When these plants are raised from seeds, they will aspire to a greater height, especially if planted in a moist soil; but those raised by cuttings generally have the appearance of shrubs. They are all, however, very beautiful, and greatly embellish those parts of the evergreen plantations where they are stationed.

3. AFRI-

3. AFRICAN CYPRESS. The branches of this ſpecies are numerous, ſlender, and ſpread themſelves all around. The leaves are narrow, awl-ſhaped, about an inch long, of a light green colour, and grow oppoſite to each other on the branches. The flowers come out from the ſides of the branches, like the Common Cypreſs, and they are ſucceeded by black fruit; but the ſeeds never ripen in England.

The method of PROPAGATING the Common Cypreſs is from ſeeds ſown in a warm border, or well ſheltered beds, of light ſandy loam, near half an inch deep, in the month of March; and by the beginning of May the plants will be up. After they have come up, if the ſummer ſhould not prove very dry, they will require little watering; and even in the greateſt drought twice a week will be ſufficient for them, provided it be done in the evenings. This is the only care they will require the firſt ſummer, except being kept clean from weeds. In the winter, if the place where they are ſown be tolerably well ſheltered, they will ſtand it very well, though it ſhould prove ſevere; but where the ſituation is not well ſheltered by plantations to break the violence of the froſty black winds, they muſt be ſcreened, otherwiſe many will be loſt. It is the black froſts, attended by high winds, which will deſtroy theſe plants; ſo that where there is not ſhelter enough to break their edge, the beds ſhould be hooped over, and covered with mats during that ſevere weather. The enſuing ſummer the plants may remain undiſturbed, when they will require no watering, and no farther care except weeding. The ſpring following, being then two years old, they ſhould be ſet out in the nurſery, exactly at two feet ſquare. In taking them out of the ſeed bed, ſome earth ſhould be taken with the roots. The latter end of March is the moſt proper time for this work; and if the weather ſhould prove dry and cold, as it often happens, the March winds blowing, the work muſt be deferred till rainy or cloudy weather; for without theſe precautions, you will find this a difficult plant to remove. After they are planted out in the nurſery, they may be now and then watered in dry weather, kept clean from weeds, and thus may ſtand till they are of a ſufficient ſize to be planted out. They will grow in almoſt any *ſoil*: but above all affect ſandy gravelly ground.

With regard to the African and the American Cypreſs, the *ſeeds* ſhould be ſown in pots or boxes. We receive them from abroad: They are very ſmall, and ſeldom come up before the ſecond ſpring; ſo that there will be leſs danger of their being loſt if they are ſown in pots or boxes, which may be ſet in the ſhade in ſummer, and removed into well ſheltered places during the winter. In the ſpring the plants will come up; and after that the Blue-berried Cypreſs may have the ſame treatment as the young ſeedlings of the Common ſort. With reſpect to the Cape Cypreſs, the plants muſt be ſet in pots, to be houſed in winter, until they are grown to be a yard high. When they are turned out into the open air, they ſhould have a dry, warm ſoil, and a well ſheltered place, and even theſe will not enſure their ſafety; ſo that whoever is deſirous of having theſe trees in his plantations, ſhould have ſome wooden ſconces made, to cover them in froſty weather; and if this is obſerved until they are grown of a tolerable ſize, there is no doubt but they will live, in a warm well ſheltered place, through our common winters.

4. The DECIDUOUS CYPRESS will grow to be near ſixty feet high, if ſtationed in a place ſuitable to its nature. It is very hardy in reſpect to cold; and a ſhare of the moiſteſt part of the plantation muſt be allotted it. In Virginia and ſeveral parts of America, where this tree is a native, it is a real aquatic; being found growing to a very large ſize in places wholly covered with water; and with us, if planted in watery places, by the edges of rivers, ponds, ſprings, &c. it will be more luxuriant, and will proportionally riſe to a greater height and bulk than if planted in a dry ſoil. This tree in the ſummer has a little the reſemblance of an evergreen, and the leaves have a pleaſing effect, appearing in ſome reſpect like ſome ſorts of the Acacias; and theſe are the chief inducements for its admiſſion into the pleaſure ground.

This ſpecies may be PROPAGATED from *ſeeds* in the ſame manner as the Common Cypreſs; alſo from *cuttings* planted in October in a moiſt ſandy ſoil. Many of of them will grow; though a general crop can ſeldom be obtained; and they ſhould be kept clean from weeds the ſummer following, as well as the ſummer after that. In the autumn, or any part of the winter they ſhould be planted out in the nurſery; and, if they are to ſtand there a conſiderable time, they ſhould be allowed a good diſtance; for they will grow, with proper care, when removed at a large ſize. If any part of the nurſery ground is moiſter than the other, they muſt have a ſhare of it. The ground ſhould be conſtantly dug between the rows every winter, the weeds hoed down in ſummer, and when planted out, theſe trees ſhould have moiſt places, in conſequence of what has been before obſerved.

CYNANCHUM.

LINNEAN Claſs and Order, *Pentandria Digynia*: Each flower contains five males and two females: There are fourteen SPECIES; moſt of them climbing: Three are ſufficiently hardy for this climate.

1. *Cyna'nchum Acu'tum*: The ACUTE-LEAVED CYNANCHUM, or the ACUTE-LEAVED MONTPELIER SCAMMONY; *an herbaceous climber*; native of the South of Europe.

2. *Cyna'nchum Monſpeli'acum*: The ROUND-LEAVED CYNANCHUM, or ROUND LEAVED MONTPELIER SCAMMONY; *an herbaceous climber*; native of Spain and the South of France.

3. *Cyna'nchum Subero'ſum*: The CAROLINA CYNANCHUM, or the CAROLINA PERI'PLOCA; *a ligneous climber*; native of Carolina and other parts of America.

1. The ACUTE-LEAVED CYNANCHUM. The root is ſtrong, creeping and ſpreads itſelf to a conſiderable diſtance. The ſtalks are herbaceous, twiſt about every thing that is near them, will grow to be ſix feet long, but always die to the ground in the autumn, and freſh ones are put forth from the roots in the ſpring. The leaves are oblong, heart-ſhaped, acute-pointed, ſmooth,

 and

and grow oppoſite by pairs on long footſtalks. The flowers come out from the wings of the leaves in ſmall bunches; they are of a dirty white colour, appear in June and July, but are not ſucceeded by good ſeeds in our gardens. This plant, on being wounded, emits a milky juice.

2. ROUND-LEAVED CYNANCHUM. The root of this ſpecies is large, thin, juicy, and ſpreads itſelf to a conſiderable diſtance. The ſtalks are herbaceous, and twine to ſix or ſeven feet high about whatever is near them. The leaves are broad, reniform, roundiſh, and grow oppoſite, on long footſtalks. The flowers come out from the wings of the leaves, in ſmall bunches; they are of a bad white colour, appear in June and July, and are rarely ſucceeded by good ſeeds in our gardens. The ſtalks die to the ground in the autumn, and freſh ones ariſe again in the ſpring. On wounding any part of this plant, a milky juice immediately flows.

3. CAROLINA CYNANCHUM. The ſtalks of this ſpecies are ſlender, ligneous, ſhrubby, and will twiſt about any thing to the height of about ſeven feet. They are hairy, and their lower part is covered with a thick, fungous, cloven, cork-like bark. The leaves are oval, heart-ſhaped, pointed, and grow oppoſite at the joints, on long hairy footſtalks. The flowers come out from the wings of the leaves, in ſmall bunches. They are greeniſh on their firſt appearance, but die away to a bad purple. They exhibit themſelves in July and Auguſt; but are not ſucceeded by good ſeeds in our gardens.

This ſort is PROPAGATED by laying down the young ſhoots as they advance in the ſummer, and covering them over with ſome fine mould. Theſe will ſoon put out roots, by the autumn will be good plants, and may then be removed to the places where they are deſigned to remain. This ſpecies is rather tender; and the ſoil in which it is planted ſhould be naturally dry, warm, light, and ſandy, and the ſituation well defended. Being thus ſtationed, it will live abroad, and continue for many years; but if the ſoil is moiſt, rich, and ill defended, the chance will be very great but it will be deſtroyed the firſt winter.

The firſt two ſorts are exceedingly hardy, will grow in

in any ſoil or ſituation, and will overrun any ſmall plants that are near them. Their ſituation, therefore, ſhould be among ſuch trees as have ſtrength enough to admit their embraces; and their propagation is by cutting the roots in the autumn. Every cut will grow; and when planted, will call for no trouble except keeping them clear from weeds, when they firſt ſhoot up in the ſpring.

CYTISUS,

LINNEAN Claſs and Order. *Diadelphia Decandria*: Each flower contains ten males and one female; the males riſing in two diviſions: There are fourteen SPECIES; five of which afford conſiderable ornament to the Engliſh garden

1. *Cy'tiſus Seſſilifo'lius*: The SESSILE-LEAVED CYTISUS (or TREFOIL TREE, or BASE TREE TREFOIL), or CYTISUS SECUNDUS CLUSII, or the SMOOTH ROUND-LEAVED CYTISUS; *a deciduous ſhrub*; native of France, Italy, and Spain.

2. *Cy'tiſus Ni'gricans*: The BLACK CYTISUS; or the BLACKISH SMOOTH CYTISUS; *a deciduous ſhrub*; native of Auſtria, Bohemia, Italy, and Spain.

3. *Cy'tiſus Auſtri'acus*: The TARTARIAN CYTISUS; or the AUSTRIAN CYTISUS; *a low deciduous ſhrub*; native of Auſtria, Siberia, and Italy.

4. *Cy'tiſus Labu'rnum*: The LABURNUM; *a deciduous tree*; native of Switzerland, Savoy, and moſt parts of Europe.

5. *Cy'tiſus Hirſu'tus*: The EVERGREEN CYTISUS; or the EVERGREEN CYTISUS OF NAPLES; or the ITALIAN CYTISUS WITH HAIRY LEAVES, *an evergreen ſhrub*; native of Italy, Spain, Auſtria, and Siberia.

1 The SESSILE-LEAVED CYTISUS will grow to the height of about five or ſix feet. The branches are numerous, erect, very brittle, and covered over with a ſmooth brown bark. The leaves are ſmall, and of a fine green: They are nearly of an oval figure, and grow

grow by threes on the twigs; on some branches they sit quite close, on others they grow on very short footstalks. The flowers grow at the ends of the branches, in short spikes: They are of a fine yellow, come out the beginning of June, and when in full blow the shrub will appear almost covered with them. The seeds usually ripen in August.

2. BLACK CYTISUS will arrive to about the height of the former, and naturally divides into many branches. The bark is brown, and the young shoots are of a greenish red. The leaves resemble Trefoil: They are smooth, and grow three together on brownish footstalks; the folioles are of an oblong oval figure, and their upper surface is of a dark green, but they are paler underneath. The flowers are produced in long, erect, close spikes, at the ends of the branches: They are of a beautiful yellow colour, come out in July, and when in full blow make a fine appearance. The seeds ripen in the autumn.

3. TARTARIAN CYTISUS. The stalks are shrubby, branching, green, and grow to three or four feet high. The leaves are oval, oblong, smooth, and of a whitish green colour. The flowers come out in close heads from the ends of the branches, in May: They are of a light yellow colour, and have a cluster of leaves under them; they are sometimes succeeded by short woolly pods, containing the seeds.

There is a *Variety* of this species, with naked stalks, smaller leaves and flowers, rather earlier in the spring, usually called the *Siberian Cytisus*.

4. The LABURNUM is a large growing plant: It will aspire to the height of near forty feet, and is one of the most beautiful trees our gardens afford. It will form itself into a fine head; its branches are smooth, of a pale green colour, and possessed of a few grayish spots. The leaves stand by threes on long slender footstalks: Each of these is oblong and entire; their upper surface is smooth, and of a shining green, but their under surface is more inclined to be downy. The time of this tree's flowering is May; and the effect can hardly be conceived which it will have, when it appears covered with its long pendulent bunches of flowers, of a delightful yellow. Each flower that helps

to compoſe one ſet is tolerably large of itſelf, and the common ſtalk to which they adhere by their own ſeparate footſtalks is often a foot or more in length; ſo that the appearance muſt be moſt noble, when it exhibits theſe long ſeries of flowers hanging down from almoſt every part of the whole head: HANBURY continues, "But this is not all; the timber when felled is exceedingly valuable. It will arrive in bulk in proportion to its height; and the timber is both heavy and hard, and of a fine colour, inclined to yellow. The very branches of this tree are ſo ponderous as to ſink in water. It poliſhes extremely well, and is ſo much like to green ebony, that it is called by the French, *Ebony of the Alps*, where the tree grows naturally. And as the timber is ſo valuable for many ſorts of rich furniture, this ſhould arouſe the timber planter's attention; for it will grow to be a timber tree of more than a yard in girt, in almoſt any poor and ſorry ſoil, where other trees will hardly grow, let the ſituation be what it will: And how enchantingly ornamental muſt large quarters or clumps of theſe trees appear, either by the borders of other woods, or in parks, and at the ſame time the expectation of the timber crop retained!"

There are ſome other ſorts of LABURNUMS, of equal or more beauty than the preceding: One is called the *Scotch Laburnum* *, another the *Italian*. The leaves of theſe are larger, and the bunches of flowers longer; and the individual flowers of which the bunches are compoſed proportionally larger. There is alſo another ſort, with ſmaller leaves, and bunches longer than the common, which difference it always preſerves from ſeeds; and theſe being planted among the common ſort, will afford the greater variety.

One method of PROPAGATION is common to all theſe ſorts: It is to be performed both by ſeeds and cuttings. 1. When by ſeeds, common garden mould, when dug, and cleared from the roots of all weeds, will do for their reception. They ſhould be ſown in the ſpring, in beds neated up, about half an inch deep, and in about ſix weeks the young plants will appear. Nothing more

* This reaches a timber ſize in Scotland. The heart is of a beautiful brown colour, and cloſe texture.

will be neceſſary than keeping them clean from weeds during the ſummer, unleſs the weather proves very dry; if it does, a little watering ſometimes will be proper. The ſpring following, the *Laburnums* ſhould be planted out in the nurſery; but the other ſorts ſhould ſtand in the ſeed bed two years, to gain ſtrength, before they are taken up. Theſe ſhould be planted a foot aſunder, and two feet diſtant in the rows; but the *Laburnums* ought to have a rather greater diſtance, eſpecially if they are deſigned to be trained up for ſtandards. 2. Another method of increaſing theſe ſorts is by cuttings. October is the beſt month for the work; and the cuttings may be planted either a foot aſunder, and two feet diſtant in the rows, ſo that they need not be removed till they are taken up for good; or they may be ſet very thick, and thoſe which live taken up the winter following, and planted out in the nurſery way, at diſtances wide in proportion to the time they are to ſtand. It will bear a very moiſt ſituation.

5. The Evergreen Cytisus. This ſhrub is naturally of an upright growth, and its common height is about ſix or ſeven feet. It may be trained up to a ſingle ſtem, for two, three, or four feet high, and will naturally ſend out many branches, which will form themſelves into a fine head. The bark on the ſtem is of a gray colour; the branches alſo are gray, with a green caſt at a diſtance; and many of them will have the appearance of being channelled, the bottom of the grooves being of a duſky green, but their upper edges white. The younger ſhoots are green and ſtreaked, and their ſurface is hairy. The leaves alſo have this property, and ſtand three upon a ſhort footſtalk. They are nearly of an oval figure, and have a ſtrong midrib running the whole length. They are of a fine green colour, and clothe the ſhrub with great beauty. The flowers are of a clear yellow colour, and are ſhaped like thoſe of the other ſorts: They appear in June, and are produced from the ſides of the branches, all over the ſhrub, in ſhort bunches; ſo that its golden head at that time is both beautiful and ſtriking. Neither is June the only time of its flowering; for it will often flower again in October, and if the winter continues

tinues open and mild, it will ſometimes ſhew its bloſſoms in November and December. The flowers that appeared in June, which is its regular time of blow, will be ſucceeded by ſmall hairy pods, in which the ſeeds are contained, and which ripen with us very well in the autumn.

This ſort ſhould be PROPAGATED by ſeeds, which ſhould be ſown in the ſpring, and managed as directed for the deciduous ſorts; only it may not be amiſs to obſerve, that it will be neceſſary to plant the ſeedlings in the nurſery when they have ſtood one year in the ſeed bed. They ſhould be ſet about a foot aſunder, in rows at two feet diſtance; and here they may ſtand for about two years, when they ſhould be planted out.

D A P H N E.

LINNEAN Claſs and Order, *Octandria Monogynia*: Each flower contains eight males and one female: There are fifteen SPECIES; eight of which are proper for our collection.

1. *Da'phne Meze'reum*: The MEZEREON, or SPURGE OLIVE; *a low deciduous ſhrub*; native of Germany; and has been diſcovered in this country in ſome woods near Andover, in Hampſhire.

2. *Da'phne Gni'dium*: The FLAX-LEAVED DAPHNE, or FLAX-LEAVED THYMELÆ'A; *a low deciduous ſhrub*; native of Italy, Spain, and about Montpelier.

3. *Daphne Cneo'rum*: The SPEAR-LEAVED DAPHNE, or the CNEORUM; or the CLUSTER FLOWERING SPEAR-LEAVED DAPHNE; *a very low deciduous ſhrub*; native of Switzerland, Hungary, the Alps, and the Pyrenean Mountains

4. *Da'phne Tartonra'ira*: The OVAL-LEAVED DAPHNE, or the TARTONRAIR; or CLUSTER FLOWERING OVAL-LEAVED DAPHNE; *a very low deciduous ſhrub*; native of France and Italy.

5. *Da'phne*

5. *Da'phne Alpi'na:* The ALPINE DAPHNE; or the ALPINE CHAMELÆ'A; *a low deciduous ſhrub*; native of the Alps, Geneva, Italy, and Auſtria.

6. *Da'phne Thymelæ'a:* The MILKWORT-LEAVED DAPHNE, or the THYMELÆA; *a low deciduous ſhrub*; native of Spain and the South of France.

7. *Da'phne Villo'ſa:* The HAIRY-LEAVED DAPHNE, or the SMALL HAIRY PORTUGAL DAPHNE, *a very low deciduous ſhrub*; native of Spain and Portugal.

8. *Da'phne Laure'ola:* The SPURGE LAUREL, or the EVERGREEN DAPHNE; *a low evergreen ſhrub*; common in ſome parts of this kingdom, alſo in Switzerland and France.

1. The MEZEREON. Of this elegant plant there are four *Varieties:* 1. The *White.* 2. The *Pale red.* 3. The *Crimſon.* And, 4. The *Purple flowering.*—HANBURY is very laviſh of his praiſe of theſe ſhrubs; he ſays, "They have each every perfection to recommend them as flowering ſhrubs. In the firſt place, they are of low growth, ſeldom ariſing to more than three or four feet in height, and therefore are proper even for the ſmalleſt gardens. In the next place, they will be in bloom when few trees, eſpecially of the ſhrubby tribe, preſent their honours. It will be in February, nay, ſometimes in January; then will the twigs be garniſhed with flowers, all around, from one end to the other. Each twig has the appearance of a ſpike of flowers of the moſt conſummate luſtre; and as the leaves are not yet out, whether you behold this tree near or at a diſtance, it has a moſt enchanting appearance. But this is not all; the ſenſe of ſmelling is peculiarly regaled by the flowers; their ſpicy ſweetneſs is diffuſed around, and the air is perfumed with their odours to a conſiderable diſtance. Many flowers, deemed ſweet, are not liked by all; but the agreeable inoffenſive ſweetneſs of the *Mezereon* has ever delighted the ſenſe of ſmelling, whilſt the luſtre of its blow has feaſted the eye. Neither is this the only pleaſure the tree beſtows; for beſides the beauty of the leaves, which come out after the flowers are fallen, and which are of a pleaſant green colour and an oblong figure, it will be full of red berries in June, which will continue growing till the autumn. Of theſe berries the birds are very fond; ſo that whoever is delighted with

with thoſe ſongſters, ſhould have a quantity of them planted all over the outſides of his wilderneſs quarters."

PROPAGATION. This ſort ripens its ſeeds with us, and may at any time be eaſily obtained, if they are ſecured from birds. Previous therefore to ſowing, the healthieſt and moſt thriving trees of the White, the Pale, and the Deep Red ſorts ſhould be marked out; and as ſoon as the berries begin to alter from green, they muſt be covered with nets, to ſecure them from the birds, which would otherwiſe devour them all. The berries will be ripe in July; and due obſervance muſt be had to pick them up as they fall from the trees, and to keep the ſorts ſeparate. As ſoon as they are all fallen, or you have enough for your purpoſe, they may then be ſown. The beſt ſoil for theſe plants is a good fat black earth, ſuch as is found in kitchen gardens that have been well manured and managed for many years. In ſuch ſoil as this they will not only come up better, but will grow to a greater height than in any other. No particular regard need be paid to the ſituation; for as this tree is a native of the northern parts of Europe, it will grow in a north border and flouriſh there as well as in a ſouth; nay, if there be any difference, the north border is more eligible than the ſouth. The ground being made fine, and cleared from roots of all ſorts, the ſeeds ſhould be ſown, hardly half an inch deep. The mould being riddled over them that depth, let the beds be neated up, and they will want no other attention until the ſpring These ſeeds will ſometimes remain in the ground two years; but for the moſt part they come up the ſpring after ſowing; and the ſeedlings will require no other care during the ſummer than weeding, and gentle watering in dry weather. After they have been in the ſeed bed one year, the ſtrongeſt may be drawn out, and planted in the nurſery, to make room for the others; though if they do not come up very cloſe, it would be as well to let them remain in the ſeed bed until the ſecond autumn: when they ſhould be taken up with care, and planted in beds at a foot aſunder each way. This will be diſtance enough for theſe low growing ſhrubs October is the beſt month for planting them out finally; for although they will grow if removed any time between then and ſpring, yet that

that will certainly be a more proper ſeaſon than when they are in full blow. Such is the culture of this ſhrub. The other ſpecies of this genus require a different management.

2. Flax-leaved Daphne ſeldom grows higher than three feet. The branches are very ſlender, and ornamented with narrow, ſpear-ſhaped, pointed leaves, much like thoſe of the Common Flax. The flowers are produced in panicles, at the ends of the branches: They are ſmall, come out in June, but are rarely ſucceeded by ſeeds in England.

3. Spear-leaved Daphne, or Cneorum. This riſes with a ſhrubby, branching ſtalk, to about a foot or a foot and a half high. The leaves are narrow, ſpear-ſhaped, and grow irregularly on the branches. The flowers are produced in cluſters, at the ends of the little twigs: They make their appearance in March, are of a purple colour, and poſſeſſed of a fragrance little inferior to that of the *Mezercon*; but they are ſeldom ſucceeded by ſeeds in England.

4. Oval-leaved Daphne, or Tartonrair. This riſes with a woody ſtalk to the height of about two feet. The branches are numerous, irregular, tough, and covered with a light brown-coloured bark. The leaves are oval, very ſmall, ſoft to the touch, and ſhining. The flowers are produced in cluſters from the ſides of the ſtalks: They are white, come out in June, and are ſucceeded by roundiſh berries, which ſeldom ripen in England. This ſort ſhould have a dry ſoil and a warm ſituation.

5. The Alpine Daphne, or Chamelæa, will grow to the height of about a yard. The leaves are ſpear-ſhaped, obtuſe, and hoary underneath. The flowers come out in cluſters from the ſides of the branches, and are very fragrant: They appear in March, and are ſucceeded by red berries, that ripen in September.

6. Milkwort-leaved Daphne, or Thymelæa, will grow to the height of a yard. The ſtalks of this ſpecies are upright, branched, and covered with a light brown bark. The leaves are ſpear-ſhaped, ſmooth, and in ſome reſpect reſemble thoſe of Milkwort. The flowers are produced in cluſters from the ſides of the ſtalks:

ſtalks: They are of a greeniſh colour, have no footſtalks, appear in March, and are ſucceeded by ſmall yellowiſh berries, which will be ripe in Auguſt. This ſort requires a dry ſoil and a warm ſituation.

7. HAIRY-LEAVED DAPHNE. The ſtalks are ligneous, about two feet high, and ſend forth branches alternately from the ſides. The leaves are ſpear-ſhaped, plane, hairy on both ſides, and grow on very ſhort footſtalks. The flowers have very narrow tubes, are ſmall, and make no great ſhow: They come out in June, and are not ſucceeded by ripe ſeeds in England. This ſhrub, in ſome ſituations, retains its leaves all winter in ſuch beauty as to cauſe it to be ranked among the low-growing evergreens; but as in others it is ſometimes ſhattered with the firſt black winds, it is left to the Gardener whether to place this ſhrub among the Deciduous Trees or Evergreens.

All theſe ſorts are with ſome difficulty PROPAGATED and retained. They will by no means bear removing, even when ſeedlings; and if ever this is attempted, not one in a hundred muſt be expected to grow. They are raiſed by ſeeds, which we receive from the places where they grow naturally; and he who is deſirous of having theſe plants, muſt manage them in the following manner: Let a compoſt be prepared of theſe equal diviſions; one fourth part of lime rubbiſh; one fourth part of drift or ſea ſand; another of ſplinters of rocks, ſome broad and others ſmaller; and the other part of maiden earth, from a rich paſture. Let theſe be mixed all together, and filled into largiſh pots. In each of theſe pots put a ſeed or two, about half an inch deep, in the fineſt of the mould. We receive the ſeeds in the ſpring; ſo that there is little hope of their coming up until the ſpring following: Let, therefore, the pots be ſet in the ſhade all the ſummer, and in the autumn removed into a warm ſituation, where they may enjoy every influence of the ſun's rays all winter. In March let them be plunged into a moderate hotbed, and the plants will ſoon after appear. This bed will cauſe them to be ſtrong plants by the autumn; and when all danger of froſt is over, they may be uncovered wholly, and permitted to enjoy the open air. In the autumn, they ſhould be removed into the greenhouſe, or ſet under a

 hot-

hotbed frame all winter; and in ſpring they ſhould be placed where they are to continue, moulding them up the height of the pot; the pots being ſufficiently broken to make way for their roots, as they ſhoot, and then left to Nature. The ſituation of the four tenderer ſorts muſt be well ſheltered, and if it be naturally rocky, ſandy, and dry, it will be the better; for in the places where they grow naturally, they ſtrike into the crevices of rocks, and flouriſh where there is hardly any appearance of ſoil.

This is one method of obtaining theſe ſhrubs. Another way is, by ſowing the ſeeds in the places where they are to remain. The ſituation and nature of the ſoil ſhould be as near that above deſcribed as poſſible; and the mould ſhould be made fine in ſome places, and a ſeed or two ſown in each. After this, pegs ſhould be ſtuck down on each ſide of them, to direct to the places where they are ſown. The exacteſt care muſt be obſerved, all ſummer, to pull up the weeds as often as they appear; for if they are permitted to get ſtrong, and have great roots, they will pull up the ſeeds with them. In the ſpring following, if the ſeeds are good, the plants will appear. During the ſummer, they ſhould be watered in dry weather; and, for the firſt winter or two, ſhould have ſome furze buſhes pricked all round them, at a proper diſtance, which will break the keen edge of the froſty winds, and preſerve the young plants until they are ſtrong enough to defend themſelves.

The Cneorum and the Alpine Chamelæa are very hardy, and will grow in the coldeſt ſituation; but the other ſorts ſhould have a warm ſoil and a well ſheltered ſite, or they will be ſubject to be deſtroyed in bad weather.

8. The Spurge Laurel, or Evergreen Daphne, is a low ſhrub, ſeldom growing more than a yard or four feet high; it ſends out many branches from the bottom, and theſe are covered with a ſmooth light brown bark, that is very thick. The bark on the younger branches is ſmooth and green; and theſe are very cloſely garniſhed with leaves of a delightful ſtrong lucid green colour. Theſe leaves ſit cloſe to the branches, and are produced in ſuch plenty, that they have the appearance, at a ſmall diſtance, of cluſters at the

the ends of the branches. They are ſpear-ſhaped, ſhining, ſmooth, and thick; their edges are entire. HANBURY extols this plant with a degree of enthuſiaſm; continuing, "and this is another excellent property of this tree, that it is thus poſſeſſed of ſuch delightful leaves for its ornament. Theſe leaves, when growing under the drip of trees, ſpread open, and exhibit their green pure and untarniſhed, in its natural colour: when planted ſingly in expoſed places, they naturally turn back with a kind of twiſt, and the natural green of the leaf is often alloyed with a browniſh tinge. This ſhrub is alſo valuable on account of its flowers; not becauſe they make any great ſhow, but from their fragrance, and the time they appear; for it will be in blow the beginning of January, and will continue ſo until the middle or latter end of April before the flowers fall off; during which time they never fail to diffuſe abroad their agreeable odours, which are refreſhing and inoffenſive. In the evenings eſpecially, they are more than commonly liberal; inſomuch that a few plants will often perfume the whole end of a garden; and when this happens early, before many flowers appear, the unſkilful in flowers, perceiving an uncommon fragrancy, are at once ſtruck with ſurprize, and immediately begin enquiring from whence it can proceed. Neither are its odours confined to a garden only; but, when planted near windows, they will enter parlours, and aſcend even into bedchambers, to the great comfort of the poſſeſſor, and ſurprize of every freſh viſitor." Theſe flowers make but little ſhow; for they are ſmall, and of a greeniſh yellow. They are produced amongſt the leaves from the ſides of the ſtalks, in ſmall cluſters, and will often be ſo hid by them, as to be unnoticed by any but the curious. They are ſucceeded by oval berries, which are firſt green, and afterwards black when ripe. Theſe berries will be in ſuch plenty as to be very ornamental; but will ſoon be eaten up by the birds; which is another good property of this tree, as it invites the different ſorts of whiſtling birds to flock where it is planted in great plenty.

This ſhrub is PROPAGATED by ſeeds, in the ſame manner as the Common *Mezereon*. The ſeeds muſt be preserved from the birds by nets, until they are ripe. Soon after, they must be sown as is directed for the *Mezereon.* They will often be two years before they come up; during which time, and afterwards, they may have the same management as has been laid down for the Common *Mezereon,* until they be finally set out.

This shrub will grow in almost any soil or situation, but flourishes most under the shade and drip of taller plants, giving a peculiar chearfulness to the bottoms of groves and clumps in winter.

DIOSPYROS.

LINNEAN Class and Order, *Polygamia Dioecia:* Some of the plants of this genus bear hermaphrodite and female flowers upon the same individual, whilst others bear male flowers only; each of which contains eight stamina. There are five SPECIES; three of which are of late discovery: The other two are,

1. *Dio'spyros Lo'tus:* The INDIAN DATE PLUM; *a very tall deciduous shrub*; native of Africa and the South of Europe.

2. *Dio'spyros Virginia'na:* The PISHAMIN PLUM; *a very tall deciduous shrub*; native of Virginia, Carolina, and many parts of North America.

1. The INDIAN DATE PLUM will arrive at the height of more than twenty feet, and is an excellent tree for shade. It aspires with an upright stem, and the young branches are covered with a smooth whitish bark. The youngest twigs stand alternately on those of the preceding year, and the buds for the next year's shoot begin to swell soon after the fall of the leaf. The leaves are of two colours; their upper surface is of a delightful green, and their lower of a whitish cast. They are of an oblong figure, end in a point, and are in length about four inches and a half, and near two inches broad. They are placed alternately on the branches, and

and ſeveral ſtrong veins run alternately from the mid-rib to the borders, which are entire. Theſe leaves will be of a deep green, even when they fall off in the autumn. The flowers have little beauty to recommend them: they are pitcher-ſhaped, and grow ſingly on ſhort footſtalks, on the ſides of the branches: they are of a reddiſh colour, and are ſucceeded by largiſh black berries, which are eatable, like the medlar, when in a ſtate of decay.

2. The PISHAMIN PLUM will not aſpire to the height of the former ſpecies, though it will ſometimes grow to near twenty feet. The branches of this tree are whitiſh, ſmooth, and produced in an irregular manner. The leaves are very large and beautiful; about five or ſix inches long, and three broad. Their upper ſurface is ſmooth, and both ſides are of a beautiful green. They are of an oblong figure, end in a point, grow irregularly on the branches, and have ſeveral veins running from the midribs to the borders, which are entire. They fall off in the autumn, at the coming on of the firſt froſts, when their colour will be that of a purpliſh red. The flowers, like thoſe of the other ſort, make no great appearance; but are ſucceeded by a fruit, which is eatable, when, like medlars, it is in a ſtate of decay.

Both theſe ſorts are PROPAGATED from the ſeeds, which we receive from abroad, in the ſpring. The compoſt proper for their reception is maiden earth, from a rich paſture, dug up ſward and all a year before, and three or four times turned in order to rot the ſward. This being made fine, a fourth part of drift or ſea ſand ſhould be added; and being all well mixed, the ſeeds ſhould be ſown in pots or boxes, three quarters of an inch deep. The pots ſhould afterwards be placed in a ſhady place during the ſummer; for the ſeeds rarely come up until the ſecond ſpring; and in the autumn they ſhould be removed into a well ſheltered place, where they may enjoy the benefit of the ſun all winter. In the ſpring the plants will come up; and if they are aſſiſted by plunging the pots into a moderate hotbed, it will make them ſhoot ſtronger; though this is not abſolutely neceſſary. All the ſummer they ſhould ſtand in a ſhady place, where they may have free air; and, if if the weather prove dry, they ſhould be watered every other evening. At the approach of winter, they ſhould be removed into the greenhouſe, or placed under a hotbed frame, or ſome ſhelter; and, when all danger of froſt is over, they muſt be put in the ſame ſhady ſituation as in the former ſummer. In the winter alſo they ſhould be hooped as before; and in ſpring may be planted in the nurſery ground. Theſe plants, when they get tolerably ſtrong, are very hardy; though even then the ends of the branches are ſubject to be killed; ſo that when they are ſeedlings, or very young, they will be in danger of being deſtroyed by the froſts, which makes the above-directed care and protection neceſſary till they have gained ſtrength.

ELÆACNUS.

LINNEAN Claſs and Order, *Tetrandria Monogynia*. Each flower contains four males and one female. There are four SPECIES; two of which have been introduced into this country; one of them requiring a ſtove heat; the other ſufficiently hardy to bear the open air; namely,

Elæagnus Anguſtifolia: The NARROW-LEAVED ELÆAGNUS, or the OLEASTER, or the WILD OLIVE; *a tall deciduous ſhrub*; native of Bohemia, Spain, Syria, and Cappadocia.

The NARROW-LEAVED ELÆAGNUS, or the OLEASTER, will grow to be near twenty feet high. Whilſt the leaves of moſt trees are poſſeſſed of a verdure, and occaſion variety by the difference of greens they exhibit, the leaves of the plant under conſideration are white, eſpecially the under ſide, and ſtand upon white twigs. The branches are of a brown colour; but the preceding year's ſhoots are white and downy, the ſilvery leaves being placed irregularly upon them: Theſe are of a ſpear-ſhaped figure, about two, and ſometimes three inches long, and three quarters of an inch broad, and

are

are as soft as satin to the touch. Neither is summer the only time the leaves afford us pleasure: They continue on the tree great part of the winter; so that the effect they cause, when other trees are despoiled of their honours, may be easily conceived. The flowers appear in July, but make no figure: They are small, and come out at the footstalks of the leaves; their colour is white, and they are possessed of a strong scent. The fruit that succeeds them much resembles a small olive.

This shrub has a *Variety*, with yellow flowers.

The culture of both the sorts is very easy. They are PROPAGATED by cuttings, which must be of the last summer's shoot. But in order to have them proper for the purpose, a sufficient number of trees must be fixed on, from which the family is to be encreased. They must be headed near the ground in the winter; which will cause them to make strong shoots the succeeding summer, and these shoots afford the cuttings. They should be taken off in the autumn, and cut into lengths of about a foot each, three parts of which should be set in the ground. They may be planted very close, and in the autumn following removed into the nursery, where they should be set a foot asunder, and two feet distant in the rows; or, if there be ground enough, they may be planted thinner, and so will want no removing until they be finally set out. The best soil for these cuttings is a rich garden mould, inclined to be moist, and lying in a shady place; in such a soil and situation almost every cutting will grow. The tree itself is exceedingly hardy, and will afterwards shoot vigorously, in almost any soil or station.

E P H E D R A.

LINNEAN Class and Order, *Dioecia Monadelphia* Male flower, containing seven stamina connected at the base, and female flowers containing two pistils, situated upon distinct plants. There are two SPECIES; one of them of a hardy nature:

E'PHEDRA *Distá'chya:* The EPHEDRA, or SHRUBBY HORSE TAIL; *a sub-evergreen shrub*; native of rocky mountains, near the sea coast of Italy, France, and Spain.

The EPHEDRA will grow to three, four, five, or six feet high, according to the nature of the soil in which it is placed; for if it be a fat moist soil, it will arrive to double the height it will attain in that of a contrary nature, and will be more tree-like; it will also have much larger leaves, and be more beautiful. The bark on the old stem is rough, and of a dark, dirty colour. These stems or branches are few; but they have joints at short intervals. Many of them are protuberant, and send forth younger shoots and leaves in prodigious plenty, so as to cause the shrub to have a close bushy look. The older branches will have bark that is smooth, and of a brown, reddish, or yellowish colour; whilst that on the younger shoots will be of a fine green. The larger branches are jointed and hollow, though they have sometimes in them a kind of reddish pith; those send forth smaller, which are called the leaves. These leaves are jointed, grow opposite by pairs, are alternately produced at every joint in opposite directions, and will thus branch out in a singular and horse tail manner, in a suitable soil, to a great length. The leaves and shoots of this shrub being bruised in the winter, emit a very fetid disagreeable scent; but in the spring when the juices begin to flow, they are possessed of a different quality, emitting a fine odour, by many supposed or fancied to be like that of the pine apple; and on account of this scent alone, in the spring, this tree is by many much coveted and admired. The flower buds will appear in May opposite at the sides of the joints; they grow by pairs, and by the middle of June will be in full blow, each standing on very short green footstalks. Male and female flowers will be found on different plants; they are small, and of a yellow colour, and afford pleasure only to the nice observer of the wonderful structure of the minute parts of the vegetable world. This shrub should always have a moist, fat soil; and in those places it will appear more luxuriant and beautiful. It is very hardy, and, although it has been used to be preserved in pots in greenhouses, will

will bear the cold of our ſevereſt winters In the winter the leaves, or rather the young ſhoots or joints, are of a dark, duſky green; but as the ſpring approaches, that goes off, and a fine, lively, chearful green poſſeſſes the whole plant. The old leaves fall off the latter end of April, or beginning of May; at which time the tree will ſend forth young ones, and will continue to do ſo until late in the autumn.

This ſhrub is very eaſily PROPAGATED; it will, indeed, propagate itſelf in great plenty, eſpecially if planted in a light, moiſt ſoil: ſo that where a quantity is wanted, ſome plants are to be procured for breeders; and theſe being planted in good light earth, will ſoon ſpread their roots, and produce plenty of *ſuckers*, which may be taken off, and planted in the nurſery ground, to gain ſtrength, for a year or two; or they may be immediately, eſpecially the ſtrongeſt plants, finally ſet out. As theſe ſhrubs naturally ſpawn, and produce ſuckers in great plenty, after they are planted out in the ſhrubery quarters, the ſpawn ſhould be every year taken off, and the ground dug about the roots; otherwiſe they will not only appear rambling and irregular, but they will diminiſh the beauty of the mother plants, which will by no means appear to be luxuriant and healthy.

This plant merits a place amongſt evergreens, rather for the ſake of variety, or as a foil to more elegant ſpecies, than for any intrinſic beauty or elegance of its own. MILLER ſays, it rarely flowers in gardens.

EUONYMUS.

LINNEAN Claſs and Order, *Pentandria Monogynia*: Each flower contains five males and one female. There are four SPECIES: three of which are cultivated in this country; one of them, however, requires a ſtove heat: The other two are,

1. EUONYMUS *Europæus*: The COMMON or EUROPEAN EUONYMUS, or the SPINDLE TREE; *a deciduous*

eiduous ſhrub; native of ſome parts of England, and of Europe in general.

2. EUO'NYMUS *America'nus:* The EVERGREEN or AMERICAN EUONY'MUS; or the EVERGREEN SPINDLE; *an evergreen ſhrub*; native of Virginia, Carolina, and other parts of North America.

1. The COMMON EUONYMUS. There are of this ſpecies five *Varieties:*

The Deep Red-berried Narrow-leaved Spindle Tree.
—— Pale Red-berried Spindle Tree.
—— White-berried Narrow-leaved Spindle Tree.
—— Broad leaved Spindle Tree.
—— Variegated Spindle Tree *.

The *Narrow-leaved Spindle Tree* will grow to be ſixteen or eighteen feet high, will aſpire with an upright ſtem to a conſiderable height, naturally forming itſelf into a regular head. The bark of the ſtem is of a dark brown; but that of the firſt and ſecond year's ſhoots is ſmooth, and of a fine green, the White-berried ſort eſpecially, which differs from the Red-berried in this reſpect, as the ſhoots of that are browner. The leaves are ſpear-ſhaped, of a fine deep green colour, about three inches long, and an inch and a half broad, very ſlightly ſerrated, and placed nearly oppoſite on the branches. The flowers have little beauty to recommend them: They are ſmall, and of a greeniſh colour, produced in ſmall bunches from the ſides of the branches, the latter end of May, the bunches hanging on long footſtalks; and are ſucceeded by fruit, which conſtitutes the greateſt beauty of theſe plants. The ſeeds are of a delightful ſcarlet; four are contained in each veſſel; and theſe opening, expoſe them to view all over the head of the plant, ſome juſt peeping out of their cells, others quite out, and ſticking to the edge; and theſe veſſels being in bunches on long pendulent footſtalks, have a look which is ſingularly beautiful. The ſeed veſſels of the firſt-mentioned ſort are of the ſame deep ſcarlet with the ſeeds; thoſe of the ſecond,

* MILLER makes the two laſt diſtinct *Species*; but HANBURY ſays, "I have raiſed thouſands of them for ſale (there being hardly any ſhrub more called for), and ever found the ſeeds of the Broad-leaved Spindle Tree to come up the Common Narrow-leaved ſort."

of

of a paler red; those of the third are white, which, together with the twigs of the latter being of a lighter green, constitute the only difference between these sorts; for the seeds themselves of all the sorts are of a deep scarlet.

The *Broad-leaved Spindle Tree* is a *Variety* of the Common Spindle Tree, though it will grow to a greater height than either of the other sorts. It will arrive at near five and-twenty feet high; and the branches are fewer, and the leaves broader. The young shoots are smooth, and of a purplish colour; and the buds at the ends of them, by the end of October, will begin to be swelled, and be near an inch long, preparing for the next year's shoot. The leaves are much larger than those of the other sorts, being, on a thriving plant, near five inches long and two broad. Their figure is like the other, though rather inclined to an oblong oval: Some are most slightly serrated, of a light green, stand opposite by pairs, and fall off much sooner in the autumn, before which their colour will be red. The flowers make an inconsiderable figure, though they are rather larger than the other sorts: The seeds that succeed them with their vessels also are proportionably larger; and many of the common footstalks to each bunch will be four inches, which causes a more noble look in the autumn; though the others are equally pleasing, as the flowers are produced on the Narrow-leaved sorts in greater plenty: Add to this, the berries of the Broad will fall off long before the others.

The wood of the Common Spindle Tree is spoken of by MILLER and HANBURY as being very valuable. The musical instrument makers, say they, use it for keys of organs, and other purposes. Toothpicks, skewers, and *spindles* of the best kind are also made from this wood; hence *Spindle Tree*.

There is but one good method of PROPAGATING the Common Spindle Tree and that is by seeds; though it may easily be done by layers or cuttings; for if the young shoots be laid in the ground in the autumn, they will have struck root by the autumn following; and if cuttings are planted in the autumn in a moist rich earth, that is shaded, many of them will grow; but neither of these methods will produce such fine upright plants,

plants, or that will grow to ſuch a height as thoſe raiſed from ſeeds, though they will be every whit as prolific of flowers and fruit. Whoever has not the convenience of procuring the ſeeds, let him improve theſe hints, if he has got a plant or two, which will be ſufficient for his purpoſe: Whoever can get the ſeeds, had better never attempt thoſe arts. The ſeeds ſhould be ſown in the autumn, ſoon after they are ripe. They will thrive in almoſt any ſoil or ſituation, if it be made fine, and clear of the roots of all weeds, &c. though if it be a fine garden mould, it will be the better. They ſhould be ſown three fourths of an inch deep. It ſeldom happens that more than a few odd plants come up the firſt ſpring; the beds muſt, therefore, remain untouched until the ſpring twelvemonth after ſowing; only conſtant weeding muſt be obſerved. At that time the plants will come up very thick, and all the ſummer they muſt be weeded. In this ſeed bed they may ſtand two years, and be then planted out in the nurſery, where they may remain, with no other care than weeding and digging between the rows in winter, until they are finally planted out.

The Broad-leaved ſort will take very well by budding it on the Common. The ſtocks for this purpoſe ſhould be planted out when they are one year's ſeedlings, and by the ſummer twelvemonth after they will be fit for working; ſo that whoever has young plants of the Common ſort, and only one of the other, may encreaſe his number this way.

2. The Evergreen Euonymus. Beſides the genuine ſpecies, there is a *Variety* of the *Americanus*, having its leaves beautifully ſtriped with yellow. Theſe ſorts grow to the height of about ſeven feet. The branches are ſlender, covered with a ſmooth green bark, and grow oppoſite by pairs at the joints. The leaves alſo grow oppoſite, are ſpear-ſhaped, and have a ſtrong midrib running their whole length. The upper ſurface is of a fine ſtrong green colour, but their under is paler. They are ſmooth, are lightly indented, acutely pointed, and juſtly entitle this ſhrub to be called a fine evergreen. The flowers are produced in July, from the ſides and ends of the branches, in ſmall bunches. They make no great ſhow; but they

they will be ſucceeded by rough, warted, red, five-cornered capſules, containing the ſeeds.

This ſpecies is to be PROPAGATED in the ſame manner as the other ſorts. 1. The beſt way is from ſeeds, which we receive from Virginia. Theſe will be two, and ſometimes three years before they appear; ſo that a perſon ſhould not be too haſty in diſturbing the beds; and after this precaution, what has been already ſaid relating to the management of raiſing the common ſorts of Spindle Trees from ſeeds, muſt conſtantly be obſerved in this ſpecies. 2. By layers alſo, and cuttings, it may be encreaſed; but when the latter way is to be practiſed, it will be proper to plant each cutting ſeparately in a ſmall pot, and plunge them into a bark bed, otherwiſe it is very ſeldom that they will grow. After they have taken root, the pots may be ſet in the natural mould up to the rims for about two years; then the plants ſhould be turned out into the places where they are to remain, and they will be ſure of growing.

F A G U S.

LINNEAN Claſs and Order, *Monoecia Polyandria*: Male flowers and female flowers upon the ſame plant; the males containing about twelve ſtamina, and the females three piſtils each: There are three SPECIES:

1. *Fa'gus Sylva'tica:* The BEECH; a well known *tall deciduous tree*; common in England and moſt parts of Europe, alſo in Canada.

2. *Fa'gus Caſta'nea:* The CHESNUT, or the SPANISH or SWEET CHESNUT; *a tall deciduous tree*; natural to the mountainous parts of the South of Europe.

3. *Fa'gus Pu'mila:* The DWARF CHESNUT, or the CHINQUEPIN; *a deciduous ſhrub*; native of North America.

1. The BEECH. In ſtatelineſs, and grandeur of outline, the Beech vies with the Oak. Its foliage is peculiarly

liarly soft and pleasing to the eye; its branches are numerous and spreading; and its stem waxes to a great size. The bark of the Beech is remarkably smooth, and of a silvery cast; this, added to the splendor and smoothness of its foliage, gives a striking neatness and delicacy to its general appearance. The Beech therefore, standing singly, and suffered to form its own natural head, is highly *ornamental*; and its leaves varying their hue as the autumn approaches, renders it in this point of view still more desirable. In point of actual *Use* the Beech follows next to the Oak and the Ash: it is almost as necessary to the cabinet makers and turners (especially about the Metropolis), as the Oak is to the ship builder, or the Ash to the plough and cart wright. EVELYN nevertheless condemns it in pointed and general terms; because "where it lies dry, or wet and dry, it is exceedingly obnoxious to the worm:" He adds, however, "but being put ten days in water, it will exceedingly resist the worm." The natural soil and situation of the Beech is upon dry, chalky, or limestone heights: It grows to a great size upon the hills of Surry and Kent; as also upon the declivities of the Cotswold and Stroudwater hills of Gloucestershire, and flourishes exceedingly upon the bleak banks of the Wye, in Hereford and Monmouth shires, where it is much used in making charcoal. In situations like those, and where it is not already prevalent, the Beech, whether as a timber tree or as an underwood, is an object worthy the planter's attention. The Beech also thrives abundantly on the thin soiled slatestone hills of Devonshire, and not less among the granite rocks of the Highlands of Scotland. It has lately been discovered, that the wood of the Beech is very durable in water.

The method of PROPAGATING the Beech is from seeds. EVELYN is brief upon this head. For woods, he says, the Beech must be governed as the Oak:—In nurseries, as the Ash; sowing the masts "in autumn, or later, even after January, or rather nearer the spring, to preserve them from vermin, which are very great devourers of them. But they are likewise to be planted of young seedlings to be drawn out of the places where the fruitful trees abound." MILLER says, the season for sowing

ing the masts " is any time from October to February, only observing to secure the seeds from vermin when early sowed, which if carefully done, the sooner they are sown the better, after they are fully ripe." HANBURY orders a sufficient quantity of masts to be gathered about the middle of September, when they begin to fall: These are to be " spread upon a mat in an airy place six days to dry; and after that you may either proceed to sow them immediately, or you may put them up in bags in order to sow them nearer the spring; which method I would rather advise, as they will keep very well, and there will be less danger of having them destroyed by mice or other vermin, by which kinds of animals they are greatly relished." They must be sown in beds properly prepared (as directed under the article PROPAGATION FROM SEED, in the Introductory Part of this Work) about an inch deep. In the first spring many of the young plants will appear, whilst others will not come up till the spring following. Having stood two years in the seminary, they should be removed to the nursery, where they may remain till wanted. More is said of the Beech under WOODLANDS.

2. The CHESNUT This is a tree of the first magnitude; growing to a great height, and swelling to an immense size Mr. Brydone, in a Tour through Sicily and Malta, measured the ruins of a celebrated Chesnut, called *Castagno de Cento Cavalli*, standing at the foot of Mount Etna, and made it " two hundred and four feet round!" The largest we know of in this country stands at Tortworth, near Berkeley, in Gloucestershire. Sir Robert Atkins, in his History of Gloucestershire, says, " By tradition, this tree was growing in King John's reign;" and Mr. Marsham calculates it to be " not less than eleven hundred years old." Sir Robert makes it nineteen yards, and Mr. Marsham forty-six feet six inches in circumference. With great deference however to the authority and veracity of these gentlemen, we have every reason to believe that what is called the Tortworth Chesnut is not one, but two trees: supposing them to be only one, its dimensions are by no means equal to what are given above. We have the highest opinion of Mr. Marsham's ingenuousness and accuracy; and fortunately, in this case, he has furnished

nished us with a proof of his candour, in saying, "As I took the measure in a heavy rain, and did not measure the string till after I returned to the inn, I cannot so well answer for this as the other measures." We will venture to add, that had the day been fine, and Mr. Marsham had viewed the *field* side as well as the *garden* side of this venerable ruin; had he climbed upon the wall, and seen the gable of the old building, adjoining, clasped in between the two stems; and had further ascended to the top of the old stump, which is not more than twelve feet high, and, looking down its hollowness, seen its cavity tending not to the centre of the *congeries*, but to the centre of the *old Tree*, we are convinced he would not have suffered so inaccurate an account to have been published with his signature, as that which appears in page 81 of the First Volume of Papers of the Bath Agriculture Society. The leaves of the Chesnut are long, somewhat large, strongly marked by the nerves, and of a dark and somewhat glossy appearance, in summer; but, in autumn, change to a yellow hue. In open uncrouded situations, the Chesnut throws out large spreading arms, forming a magnificent strongly-featured outline; whilst in a close planted grove the stem will shoot up clean and straight as an arrow to a great height.

As an *Ornamental*, the Chesnut, though unequal to the Oak, the Beech, and the Esculus, has a degree of greatness belonging to it which recommends it strongly to the gardener's attention. Its *Uses* have been highly extolled; and it may deserve a considerable share of the praise which has been given it. As a substitute for the Oak, it is preferable to the Elm: For door jambs, window frames, and some other purposes of the house carpenter, it is nearly equal to Oak itself; but it is liable to be *shakey*, and there is a deceitful brittleness in it which renders it unsafe to be used as beams, or in any other situation where an uncertain load is required to be borne. It is universally allowed to be excellent for liquor casks; as not being liable to shrink, nor to change the colour of the liquor it contains: it is also strongly recommended as an underwood for hop poles, stakes, &c. Its fruit too is valuable, not only for swine and deer, but as a human food: Bread is said to have been

been made of it. Upon the whole, the Chesnut, whether in the light of ornament or use, is undoubtedly an object of the planter's notice.

While young and in full growth, the Chesnut affords wood of a very superior quality, for many purposes. Its shakeyness and brittle texture arise, in some measure, from its being suffered to stand too long. For hop poles no wood is equal to it: for gate posts, or any work where it is proper that timber should be placed in contact with the ground, the wood of the Chesnut is found preferable to any other, except those of the Yew and the Larch.

The PROPAGATION of the Chesnut is chiefly from seeds: EVELYN says, "Let the nuts be first spread to sweat, then cover them in sand; a month being past, plunge them in water, and reject the swimmers; being dried for thirty days more, sand them again, and to the water ordeal as before. Being thus treated until the beginning of spring, or in November, set them as you would do Beans; and, as some practise it, drenched for a night or more in new milk; but with half this preparation they need only to be put into the holes with the point upmost, as you plant tulips."—"If you design to set them in winter or autumn, I counsel you to inter them in their husks, which being every way armed, are a good protection against the mouse, and a providential integument."—"Being come up, they thrive best unremoved, making a great stand for at least two years upon every transplanting; yet if needs you must alter their station, let it be done about November."—Thus far EVELYN. MILLER cautions us against purchasing foreign nuts that have been kiln dried, which, he says, is generally done to prevent their sprouting in their passage; therefore, he adds, "if they cannot be procured fresh from the tree, it will be much better to use those of the growth of England, which are full as good to sow for timber or beauty as any of the foreign nuts, though their fruit is much smaller." He also recommends preserving them in sand, and proving them in water. In setting these seeds or nuts, he says, "The best way is to make a drill with a hoe (as is commonly practised for kidney beans) about fout inches deep, in which you ſhould place the nuts, at about four inches diſtance, with their eye uppermoſt; then draw the earth over them with a rake, and make a ſecond drill at about a foot diſtance from the former, proceeding as before, allowing three or four rows in each bed."—" In April" (he does not mention the time of ſowing) " theſe nuts will appear above ground; you muſt therefore obſerve to keep them clear from weeds, eſpecially while young: in theſe beds they may remain for two years, when you ſhould remove them into a nurſery at a wider diſtance. The beſt time for tranſplanting theſe trees is either in October, or the latter end of February, but October is the beſt ſeaſon: the diſtance theſe ſhould have in the nurſery is three feet row from row, and one foot in the rows. If theſe trees have a downright tap root, it ſhould be cut off, eſpecially if they are intended to be removed again; this will occaſion their putting out lateral ſhoots, and render them leſs ſubject to miſcarry when they are removed for good. The time generally allowed them in the nurſery is three or four years, according to their growth, but the younger they are tranſplanted the better they will ſucceed. Young trees of this ſort are very apt to have crooked ſtems; but when they are tranſplanted out and have room to grow, as they increaſe in bulk they will grow more upright, and their ſtems will become ſtraight, as I have frequently obſerved where there have been great plantations."—HANBURY follows MILLER almoſt literally; except that he mentions February as the time of ſowing; and recommends that the young plants, a year after they have been planted in the nurſery, be cut down to within an inch of the ground; which, he ſays, " will cauſe them to ſhoot vigorouſly with one ſtrong and ſtraight ſtem." There is one material objection againſt ſowing Cheſnuts in drills, which are well known to ſerve as guides or conductors to the field mouſe, who will run from one end to the other of a drill without letting a ſingle nut eſcape her: we rather recommend ſetting them with a dibble, either promiſcuouſly or a quincunx, at about ſix inches diſtance.

EVELYN

EVELYN ſays, that coppices of Cheſnuts may be thickened by layering the tender young ſhoots; but adds, that "ſuch as ſpring from the nuts and marrons are beſt of all." There is a ſtriped-leaved variegation which is continued by budding; and the French are ſaid to graft Cheſnuts for their fruit; but MILLER ſays, ſuch grafted trees are unfit for timber.

The Cheſnut will thrive upon almoſt any ſoil which lies out of the water's way; but diſaffects wet moory land *. See more of this tree under WOODLANDS.

3. The DWARF CHESNUT grows to about eight or ten feet high. The ſtem is of a brown colour, and divides into ſeveral branches near the top. The leaves are of an oval, ſpear-ſhaped figure, acutely ſerrated, with a hoary caſt on their under ſide. The flowers come out in the ſpring, in ſlender knotted catkins: They are of a greeniſh yellow colour, and are very ſeldom ſucceeded by ripe ſeeds in England. This tree is hardy, and thrives beſt in a moiſt ſoil and ſhady ſituation.

The method of PROPAGATING the Dwarf Cheſnut is from ſeeds, which we receive from America. Theſe ſhould be planted in drills, as ſoon as they arrive, in a moiſtiſh bed of rich garden mould. If the ſeeds are good, they will come up pretty ſoon in the ſpring. After they appear, they will require no trouble, except keeping them clean from weeds, and watering them in dry weather. They may ſtand in the ſeed bed two years, and be afterwards planted in the nurſery ground at a foot aſunder and two feet diſtance in the rows; and here when they are got ſtrong plants, they will be fit for any purpoſe.

* The Cheſnut flouriſhes in the Vallies of the Highlands; particularly at Taymouth and Dunkeld. It ſeems to diſlike ſtiff cold land.

FRAXINUS.

LINNEAN Claſs and Order, *Polygamia Dioecia:* Hermaphrodite flowers and female flowers upon diſtinct plants; the former containing two males and one female each; the latter one piſtillum only: There are three SPECIES.

1. *Fra'xinus Excé'lſior:* The COMMON ASH; *a well known tall deciduous tree*; common throughout England and moſt parts of Europe.

2. *Fra'xinus O'rnus:* The FLOWERING ASH; *a low deciduous tree*; native of Italy and other ſouthern parts of Europe.

3. *Fra'xinus America'na:* The AMERICAN ASH; *a low deciduous tree*; native of Carolina and Virginia.

1. The COMMON ASH is one of the loftieſt of our foreſt trees. In a cloſe grove and in a ſoil it affects it lengthens out into a beautifully clean ſtem, and riſes to an aſtoniſhing height: But ſtanding ſingly, it throws out large arms, forms a full ſpreading head, and ſwells out into a ſtem proportionable: Mr. Marſham mentions a very flouriſhing one, growing in Benel churchyard, three miles north of Dunbarton, in Scotland, which, in 1768, meaſured, at five feet high, ſixteen feet nine inches in circumference. The leaves of the Aſh, too well known to require deſcription, are amongſt the laſt which foliate in the ſpring, and amongſt the firſt which fall in autumn. This alone depreciates its value very much as an *Ornamental*, eſpecially near gardens and gravel walks: and planted ſingly or in hedges, it becomes an utter nuiſance in the neighbourhood it ſtands in: every huſbandman knows the injury it does to corn; and there are few dairy-women who are not well acquainted with the evil effects of its leaves, in autumn, upon the produce of the dairy; beſides, being large and numerous, they foul and injure the after-graſs by rotting amongſt it. Cloſe groves are the only proper ſituation for the Aſh; its uſes require a length and cleanneſs of grain; and it would be well for the occu-

piers

piers of land, and, indeed, for the community at large, if a ſevere penalty was laid upon planting it in any other ſituation. To enumerate the *Uſes* of the Aſh would require a ſeparate volume: in this point of view it undoubtedly ſtands next to the Oak. The Farmer would find it difficult to carry on his buſineſs without it: and indeed, the cooper and the coachmaker would be equally at a loſs with the wheelwright, ſhould a ſcarcity of Aſh take place; and we know of no ſpecies of timber ſo likely to be worn out in this country as the Aſh. The juſt complaints of the Huſbandman are expelling it very properly from our hedges; and we are concerned to ſee, amongſt the numerous plantations which have of late years been made, ſo few of this neceſſary tree: it is therefore more than probable that no tree will pay better for planting; not, however, in ſingle trees and hedge-rows, but in cloſe plantations, in the manner which is pointed out under WOODLANDS.

The method of PROPAGATING the Aſh is from ſeeds; which are peculiarly prone to vegetation, and frequently catch under or near the tree they are produced upon, from whence tolerable plants may ſometimes be collected; but in general they are either cropt by cattle, or are drawn up ſlender and ill rooted, and ſeldom make ſo good plants as thoſe raiſed by the gardener's aſſiſtance in a prepared ſeed bed. EVELYN directs us to gather the keys from a young thriving tree in October or November, and having laid them to dry, ſow them "any time betwixt then and Chriſtmas; but not altogether ſo deep as your former maſts" (meaning thoſe of Beech, Hornbeam, &c.). "Thus they do in Spain, from whence it were good to procure ſome of the keys from their beſt trees." He recommends the young plants' ſtanding two years in the ſeminary, and cautions us, in removing them into the nurſery, "not to cut their head at all, which being young is pithy, nor by any means the fibrous part of the roots; only that downright or tap root, which gives our huſbandmen ſo much trouble in drawing, is to be totally abated; but this work ought to be in the increaſe of October or November, and not in the ſpring. We are, as I told you, willing to ſpare his head rather than the ſide branches (which whilſt young may be cut cloſe), becauſe being yet young, it is but of a ſpungy ſubſtance; but being once fixed, you may cut him as cloſe to the earth as you pleaſe; it will cauſe him to ſhoot prodigiouſly, ſo as in a few years to be fit for pike ſtaves."—"Young aſhes are ſometimes in winter froſt-burnt, black as coals; and then to uſe the knife is ſeaſonable, though they do commonly recover of themſelves ſlowly." He adds, "You may accelerate their ſpringing by laying the keys in ſand, and ſome moiſt earth, *ſtratum ſuper ſtratum*;" but does not ſay that this preparation will cauſe them to vegetate the firſt ſpring. MILLER ſays, "the ſeeds ſhould be ſown as ſoon as they are ripe, and then the plants will come up the following ſpring; but if the ſeeds be kept out of the ground till ſpring, the plants will not come up till the year after."—"If they make good progreſs in the ſeed bed, (he ſays) they will be fit to tranſplant by the following autumn,"——"as ſoon as their leaves begin to fall." Great care (he ſays) is neceſſary in taking them up: they ſhould not be drawn, but taken up with a ſpade; clearing the whole bed at once, placing the larger together in rows, and the ſmaller by themſelves. "The rows ſhould be three feet aſunder, and the plants a foot and a half diſtance in the rows: in this nurſery they may remain two years, by which time they will be ſtrong enough to plant where they are to remain; for the younger they are planted the larger they will grow."——HANBURY is very deficient upon the ſubject of raiſing Aſhes in the nurſery way: he does not even tell us the depth at which the keys are to be ſown; nor, except in general terms, when they are to be ſown; namely, "ſoon after they are gathered."——We beg leave, however, to differ from theſe three great authorities. Inſtead of ſowing the keys in autumn, preſently after they are gathered, we venture to recommend their being ſown in the ſpring, in the firſt favourable opportunity in February or March; for being ſown in autumn ſome few may, and in general will, vegetate the firſt ſpring, whilſt much the greateſt part will lie in the ground until the ſpring following: the few that come up will be an incumbrance upon the beds, and will render the expence of clearing them the firſt ſummer unne-

unneceſſarily great; whereas, on the contrary, if the ſowing be deferred until ſpring, the hoe and rake will have free range over the beds, and the expence of cleaning them the firſt ſummer will be comparatively trifling. If the keys be well cured by ſpreading them thin in an airy place, and keeping them turned for a few days after gathering, they may be kept in a heap (moving them now and then) until ſpring, with ſafety. The depth proper for ſowing Aſhen keys is from one inch to an inch and a half or two inches, according to the ſtiffneſs or the lightneſs of the ſoil of the ſeed bed. If they be ſown too deep in a cloſe-textured ſoil, they will be apt to be ſmothered; and if too ſhallow in a porous one, the drought has too much power over them, and they are liable to be diſturbed by the hoe and rake in clearing them the firſt ſummer after ſowing. Upon the approach of the ſecond ſpring, the ſurface of the beds ſhould be made as light and pulverous as poſſible, in order to give to the embryo plants a free admiſſion of air, and to facilitate their riſing: if part of the ſtale mould be raked off, and a little freſh earth be ſifted over in its ſtead, it will add conſiderable vigour to the young plants; which may be removed into the nurſery whenever the croudedneſs of the beds, the ſtrength of the plants, or the conveniency of the planter, may render it requiſite. For raiſing groves of Aſh ſee WOODLANDS.

There are three *Varieties* of the Common Aſh: The *Silver-ſtriped*; the *Gold-ſtriped*; and the *Yellow-coloured Aſh*. Theſe *Varieties* may be continued by *budding*.

EVELYN tells us, "that Aſh may be propagated from a bough *ſlipt* off with ſome of the old wood, a little before the bud ſwells, but with difficulty by *layers*.

The Aſh will thrive in almoſt any ſoil; but delights moſt in a moiſt ſituation, ſo that it ſtand above the level of ſtagnant water; in marſhes, half-drained bogs, and by the ſides of rivers, it flouriſhes extraordinarily, outgrowing even many of the aquatics themſelves.

2. The FLOWERING ASH. Of this ſpecies there are two kinds or *Varieties*: The *Virginia Flowering Aſh*;—and the *Dwarf Aſh of Theophraſtus*.

The *Virginia Flowering Aſh* when in blow is inferior in beauty to few of our flowering trees. It will grow to near thirty feet in height. The branches of this ſort, in the winter, have nearly the ſame appearance with the Common; only they are, eſpecially the youngeſt, more inclined to a black caſt: The buds alſo, which will begin to ſwell in the autumn, are of that hue. The branches will not burn, when green, ſo well as thoſe of the Common Aſh. The leaves are of a fine green, ſmooth, ſerrated, and conſiſt of about three or four pair of folioles, placed a good way aſunder along the mid-rib; and they are uſually terminated by an odd one, The midrib is long, but not ſtraight; ſwelling where the leaves, which fall off early in the autumn, come out. The flowers are white, produced in May, in large bunches, at the ends of the branches. HANBURY ſays, "I have had this tree, the ſecond year from the bud, produce, on the leading ſhoot, a tuft of flowers; and although this is not common, yet, when it gets to be about ten feet high, almoſt every twig will be terminated with them The flowers exhibit themſelves not in a gaudy dreſs, but in a looſe eaſy manner, all over the tree, which, together with the green leaves peeping from amongſt this white bloom, makes the appearance extremely pleaſing. I have never yet known the flowers to be ſucceeded by ſeeds."

Dwarf Aſh of Theophraſtus is, as the name imports, a low tree for the Aſh tribe; about fourteen or fifteen feet is the height it generally aſpires to. The branches are ſmooth, and of a darkiſh green. The leaves are pinnated, of a dark green, and ſerrated on the edges, but proportionably ſmaller than thoſe of the Common Aſh. The flowers of this ſort make no ſhow, though they are poſſeſſed of the petals neceſſary to complete a flower, which are denied the Common Aſh.

3. AMERICAN ASH. The *Varieties* of this ſpecies are, *Manna Aſh*, *White Aſh*, *Red Aſh*, *Black Aſh*, and *New-diſcovered Aſh*.

Manna Aſh will grow to about twenty feet high. It will ſometimes ſhoot eight feet the firſt year from the bud, though it ſeldom ſhoots more than two feet in a ſummer afterwards. The bark of the young ſhoots is ſmooth, of a browniſh green, and has a few grayiſh ſpots. The leaves are compoſed of four or five pair of folioles, placed on a ſtraight midrib; they are of a fine pleaſant

pleaſant green, and more acutely and deeply ſerrated than any of the other ſorts. The flowers make no ſhow: They are partly the colour of thoſe of the Common Aſh, and are produced, like them, early in the ſpring, before the leaves appear.

White Aſh is ſo called from the whitiſh colour of the young branches in winter. They are ſpotted all over with many white ſpots, which makes their colour that of a lightiſh gray. This ſort will arrive to about thirty feet high; and the branches are ſtrong, and produced in an irregular manner. The folioles which compoſe the leaves are of a light green, and obtuſely ſawed on the edges: they ſeldom conſiſt of more than three pair, with the uſual odd one, which has a long point; and theſe are placed far aſunder, on the midrib. Theſe leaves fall off early in the autumn, when they are of a light colour: This together with the gray branches make the tree have a whitiſh look. The flowers are produced in the ſpring, and make no ſhow. This ſort is commonly called the New-England Aſh.

Red Aſh. The Red Aſh is a ſtronger ſhooting tree than any of the former, the Common Aſh excepted. The branches which are fewer, are ſmooth, and the young ſhoots are of a reddiſh colour in the autumn. The leaves of this ſort make the moſt noble figure of any of the others; for although they are ſeldom compoſed of more than three pair of folioles, beſides the odd one, yet theſe are exceedingly large, eſpecially the odd one, which will be ſometimes ſix inches long, and three and a half broad. The pair next it, alſo, will be fine and large; though they diminiſh in ſize as they get nearer the baſe of the footſtalk. Theſe folioles are diſtinctly ſawed on their edges, are of a fine light-green during the ſummer, and in the autumn die to a red colour; from which circumſtance, together with that of their red twigs, this ſort takes the denomination of the Red Aſh. It has its ſeeds very broad, and is commonly called the Carolina Aſh.

Black Aſh we receive from abroad by that name; though it is difficult to ſee the propriety of its being ſo called. The colour of the ſhoots is nearly like that of the White Aſh; but they ſhoot ſtronger, and promiſe

to

to form a larger tree. The leaves are large, and ribbed underneath; of a very dark green, and die to a ſtill darker in the autumn. The folioles are not ſo large as thoſe of the Red ſort, but they quit the tree about the ſame time. The keys are very broad, and, when we receive them, of a blackiſh colour.

"*New-diſcovered Aſh* I received from Pennſylvania, where it was diſcovered growing in the woods near Philadelphia. The keys are very ſmall and flat, and come up in a fortnight after being ſown. The young ſhoots of this ſort are covered with the ſame kind of bark as the White Aſh, and the leaves nearly reſemble thoſe of the Black Aſh, though they are not quite ſo large." HANBURY.

All the ſorts of foreign Aſhes are eaſily PROPAGATED. 1. By ſeeds, if they can be procured from abroad. We often have them in February; and if they are ſown directly, they will ſometimes come up the beginning of May, though they generally lie, or at leaſt the greateſt part of them, until the ſpring following. The beds may be made in any part of the garden; and almoſt any ſort of garden mould, made fine, will do for the purpoſe. After the ſeeds are ſown, they will want no other care than weeding, until the plants are a year or two old in the ſeed bed, when they may be taken up, and planted in the nurſery, at the uſual diſtance of a foot aſunder, and two feet in the rows, which will be ſufficient for them until they are finally taken up. 2. Budding is another good method of propagating theſe trees; ſo that thoſe who have not the convenience of a correſpondence in the countries where they grow naturally, ſhould procure a plant or two of a ſort, and raiſe young Aſhes of the Common ſort for ſtocks. Theſe ſtocks ſhould be planted out in the nurſery, a foot aſunder, and two feet diſtant in the rows. When they are one year old, and grown to be about the thickneſs of a bean ſtraw, they will be of a proper ſize for working. A little after Midſummer is the time for the operation; and care muſt be obſerved not to bind the eye too tight. They need not be unlooſed before the latter end of September. In March, the head of the ſtock ſhould be taken off, a little above the eye; and by the

the end of the ſummer following, if the land be good, they will have made ſurpriſing ſtrong ſhoots, many of them ſix feet or more.

GENISTA.

LINNEAN Claſs and Order, *Diadelphia Decandria*: Each flower contains ten males and one female; the males ſtanding in two diviſions: There are fourteen SPECIES; ſeven of which come under our notice:

1. GENISTA *Tridentata*: The PORTUGAL BROOM; *a deciduous ſhrub*; native of Portugal and Spain.

2. GENISTA *Tinctoria*: The DYER's BROOM, or WOODWAXEN; *a low deciduous ſhrub*; native of England and Germany.

3. GENISTA *Piloſa*: The BRANCHING BROOM; *a deciduous ſhrub*; native of Hungary, Germany, and France.

4. GENISTA *Anglica*: The DWARF ENGLISH BROOM, or PETTY WHIN; *a deciduous ſhrub*; natural to moiſt, heathy grounds in ſeveral parts of England.

5. GENISTA *Germanica*: The PRICKLY GERMAN BROOM; *a low deciduous ſhrub*; native of Germany.

6. GENISTA *Hiſpanica*: The PRICKLY SPANISH BROOM; *a deciduous ſhrub*; native of Spain and France.

7. GENISTA *Candicans*: The ITALIAN BROOM, or The CYTISUS OF MONTPELIER; *a low deciduous ſhrub*; native of Italy and about Montpelier in France.

*** For another Claſs of BROOMS, ſee SPARTIUM.

1. The PORTUGAL BROOM is one of the larger growers: It will arrive to be five or ſix feet high: the branches are very ſlender, tough, and for the moſt part three-cornered and jointed. The leaves end in three points, and are ſmall; though ſome of them will be produced by threes, in ſuch a manner as to be entirely trifoliate leaves; whilſt others again are often found ſingle. By the beginning of May, this ſhrub will be in blow. The flowers, which are yellow and of the butterfly

terfly kind, are each very large: They grow from the ſides of the branches, and wings of the leaves, ſingly, on ſhort footſtalks, and are produced in ſo free and eaſy a manner, that they may not improperly be ſaid to have a genteel appearance. They are ſucceeded by pods, in which are contained kidney-ſhaped ſeeds, that will be ripe in autumn.

There are two *Varieties* of this ſpecies of Broom, one with larger, the other with narrower leaves, both of which are ſought after by thoſe who are fond of having great varieties. Theſe ſorts are the leaſt kinds, and require a ſheltered ſituation.

2. The DYER's BROOM. Of this ſpecies there are two varieties, one of which has a narrower leaf, and grows more upright; the other is more ſpreading in its branches. Their natural growth is about two or three feet high, and their branches are taper and channelled, The leaves are of a lance-like figure, and placed alternately on the branches. Theſe branches will produce ſpikes of yellow flowers in June, in ſuch a manner, that though each individual flower is but ſmall for thoſe of the butterfly kind, the whole ſhrub will appear covered with them to the pleaſure of all beholders. Theſe flowers are ſucceeded by pods, which will have ripe ſeeds in the autumn.

3. BRANCHING BROOM, as the name indicates, is a plant whoſe branches ſpread abroad, and decline towards the earth's ſurface. The main ſtalk is beſet all over with tubercles, and the leaves that ornament the ſlender branches are obtuſe and ſpear-ſhaped. The flowers, which are yellow, are produced at the ends of the branches, in ſpikes, in June; and they are exhibited in ſuch profuſion as to make a delightful ſhow. They are ſucceeded by pods that ripen their ſeeds in autumn.

4. DWARF ENGLISH BROOM has many beauties to recommend it to the gardener, though it grows common on many of our barren heaths. In theſe places, it goes by the cant name of *Petty Whin*. All the ſorts of our choiceſt cultivated plants grow wild in ſome parts of the globe, but loſe nothing of their value becauſe they appear thus ſpontaneouſly: Why then ſhould this, becauſe it is common in ſome parts of England, be denied admittance into gardens, eſpecially thoſe that

are

are at a remote diſtance from ſuch places, as it has many natural beauties to recommend it? It is a low plant, ſeldom growing to be more than two feet high; on which account no garden is ſo ſmall but it may be there planted, if the commonneſs of it be no objection to the owner. This ſhrub has ſome ſingle, long ſpines, though the flower branches are entirely free from them. The leaves, like the ſhrub, are proportionally ſmall, of a lanceolated figure, and grow alternately on the branches. The flowers, which are of a fine yellow, are produced the beginning of May, in cluſters, at the ends of the branches; and are ſucceeded by thick ſhort pods, in which the ſeeds are contained.

5. German Prickly Broom will grow to be about a yard high. This ſhrub is armed with many compound ſpines; the branches are ſlender and numerous, though thoſe that produce the flowers are entirely free from ſpines. The leaves of this ſort, alſo, are ſmall, and of a lanceolate figure, and grow alternately on the branches. The flowers are produced in plenty at the ends of the branches, in June: They are of the colour and figure of the others, and are ſucceeded by pods, in which the ſeeds are contained.

6. Prickly Spanish Broom will grow to be five or ſix feet high. This ſhrub is poſſeſſed of many compound ſpines; though the branches that produce the flowers are entirely free from them. The leaves are exceedingly narrow, many of them being no wider than a thread, but very hairy. The flowers are yellow, produced in May, in cluſters, at the ends of the branches; and are ſucceeded by hairy compreſſed pods, in which the ſeeds are contained.

7. Italian Broom riſes, with an erect, ſhrubby, branching, ſtriated ſtalk, to the height of about a yard. The leaves are trifoliate, oval, and hairy underneath. The flowers come out on leafy footſtalks, from the ſides of the branches: They are of a bright yellow colour, appear in June, and are ſucceeded by hairy pods, containing ripe ſeeds, in September.

The beſt way of propagating all theſe ſorts is by ſeeds; and if theſe are ſown ſoon after they are ripe, they will come up earlier in the ſpring, and make better plants by the autumn. They ſhould only ſtand one year

year in the ſeed bed before they are tranſplanted. They ſhould be taken up in the ſpring, and planted out finally, in ground properly prepared for ſuch ſmall plants; for the leſs they are removed, ſo much the faſter will they thrive; as they naturally grow with long ſtrong ſtringy roots, that do not love to be diſturbed; on which account, if places in the plantations were to be marked out, the mould made fine, a few ſeeds of the different ſorts ſown, and ſticks ſet as guides to prevent their being hoed or dug up; plants that have been thus raiſed, without removing, will ſhoot ſtronger, and flower better, than any that have been brought from the ſeed bed or nurſery. After they are come up, if there be too many in a place, the weakeſt may be drawn out, and only two or three of the ſtrongeſt left, which will cauſe them to flower better and ſtronger.

GLEDITSIA.

LINNEAN Claſs and Order, *Polygamia Dioecia*: Hermaphrodite flowers and male flowers upon one plant, and female flowers upon a different plant. There are two SPECIES: one of them a Stove plant; the other

GLEDI'TSIA *Triaca'nthos*: The GLEDITSIA, or TRIPLE-THORNED ACACIA; *a deciduous tree*; native of Virginia and Pennſylvania.

The GLEDITSIA. Its growth is naturally upright, and its trunk is guarded by thorns of three or four inches in length, in a remarkable manner. Theſe thorns have alſo others coming out of their ſides at nearly right angles: Their colour is red. The branches are ſmooth, and of a white colour. Theſe are likewiſe armed with red thorns, that are proportionally ſmaller: They are of ſeveral directions, and at the ends of the branches often ſtand ſingle. The young ſhoots of the preceding ſummer are perfectly ſmooth, of a reddiſh green, and retain their leaves often until the middle of Novem-

November. Although there is a peculiar oddity in the nature and poſition of the ſpines, yet the leaves conſtitute the greateſt beauty of theſe trees: They are doubly pinnated, and of a delightful ſhining green. The pinnated leaves that form the duplication do not always ſtand oppoſite by pairs on the middle rib; the pinnæ of which they are compoſed are ſmall and numerous; no leſs than ten or eleven pair belong to each of them; and as no leſs than four or five pair of ſmall leaves are arranged along the middle rib, the whole compound leaf conſiſts often of more than two hundred pinnæ of this fine green colour: They ſit cloſe, and ſpread open in fine weather; though during bad weather they will droop, and their upper ſurfaces nearly join, as if in a ſleeping ſtate. The flowers are produced from the ſides of the young branches, in July: They are a greeniſh catkin, and make little ſhow; though many are ſucceeded by pods, that have a wonderful effect; for theſe are exceedingly large, more than a foot, ſometimes a foot and a half in length, and two inches in breadth, and of a nutbrown colour when ripe; ſo that the effect they occaſion, when hanging on the ſides of the branches, may eaſily be gueſſed.

There is a *Variety* of this ſpecies, with fewer thorns, ſmaller leaves, and oval pods. It has nearly the reſemblance of the other; though the thorns being not ſo frequent, and the pods being ſmaller, each containing only one ſeed, this ſort loſes that ſingular effect which the other produces by them.

The PROPAGATION of theſe trees is not very difficult. We receive the ſeeds from America in the ſpring, which keep well in the pods, and are for the moſt part good. They generally arrive in February; and, as ſoon as poſſible after, they ſhould be ſown in a well ſheltered warm border of light ſandy earth. If no border is to be found that is naturally ſo, it may be improved by applying drift ſand and making it fine. The ſeeds ſhould be ſown about half an inch deep; and they will for the moſt part come up the firſt ſpring. If the ſummer ſhould prove dry, they muſt be conſtantly watered; and if ſhade could be afforded them in the heat of the day, they would make ſtronger plants by the autumn. A careful attention to this article is peculi-

arly

arly requiſite; for as the ends of the branches are often killed, if the young plant has not made ſome progreſs, it will be liable to be wholly deſtroyed by the winter's froſt, without protection: And this renders the ſowing the ſeeds in a warm border, under a hedge, in a well ſheltered place, neceſſary; for there theſe ſhrubs will endure our winters, even when ſeedlings, and ſo will require no farther trouble; nay, though the tops ſhould be nipped, they will ſhoot out again lower, and will ſoon overcome it. It will be proper to let them remain two years in the ſeed bed, before they are planted out in the nurſery. The ſpring is the beſt time for the work. Their diſtances ſhould be one foot by two; the rows ſhould be dug between every winter; and, being weeded in ſummer, here they may continue with no other particular care, until they are ſet out to remain. Theſe trees are late in the ſpring before they exhibit their leaves, but keep ſhooting long in the autumn.

G L Y C I N E.

Linnean Claſs and Order, *Diadelphia Decandria*: Each flower contains ten males riſing in two diviſions, and one female. There are fourteen Species; five of which have been enured to the open air of this country.

1. Glyci'ne *Frute'ſcens*: The Carolina Kidney Bean; *a climber*; native of Carolina and Virginia.

2. Glyci'ne *A'pios*: The Ash-leaved Milk Vetch; *a climber*; native of Virginia.

3. Glyci'ne *Tomento'ſa*: The Climbing Rest-harrow; *a climber*; native of Virginia.

4. Glyci'ne *Como'ſa*: The Maryland Kidney Bean; *a climber*; natural to moiſt ſhady places in Virginia.

5. Glyci'ne *Mono'ica*: The Virginia Glycine; *a climber*; native of moiſt ſhady places in Virginia.

1. Carolina Kidney Bean does not riſe by the aſſiſtance of claſpers, but by the twining branches, which

which naturally twiſt round any adjacent tree; nay, if trees are ten feet or more diſtance from the root of the plant, its branches, being too weak to ſupport themſelves, will trail along the ground until they reach theſe trees, and then they will twine their branches with theirs, and arrive to a great height: Indeed, where trees are near at hand, and they begin by the firſt ſpring ſhoot to twiſt about them, they will twine up to them to the height of near twenty feet. This climber is poſſeſſed of noble large pinnated leaves, very much like thoſe of liquorice. The folioles are about three pair in number, arranged on their common midrib, and they always end with an odd one. Their colour is for the moſt part of a lightiſh hoary caſt, with a blueiſh tinge. The flowers are very large and ornamental: Their colour is that of a blueiſh purple, and their general characters indicate their ſtructure. They are produced from the wings of the leaves, in July and Auguſt; and are ſucceeded by long pods, like Kidney Beans.

This fine climber is eaſily PROPAGATED, 1. By ſeeds, if there is a conveniency of procuring them from abroad; for they never ripen with us. In the ſpring, as ſoon as we receive them, they ſhould be ſown in fine beds of light ſandy earth, half an inch deep. They will readily come up, and all ſummer muſt have frequent waterings; and if the beds be ſhaded in hot weather, it will be the better. In winter the beds ſhould be hooped, and covered with mats in froſty weather: And in ſpring the ſtrongeſt may be drawn out, which will thin the bed, and make way for the others, which ſhould ſtand until the next ſpring. Plants thus drawn ſhould be ſet in the nurſery, at ſmall diſtances, and in a year or two after they will be good plants for any place where they are wanted. 2. This plant is alſo eaſily encreaſed by layers; for if the young ſhoots of the preceding ſummer be laid in the ground in the autumn, by the autumn following they will have ſtruck root; when the beſt-rooted and ſtrongeſt layers may be planted out to ſtand where they are wanted, whilſt the weaker, or thoſe with hardly any root, may be ſet in the nurſery, like the ſeedlings, to gain ſtrength.

2. The ASH-LEAVED MILK VETCH will twine from ſix to twelve feet high, according to the nature of the ſoil;

ſoil; for in a rich fat mould it will grow near double the length it will in a ſoil of an oppoſite nature. The ſtalks die to the ground every autumn: and in the ſpring new ones are iſſued forth from the roots, which are compoſed of many knobs, that encreaſe in number, the longer the plant is ſuffered to remain. The leaves ſomewhat reſemble thoſe of the Aſh-tree, being pinnated almoſt in the ſame manner. The folioles, which conſiſt of three pair beſides the odd one, are of an oval lanceolate ſhape; and being arranged oppoſite along the midrib, and terminated with a ſingle one form a fine leaf. The flowers are produced from the ſides of its twining ſtalks, in Auguſt. They grow in ſmall ſpikes, are of a reddiſh colour, and being of the butterfly or pea-bloſſomed kind, make a pretty good ſhow. Theſe flowers are ſometimes ſucceeded by pods, which never perfect their ſeeds with us.

3. Climbing Restharrow is but a low plant for a climber, ſeldom ariſing higher than five feet. The ſtalk dies to the ground every autumn; and the loſs is repaired by a natural ſucceſſion preſented from the root every ſpring. The leaves are trifoliate, and very downy. Every one knows the beauty that ariſes from leaves of a hoary nature, amongſt the variety of greens of different tinges. The flowers are of the pea-bloom kind, and are produced in ſhort bunches, in June and July, from the ſides of the ſtalks. They are of a yellow colour, and, though they are rather ſmall in proportion, are very beautiful. They are ſucceeded by pods, in which two ſeeds only are contained, and which will be ripe with us in September.

4. The Kidney Bean Plant of Maryland has a ſlender, annual, twining ſtalk, which will ariſe to be three or four feet high. The leaves are trifoliate, and ſit cloſe to the ſtalks. They are hairy, and the folioles are of an oval lanceolate ſhape; and being of a good green, make the whole ornamental enough. But the greateſt ornament this plant receives is from the flowers, which are alſo of the pea-bloom kind, and are of a clear blue. They are produced in June, from the ſides of the ſtalks, in fine recurved bunches; and theſe are ſucceeded by pods, which will have ripe ſeeds in Auguſt or September.

5. The

5. The VIRGINIAN GLYCINE will ariſe with its ſlender branches to a degree higher than the other. The ſtalks are hairy, and the leaves with which they are ornamented are trifoliate and naked. The flowers are produced from the ſides of the ſtalks, in June and July. They grow in pendulent bunches, and are alſo of the butterfly kind. They are very beautiful, and each exhibits a variety of colours; for the wings and the keel are white, whilſt the ſtandard is of a pale violet colour. Theſe flowers are ſucceeded by compreſſed half-rounded pods, hanging by lengthened peduncles; and the ſeeds will often be ripe in September.

All theſe ſorts are PROPAGATED by the ſeeds; and this may be in the places where they are to remain, or in warm well ſheltered beds or in pots, to be houſed for the firſt winter, if it ſhould prove ſevere. They will very readily come up; and if they are ſown in the open ground, the beds ſhould be hooped at the approach of winter, to be covered with mats, in caſe it ſhould prove bad. It will be proper to plunge thoſe ſown in pots, immediately after, up to the rims in the natural mould; this will keep them cool and moiſt: At the approach of hard froſts, they may be removed into the greenhouſe; and in ſpring may be turned out into the places where they are deſigned to remain. Thoſe in the beds, alſo, ſhould be tranſplanted to ſuch places: Their after-management will be only to part the roots about every three or four years; and by this method alſo they may be all encreaſed. The ſpring is the beſt time for parting the roots; and by this way they may be multiplied faſt enough. As to the firſt ſort, this method is chiefly practiſed for its propagation, as it does not ripen its ſeeds here, unleſs there is a conveniency of procuring them from abroad. The roots of this ſort are compoſed of ſeveral knobs; and theſe being taken up and divided readily grow, and become good plants.

Theſe perennials are all proper to be planted amongſt ſhrubs in warm and well ſheltered places; for they are rather of a tender nature, and are often deſtroyed by ſevere froſts. As the ſtalks are all annual, as ſoon as they decay at the approach of winter, they ſhould be cut up cloſe to the ground, and cleared off ſuch plants as are near them, by which they have aſpired, otherwiſe they will have a dead paltry look, and render the place inelegant; for, even in the dead of winter, neatneſs and elegance muſt be obſerved, which will not only ſhew a more promiſing expectation of a reſurrection, but the clearing away old ſtalks, &c. will be better for the plants themſelves, as they would in ſome degree hinder and choke the young ſhoots as they advance in the ſpring.

GUILANDINA.

Linnean Claſs and Order, *Decandria Monogynia*: Each flower contains ten males and one female: There are five Species; one of which will ſtand our winter.

Guilandi'na *Dio'ica*: The Canada Nickar Tree; *a deciduous tree or ſhrub*; native of Canada.

The Canada Nickar Tree. The ſtem is erect, firm, often twenty feet high, and ſends forth ſeveral branches, which are covered with a ſmooth, bluiſh, aſh-coloured bark. The leaves are bipinnated; and the folioles are large, ſmooth, entire, and ranged alternately on the midrib. The flowers appear in July or Auguſt; but are very rarely ſucceeded by ſeeds in England.

This ſpecies is propagated, 1. By ſeeds, which muſt be procured from the places where the tree naturally grows. The ſeeds are very hard, and often lie two years before they make their appearance; ſo that if they are ſown in common ground, the beds muſt all the time be kept clean from weeds. In the autumn it will be proper to ſtir the ſurface of the mould, but not ſo deep as to diſturb the ſeeds. In the ſpring the plants will come up: All ſummer they muſt be kept clean from weeds, watered in dry weather, and in the autumn the ſtrongeſt may be planted out in the nurſery, at the uſual diſtance, while the weakeſt may remain another year in the ſeed bed to gain ſtrength. The ſeeds alſo may be ſown in pots, and plunged into a hotbed: This will

will bring the plants up the firſt ſpring. After they make their appearance, they muſt be hardened by degrees to the open air. 2. This tree may likewiſe be propagated by layers. Theſe muſt be the ſmaller ſhoots of the laſt year's wood. The operation muſt be performed by making a ſlit, as is practiſed for carnations; and the beſt time for the buſineſs is the autumn. 3. By cutting the root, alſo, this tree may be encreaſed. In order to this, bare away the earth from the top of the root; then with the knife cut off ſome parts of it, leaving them ſtill in the ground, and only directing their ends upwards; Then cover the whole down lightly with mould. The parts that have been ſeparated will ſhoot out from the ends, and come up as ſuckers all round the tree. If dry weather ſhould happen, you will do well to water them all the ſummer; and in the autumn they may be removed to the place where they are deſigned to remain; which ought always to be in a light dry ſoil, in a well ſheltered place.

HAMAMELIS.

LINNEAN Claſs and Order, *Tetrandria Digynia*: Each flower contains four males and two females: There is only one SPECIES:

HAMAME'LIS *Virgi'nica*: The DWARF HAZEL; *a low deciduous ſhrub*; native of Virginia.

The DWARF HAZEL is a ſhrub of about four feet in growth, and will conſtitute a variety among other trees, though there is no great beauty in it, except what is afforded by the leaves. Theſe are placed on the branches, which are numerous and ſlender, in an alternate manner, and much reſemble thoſe of our Common Hazel, that are known to all. The flowers make no ſhow; but perhaps the time of their appearing, which happens in winter, in November or December, when they will be produced in cluſters from the joints of the young ſhoots, may make the plant deſirable to ſome persons. Nothing farther need be said to the gardener concerning this shrub, which Nature seems to have designed for the stricter eye of the botanist; so that we shall proceed to its culture.

It is PROPAGATED, 1. By seeds, which must be procured from America, for they do not ripen here. An easterly border, well defended from the north and westerly winds, is best for their reception; for these plants, when seedlings, are rather tender; when older, they are hardy enough. They will grow in almost any kind of good garden mould, made fine; and they should be covered about half an inch deep. They will not come up before the second, and sometimes the third spring. 2. This tree may also be PROPAGATED by layers; so that whoever has not the conveniency of procuring the seeds from abroad, having obtained a plant or two, may encrease them this way. The operation should be performed on the twigs of the preceding summer's shoot: These should be slit at the joint, and a bit of chip, or something, put in to keep the slit open. If these stools stand in a moistish place, which these shrubs naturally love, and are layered in the autumn, they will have shot root by the autumn following; and may be then either planted out in the nursery, or where they are to remain.

HEDERA.

LINNEAN Class and Order, *Pentandria Monogynia*: Each flower contains five males and one female: There are only two SPECIES:

1. HEDERA *Hélix*: The COMMON IVY; a well known *evergreen climber*; native of England and most parts of Europe.

HEDERA *Quinquefólia*: The DECIDUOUS IVY, or the VIRGINIAN CREEPER; *a deciduous climber*; native of Virginia and Canada.

1. The COMMON EVERGREEN IVY. Besides the genuine

genuine ſpecies there are three *Varieties:* namely, the Yellow-berried Ivy; the Gold-ſtriped Ivy; and the Silver-ſtriped Ivy.

The *Common Ivy* is well known all over England, and how naturally it either trails on the ground, or riſes with walls or trees, ſtriking its roots all along the ſides of the branches for its ſupport. It chiefly delights in old houſes or walls; and when it has taken poſſeſſion of any outſide of the outer buildings, will ſoon cover the whole. It will make ſurpriſing progreſs when it reaches old thatch; and will ſoon, if unmoleſted, climb above the chimney itſelf. Neither are old houſes or walls what it chiefly likes to grow on; for it will ſtrike its roots even into the bark of trees. But above all, it chiefly affects old rotten trees or dodderels; for theſe it will almoſt cover, and rear its head with a woody ſtem above the trunk, and will produce flowers and fruit in great plenty. There, as well as on the ſides of old walls and buildings, it becomes a habitation for owls and other birds. The uſefulneſs of Ivy, then, in gardening, is to overrun caves, grottos, old ruins, &c. to which purpoſe this plant is excellently adapted; and were it not for its commonneſs, it would be reckoned inferior to few evergreens; for the older gray ſtalks look well, whilſt the younger branches, which are covered with a ſmooth bark of a fine green, are very beautiful. The leaves, alſo, are of a fine ſtrong green, are large and bold, and make a variety among themſelves; for ſome are compoſed of lobes, whilſt others are large, and of an oval figure. The flowers are nothing extraordinary, unleſs it be for the figure in which they grow. This is ſtrictly the *Corymbus*; and all flowers growing in ſuch bunches are called by Botaniſts Corymboſe Flowers. The fruit that ſucceeds them, however, is very beautiful; for being black, and growing in this round regular order, and alſo continuing on all winter, it makes the tree ſingular, and, were it not for its commonneſs, deſirable. It is obſervable, if Ivy has no ſupport, but is left to creep along the ground only, it ſeldom flowers; but having taken poſſeſſion of rails, hedges, trees, or buildings, from theſe it ſends out woody branches, which produce the flowers and fruit.

The *Yellow-berried Ivy* differs from the Common Ivy in that its berries are yellow. It grows common in the iſlands of the Archipelago; and is at preſent rare with us. This is the *Hedera Poetica* of old authors.

The *Gold-ſtriped Ivy* is the Common Ivy with yellow blotched leaves; though it is obſervable, that this ſort has very little inclination to trail along the ground, or up trees or buildings, as it naturally riſes with woody branches, and forms itſelf into a buſhy head: So that this ſort may be planted amongſt variegated trees, or evergreens, as a ſhrub. Let it be ſet where it will, it is very beautiful; for the leaves will be a mixture of yellow and green; and ſometimes they will have the appearance of being all yellow, thereby cauſing a very ſingular and ſtriking look at a diſtance

The *Silver-ſtriped Ivy* is a variety of our Common ſort, though the branches are naturally more ſlender. The leaves alſo are ſmaller; and of all the ſorts, this creeps the cloſeſt to walls or buildings, or is of ſtrength ſufficient to form its ligneous branches, when got to the top, to any head. "This plant (continues HANBURY, with whom it ſeems to be a very great favourite) is of all others to be planted againſt walls for ornament; for its leaves are very finely ſtriped with ſtreaks of ſilver, and the ſets being firſt planted at ſmall diſtances, will ſoon cover them all over, ſo as to have a delightful look. A more beautiful ornament to a wall cannot be conceived, than what belongs to a wall of Charles Morris, Eſq of Loddington. It conſiſts of theſe plants, which having firſt taken properly to the ground, and afterwards to the mortar-joints, have ſo overſpread the ſurface as to be a ſight, of the kind, ſuperior to any I ever beheld; and I am perſuaded there are few people of taſte, who had ſeen anything of this nature, but would be induced to have the like, even againſt their choiceſt walls. And here let it always be remembered, that whereas our Common Green Ivy is to hide and keep from view all old and unſightly walls, ſo the Silver-ſtriped Ivy is to ornament all walls, even thoſe of the fineſt ſurface."

2. The DECIDUOUS AMERICAN IVY is a real ſpecies of *Hedera*. It ſheds its leaves in the autumn; and will ſpread itſelf over pales, walls, buildings, &c. in a very little

little time. It puts forth roots at the joints, which faften into mortar of all forts; fo that no plant is more proper than this to hide the unfightly furface of an old barn end, or any other building which cannot be concealed from the view by trees being planted at fome diftance; as in one year it will fhoot often near twenty feet, and, let the building be ever fo high, will foon be at the top of it. The bark on the fhoots is fmooth, and of a brown colour; and the buds in the fpring, as they are beginning to open, will be of a fine red. The leaves are large and well-looking. Each is compofed of five fmaller, which are ferrated at their edges. Their common footftalk is proportionably ftrong, and they die to a fine red in the autumn. It grows with great luxuriance in the atmofphere of the metropolis.

All the forts are to be PROPAGATED by cuttings; for thefe being fet any time in the winter, in almoft any foil, will ftrike root by the autumn following; and if they are permitted to remain another year, they will then be ftrong plants, fit to be fet out for good. The Common Ivy is alfo to be raifed from feeds.

H I B I S C U S.

LINNEAN Clafs and Order, *Monadelphia Polyandria*: Each flower contains numerous males and one female; the males being joined together at the bafe: There are thirty-feven SPECIES; one of which adds great beauty to our grounds and fhruberies, in autumn.

HIBI'SCUS *Syri'acus*: The ALTHEA FRU'TEX, or the SYRIAN MALLOW, or the SYRIAN HIBISCUS; *a deciduous fhrub*; native of Syria.

The ALTHEA FRUTEX, or the SYRIAN HIBISCUS. Of this fpecies there are feveral *Varieties*:

The *White Althea Frutex.*
Red-flowering Althea Frutex.
Yellow-flowering Althea Frutex.
Pale Purple-flowering Althea Frutex.
Deep Purple Althea Frutex.

All

All thefe, though fuppofed to be only forts of one fpecies of *Hibifcus*, afford wonderful varieties to the gardener. They will grow to the height of about fix feet. Their branches are not very numerous; they are fmooth, and of a whitifh colour. The leaves are of a pleafant green, and grow on fhort footftalks, irregularly on the branches. They are of an oval, fpear fhaped figure, ferrated at the edges, and many of them are divided at the top into three diftinct lobes. The flowers have longer footftalks than the leaves, and come out from the fides of the young fhoots with them; infomuch that the young fhoots are often garnifhed with them their whole length. The Common Mallow produces not a bad flower, did not its commonnefs render it unnoticed. The flowers of thefe fpecies fomewhat refemble it in fhape, but by far exceed it both in fize and fplendor of colour; and each has a greater variety; infomuch that though they are termed Red, White, Purple, &c. from the colour of the upper part of the petals, yet the lower part of all of them is very dark, and feems to fhoot out in rays in directions towards the extremity of each petal. Auguft is the month we may expect to be entertained with this bloom; though in ftarved cold foils, the flowers rarely ever appear before September. It will bear a moift fituation.

This beautiful fhrub may be PROPAGATED by two methods. 1. By feeds, which we receive from abroad. Thefe fhould be fown in a bed of light fandy earth; and if it is not naturally fo, drift fand muft be added; and if fome old lime rubbifh, beat to powder, be alfo mixed with it, it will be the better. Having worked them all together, and made the bed fmooth and fine, the feeds fhould be covered about a quarter of an inch deep. The fituation of this bed muft be in a warm well fheltered place, that the young plants may not fuffer by frofts the firft winter. Any time in March will do for the work; and in about fix weeks the young plants will come up. In the heat of fummer it will be proper to fhade them; and if conftant waterings are afforded them in dry weather, they will acquire greater ftrength and vigour by the autumn. At the beginning of November, befides the natural fhelter of thefe beds, it will be proper to prick furze bufhes at a little diftance

all

all around, to break the keen edge of the black frofts, which otherwife would deftroy many of them the firft winter: After that, they will be hardy enough for our fevereft weather. They fhould ftand in thefe feed beds two years, and all the while be weeded and watered in dry weather. The fpring is the beft time for planting them out in the nurfery, where no more diftance need be allowed them than one foot. 2. Thefe plants may be propagated by layers; for which purpofe the ftools fhould be headed near the ground, to throw out fome good ftrong fhoots the following fummer. Thefe fhould be laid in the ground, the bark being broken, or cut at one or two of the joints, and they will have ftruck root by the autumn following, when they may be taken up and planted in the nurfery, like the feedlings; and a fecond operation performed on the ftools. 3. Thefe plants may be raifed alfo by cuttings; for by planting them in a fhady border, many of them will grow; though this is not a certain method.

H I P P O P H Æ.

LINNEAN Clafs and Order, *Dioecia Tetrandria*: Male flowers containing four ftamina, and female flowers containing one piftil, upon diftinct plants: There are only two SPECIES:

1. HIPPOPHÆ *Rhamnoides*: The EUROPEAN SEA BUCKTHORN; *a tall deciduous fhrub*; native of the fea fhores of this country and moft parts of Europe.

2. HIPPOPHÆ *Canadenfis*: The AMERICAN SEA BUCKTHORN; *a tall deciduous fhrub*; native of Canada.

1. The EUROPEAN SEA BUCKTHORN will grow to the height of about twelve feet, and fends forth numerous branches in an irregular manner. Their colour is that of a dark brown; and on them a few ftrong and long fharp fpines are found, nearly like thofe of the Common Buckthorn. This tree is chiefly admired for its fingular appearance in winter; for the young fhoots of

of the preceding ſummer are then found thickly ſet on all ſides with large, turgid, uneven, ſcaly buds, of a darker brown, or rather a chocolate colour, than the branches themſelves: Theſe give the tree ſuch a particular look, that it catches the attention, and occaſions it to be enquired after, as much as any ſhrub in the plantation. About the end of February theſe turgid buds will be much larger; and a little before their opening, upon ſtriking the tree with a ſtick, a yellow duſt, like brimſtone, will fall from them. Though ſome think the beauty of this ſhrub to be diminiſhed after the leaves are opened, yet theſe have their good effect; for they are of two colours: Their upper ſurface is of a dark green, their under hoary; they are long and narrow, entire, have no footſtalks, nearly like thoſe of the roſemary, though rather longer and broader; and they are placed alternately all around, without any footſtalks, on the branches. They continue on the tree green and hoary late, ſometimes until the beginning of December, and at length die away to a light brown. The flowers are of no conſequence to any but Nature's ſtrict obſervers. They are produced in July, by the ſides of the young ſhoots; the male flowers appear in little cluſters, but the females come out ſingly. They are ſucceeded by berries, which, in the autumn, when ripe, are either of a red or yellow colour, for there are both thoſe ſorts. The wood is of a brittle texture, and a bright brown colour.

2. CANADA SEA BUCKTHORN will grow to about the ſame height as the other ſpecies; nearly the ſame dark brown bark covers their branches; and, except the figure of their leaves, which are oval, this plant differs in few reſpects from the European Sea Buckthorn.

Both theſe ſorts may be PROPAGATED, 1. By cuttings of the young ſhoots, planted in a ſhady border, in October; though the moſt certain method is by layers. If the trees to be encreaſed are of ſome years growth, the ground ſhould be dug and made fine, as well as cleared of the roots of bad weeds, &c. all round. The main branches may be plaſhed, and the young twigs that form the head laid in the ground; taking off their ends with a knife, that th y may only juſt peep. If this work be performed in the autumn, they will be good

good rooted plants by the autumn following, when they may be taken off, and either planted in the nursery, or where they are to remain. 2. Both these sorts are subject to spawn, and throw out many suckers, sometimes at a good distance from the plants; so that by this method they propagate themselves.

HYDRANGEA.

LINNEAN Class and Order, *Decandria Digynia*: Each flower contains ten males and two females: There is only one SPECIES:

HYDRANGE'A *Arbore'scens*: The HYDRANGEA; *a deciduous shrub*; native of Virginia.

The HYDRANGEA seldom grows to more than a yard or four feet high, and affords as much pleasure to those who delight in fine flowers as it does to the botanist. It forms itself into no regular head; but the branches of which it is composed shoot chiefly from the root. These, when young, are four-cornered and green; when old, of a fine brown colour: They are very large for their height, as well as very full of pith. The leaves are a great ornament to these plants; being also very large, and having their upper surface of a fine green, and their under rather downy. Their figure is nearly shaped like a heart, but ends in an acute point; and their size will prove according to the nature of the soil they grow in. On a dry soil, they will often be no more than two inches long, and scarcely an inch and a half broad; but, in a moist rich soil, they will frequently grow to near four inches long, and two and three quarters broad in the widest part. They are serrated at their edges, and are placed on long footstalks, opposite to each other, on the branches. But the flowers constitute the greatest beauty of these plants; for they are produced in very large bunches, in August: Their colour is white, and the end of every branch will be ornamented with them. They have an agreeable odour,

odour, and make ſuch a ſhow all together as to diſtinguiſh themſelves even at a conſiderable diſtance. With us, however, they are ſeldom ſucceeded by any ſeeds.

The PROPAGATION of this plant is more eaſy than to keep it within bounds: for the roots creep to a conſiderable diſtance, and ſend up ſtalks which produce flowers; ſo that theſe being taken off, will be proper plants for any place. It likes a moiſt ſoil.

HYPERICUM.

LINNEAN Claſs and Order, *Polyadelphia Polyandria*: Each flower contains many males and about three females; the males being joined at the baſe in many ſets. There are forty SPECIES; two of which are proper for ornamental ſhruberies.

1. HYPE'RICUM *Hirci'num*: The SHRUBBY ST. JOHN'S WORT, or the STINKING or GOAT-SCENTED ST. JOHN'S WORT; *a deciduous ſhrub*; grows naturally by the ſides of rivers in Sicily, Calabria, and Crete.

2. HYPERICUM *Canarie'nſe*: The CANARY ST. JOHN'S WORT; *a deciduous ſhrub*; native of the Canaries.

1. The SHRUBBY ST. JOHN'S WORT. Of this there are ſeveral *Varieties*. The Common is a beautiful ſhrub, near four feet in height. The branches are ſmooth, of a light brown, and come out oppoſite by pairs from the ſides of the ſtrongeſt ſtalks; and theſe alſo ſend forth others, which alternately point out different directions. The leaves are of an oblong oval figure, grow oppoſite by pairs, and ſit very cloſe to the ſtalks. Theſe being bruiſed, emit a very ſtrong diſagreeable ſcent. The flowers are yellow, and make a good ſhow in June and July; for they will be produced in ſuch cluſters, at the ends of the young ſhoots, that the ſhrub will appear covered with them. They are ſucceeded by oval black-coloured capſules, containing ripe ſeeds, in the autumn.

There is a *Variety* of this ſpecies, which will grow to be eight feet high: The ſtalks are ſtrong, the leaves broad, and the flowers large; and being produced in great plenty, cauſes it to be a valuable ſhrub for the plantation. There is another Variety with variegated leaves, which is admired by thoſe who are fond of ſuch kinds of plants. There is alſo a Variety diſpoſſeſſed of the diſagreeable ſmell, which cauſes it to be preferred by many on that account.

2. The CANARY ST. JOHN'S WORT is a ſhrub of about ſix or ſeven feet high: The branches divide by pairs, and the leaves, which are of an oblong figure, grow oppoſite by pairs, without any footſtalks. The flowers come out in cluſters from the ends of the branches: They are of a bright yellow, have numerous ſtamina, which are ſhorter than the petals, and three ſtyles. They appear in July and Auguſt, and are ſucceeded by oval roundiſh capſules, containing the ſeeds.

No art need be uſed in PROPAGATING theſe ſhrubs; for, 1. Having obtained a plant or two of each, they will afford encreaſe enough by ſuckers. Having ſtood about three years, the whole of each plant ſhould be taken up, and the ſuckers and ſlips with roots that this may be divided into, may reaſonably be ſuppoſed to be twenty in number. The ſtrongeſt of theſe may be planted where they are to remain, while the weaker may be ſet out in the nurſery to gain ſtrength. 2. Theſe ſhrubs may alſo be propagated by ſeeds, which ripen well with us, and will come up with common care; nay, they will often ſhed their ſeeds, which will come up without ſowing, eſpecially the laſt ſort.

JASMINUM.

LINNEAN Claſs and Order, *Dyandria Monogynia*: Each flower contains two males and one female: There are ſix SPECIES; three of which are hardy enough for our purpoſe.

1. JAS-

1. JA'SMINUM *Officina'le:* The COMMON WHITE JASMINE; *a deciduous ſhrub or climber*; native of India.

2. JA'SMINUM *Fru'ticans:* The COMMON YELLOW JASMINE; *a deciduous ſhrub or climber*; native of the South of Europe, and of the Eaſt.

3. JA'SMINUM *Hu'mile:* The ITALIAN JASMINE; *a deciduous ſhrub or climber*; native of Italy.

1. The COMMON WHITE JASMINES have uſually been planted againſt walls, &c. for the branches being ſlender, weak, and pithy, by ſuch aſſiſtance they have arrived to a good height; though this ſhrub is not the moſt eligible for that purpoſe as its branches, which are numerous, are covered with a brown, dirty-looking bark, and afford ſhelter for ſnails, ſpiders, and other inſects, which in winter, when the leaves are fallen, will give them an unſightly look; and if they are clipped and kept up to the wall, as the flowers are produced from the ends and wings of the ſhoots, theſe muſt of courſe be ſheared off; ſo that little bloom will be found, except what is at the top of the tree. It is not meant, however, to diſſuade thoſe people who are fond of it from planting it againſt walls: It naturally requires ſupport, though attended with thoſe defects. It may, neverthelefs, be planted among ſhrubs in the ſhrubery, to appear to great advantage. It ſhould keep company with the lower kinds of ſhrubs; and whenever the branches grow too high to ſuſtain themſelves without nodding, and diſcover their ruſty ſtems, theſe ſhould be taken off from the bottom. There will always be a ſucceſſion of young wood; and theſe young ſhoots, which are covered with a ſmooth bark, of a delightful green colour, alſo exhibit the leaves and bloom. The leaves are pinnated, and very beautiful: They grow oppoſite by pairs, and the folioles are uſually three pair in number, beſides the odd one with which each leaf is terminated. They are all of a dark ſtrong green colour, are pointed, and the end one is generally the largeſt, and has its point drawn out to a greater length. The flowers are produced from the ends and joints of the branches, during moſt of the ſummer months: They are white, and very fragrant; but are ſucceeded by no fruit in England.

There

There is a *Variety* of this ſort with yellow, and another with white ſtriped leaves.

2. The YELLOW JASMINE is often planted againſt walls, pales, &c. as the branches are weak and ſlender; and it will grow to be ten or twelve feet high, if thus ſupported. It may, however, be planted in ſhrubery quarters, in the ſame manner as the other. The young ſhoots are of a fine ſtrong green colour, angular, and a little hairy. The leaves are trifoliate, though ſometimes they grow ſingly: They are placed alternately on the branches, are of a thick conſiſtence, ſmooth, and of a fine deep green colour. Theſe leaves, in well ſheltered places, remain until the ſpring before they fall off; ſo that this plant may not improperly be ranked among evergreens, eſpecially as the young ſhoots are always of a ſtrong green. The flowers are yellow, and do not poſſeſs the fragrance of the preceding ſpecies: They are produced in June, and the blow is ſoon over; but they are ſucceeded by berries, which, when ripe, are black. Theſe have occaſioned this ſort to be called by ſome perſons the Berry-bearing Jaſmine.

3. The ITALIAN JASMINE is, of all the ſorts, beſt adapted to a ſhrubery, becauſe it loſes part of its beauty if nailed to a wall. It is naturally of lower growth, and the branches are ſtronger, fewer in number, able to ſupport themſelves in an upright poſition, and are angular. The bark is ſmooth, and of a fine deep green colour. The leaves grow alternately: They are chiefly trifoliate, though ſome pinnated ones are found upon this ſhrub. The folioles are ſmooth, and of a fine ſtrong green: They are much broader than the preceding ſorts, and often continue till ſpring before they drop off; ſo that this ſhrub, on account of the beautiful green colour of the young ſhoots, might have a place among evergreens. The flowers are yellow, and much larger than thoſe of the other ſorts: They are produced in July, and are ſometimes ſucceeded by berries; but theſe ſeldom if ever come to perfection. This ſpecies is very hardy, and has grown in the moſt expoſed places, reſiſting the ſevereſt froſts for many years.

Little need be ſaid concerning the PROPAGATION of theſe plants; for they will all grow by layers or cuttings; ſo that if either way be purſued in the winter,

 you

you will have plenty of plants by the autumn following. The cuttings, however, muſt have a moiſt good ſoil, and ſhould be ſhaded and watered, as the hot weather comes on, the beginning of ſummer. The Common Yellow Jaſmine may be propagated by the ſeeds; but it naturally ſends forth ſuch plenty of ſuckers as to render it needleſs to take any other method for its encreaſe; for theſe being taken off, will be good plants; nay, if it is planted in borders, they muſt be annually taken for uſe, or thrown away, or they will overſpread every thing that grows near them. The Yellow and White ſtriped-leaved Jaſmines are propagated by grafting, budding, or inarching into ſtocks of the Common White: They are rather tender, eſpecially the White, therefore muſt have a warm ſituation. The Yellow-ſtriped is the moſt common and leaſt beautiful, and may be encreaſed by layers and cuttings, like the plain ſort.

I L E X.

LINNEAN Claſs and Order, *Tetrandria Tetragynia*: Each flower contains four males and four females: There are three SPECIES; one of Europe, one of Aſia, and one of America: the laſt has been introduced into our gardens and ſhruberies, making with our own Holly two ſpecies.

1. I'LEX *Aquifo'lium*: The COMMON HOLLY, or the EUROPEAN HOLLY; a well known *evergreen tree or ſhrub*; native of moſt parts of Europe, particularly of England.

2. I'LEX *Caſſi'ne*: The AMERICAN HOLLY, or the DOHOON HOLLY; *an evergreen tree or ſhrub*; native of Carolina.

1. The EUROPEAN or COMMON HOLLY will grow to thirty or forty feet high, with a proportionable ſtem. In its natural ſoil and ſituation, namely, a high chalky, marly, or limeſtone loam, the ſtem frequently ſhoots

ſhoots up naked and ſilvery, ſix or eight feet high, ſupporting a cloſe, ſnug, elliptical head: This may be called its tree ſtate. But the Holly, almoſt as frequently, puts on a very different appearance; feathering from the ground and riſing with an irregular, looſe, elegant outline; forming one of the moſt *Ornamental* evergreens which Nature has furniſhed us with. What renders it in this point of view peculiarly valuable,—it is not only highly ornamental in ſinglets or groups ſtanding in the open air, but will flouriſh with great beauty under the ſhade and drip of the more lofty deciduous tribes. Beſides, the bluſhing fruit of the Holly renders it moſt ornamental at a time when the face of nature is in a manner diveſted of every other ornament: In this light it is ſuperior to the Box; and, indeed, taken all in all, the Holly is undoubtedly entitled to take rank amongſt the firſt claſs of Ornamentals. In reſpect of *Utility*, the Holly gives place to the Box; except for the purpoſe of Hedges, and for this purpoſe it ſtands unrivalled;—but of this the reader will find more under the Article HEDGES. Its wood, however, is in good eſteem among the inlayers and turners; it is the *whiteſt* of all woods; its colour approaching towards that of Ivory.

The PROPAGATION of the Holly is principally from ſeeds. EVELYN tells us, that young ſeedlings, collected from the woods, and planted in a nurſery, in a few years will make tolerable plants. MILLER ſays, the ſeeds "never come up the firſt year; but lie in the ground as the Haws do: therefore the berries ſhould be buried in the ground one year, and then taken up and ſown at Michaelmas, upon a bed expoſed only to the morning ſun; the following ſpring the plants will appear."---"In this ſeed bed, he ſays, the plants may remain two years; and then ſhould be tranſplanted in the autumn, into beds at about ſix inches aſunder, where they may ſtand two years longer, during which time they muſt be conſtantly kept clean from weeds; and if the plants have thriven well, they will be ſtrong enough to tranſplant where they are deſigned to remain; for when they are tranſplanted at that age, there will be leſs danger of their failing, and they will grow to a larger ſize than thoſe which are removed when they are much larger."---He alſo tells us, " the beſt time for removing Hollies is in autumn, eſpecially in dry land; but where the ſoil is cold and moiſt, they may be tranſplanted with great ſafety in the ſpring; if the plants are not too old, or have not ſtood long unremoved; for if they have, it is great odds of their growing when removed." HANBURY differs with MILLER in regard to ſowing: he thinks the beſt way is to ſow them as ſoon as they are ripe, and then, he ſays, " they will undoubtedly come up the ſpring twelvemonth following." ---" However, he adds, if the ſeeds have been buried, let them be taken up in October; and having ſome fine light ſoil for the ſeminary, let them be ſown half an inch deep, and carefully covered from the mice." He recommends, when the ſeedlings are two years old, that " in the ſpring they ſhould be taken out of the beds and planted in the nurſery in rows, a foot aſunder, and two feet diſtance between the rows. Here they may ſtand until they are of a ſufficient ſize to be finally planted out." He follows MILLER as to the time of tranſplanting; recommending autumn if the land be naturally dry: but if of a moiſt nature, he ſays, " the planter need not be very anxious about the time of the winter in which he makes his plantations of Hollies." It is ſomewhat extraordinary that men practical as MILLER and HANBURY undoubtedly were, ſhould not have been acquainted with the proper time of removing ſo prevalent and ſo uſeful a plant as the Holly: and it is ſtill more remarkable, that the profeſſional nurſerymen of the preſent day ſhould, in general, be involved in the ſame darkneſs. Spring is the very worſt time for performing this buſineſs; winter and autumn may be ſomewhat more eligible; but SUMMER is of all others the moſt proper ſeaſon for tranſplanting the Holly. At this time of the year, if the plants be young and well rooted, it matters not much how dry the ſoil is; for they will, notwithſtanding, ſucceed with great certainty. MILLER neverthelеſs is right in ſaying that large Hollies which have not lately been moved are difficult to tranſplant; more eſpecially ſuch as have ſtood in a thicket, or under the ſhade of other trees. If, however, even theſe be taken up with good roots, together with a large quantity of native mould adhering to them; their

heads

heads lessened by pruning them in the conoidic manner, and be planted during the summer months in a well tempered paste, agreeably to the directions given under the Article TRANSPLANTING, success, though it cannot be insured, may with great probability be expected. Thus far the Common Holly;—which has been already treated of under the Article HEDGES.

But besides the genuine species, there are of the European Holly almost endless *Varieties*. MILLER mentions some eight or ten; and HANBURY enumerates upwards of forty; Five of them are sufficiently distinct to merit separate descriptions; the rest, distinguished chiefly by the variegation or mottled appearance of their respective leaves, would only form a long list of uncouth names, wholly uninteresting upon paper, how elegant soever they themselves may appear in a group of Evergreens.

The Smooth-leaved Holly.
The Green-leaved Yellow-berried Holly.
The Box-leaved Holly.
The Hedge-Hog Holly.
The Saw-leaved Holly.

The *Smooth-leaved Holly* resembles in general appearance the Common sort. Of the two it seems to be the strongest shooter, and bids fair for the largest growing tree. The leaves are nearly oval, and most of them are entirely free from prickles, only they end in acute points. This sort is commonly called the Carolina Smooth-leaved Holly: But it is a native of England, and is found growing amongst the others in many parts.

The *Green-leaved Yellow-berried Holly* differs in no respect from the Common Holly, only the berries are yellow; and as this tree produces berries in plenty, which are thought by most people to be uncommon and curious, this sort, on their account, is deserving of a place, either in small or large gardens, in shrubery-quarters or plantations of any kind.

The *Box-leaved Holly* has but little claim to be so called; for though some of the leaves be small, pretty free from prickles, and nearly oval, yet there will be so many nearly as prickly as the Common Holly as to merit no claim to that appellation. The leaves, however, are ſmall; and by them, on that account, the chief variety is occaſioned.

The *Hedge-Hog Holly* has the borders of the leaves armed with ſtrong thorns, and the ſurface beſet with acute prickles, a little reſembling thoſe of a hedge-hog, which gave occaſion to this ſort being ſo called by the Gardeners. This, together with the Striped ſorts of it, is juſtly ranked among our Hollies of the firſt rate.

The *Saw-leaved Holly* is a kind very different from any of the other ſorts. The leaves are of the ordinary length, but very narrow and of a thick ſubſtance. Their edges are formed into the likeneſs of a ſaw; though they are not very ſharp and prickly. This is a very ſcarce and valuable Holly, and is by all admired.

Theſe ſix ſorts of themſelves form a Collection truly valuable to our evergreen ornamental plantations: if the variegated ſorts are alſo to have a place, which they may properly enough, we introduce then a freſh Collection, which for variety and beauty far exceeds not only the variegated ſorts belonging to any one genus, but perhaps all the variegated ſorts of trees and ſhrubs put together.

But let us proceed to the CULTURE of theſe ſorts. We have already ſhewn how the Common Engliſh Holly may be raiſed from the berry. That method is to be practiſed, and plenty of that ſort may be raiſed. Theſe are to be ſtocks, on which the others are to be budded or grafted: for though they will take by layers, yet plants raiſed that way are of little or no value; and if the berries of the variegated ſorts be ſown, the plants will come up plain, and be our Common Engliſh Holly (though from Hedge-Hog berries plants of the Hedge-Hog Holly are frequently raiſed). By grafting or budding, then, theſe ſorts muſt be propagated; and for this purpoſe young ſtocks muſt be raiſed of the Common Holly, as has been already directed. After theſe have ſtood two years in the ſeed bed, they ſhould be taken up, have their roots ſhortened, and be planted out in the nurſery, a foot aſunder, in rows at two feet diſtance. The ſummer following they will probably make few ſhoots; but the ſummer after that, they will

ſhoot

ſhoot ſtrongly; and when the operation is to be performed by grafting, theſe will be proper ſtocks for the purpoſe by the ſpring following. The firſt week in March is a good time for the work. Whip-grafting is the method to be practiſed; and it muſt be performed on the young wood, namely, on that of the preceding ſummer's ſhoot. The cions being cut true and even, and well jointed to the ſtock, many of them will grow; and this is a very good method of encreaſing theſe trees. They may alſo be multiplied at pleaſure by inoculation. This operation is beſt performed about ten days after Midſummer, in cloudy weather; and for want of this, evening ſhould be the time: and if much work is to be done, morning too may be added; nay, it may be practiſed all day in the hotteſt ſeaſons, with tolerable ſucceſs; but this is never ſo eligible, unleſs when the multiplicity of work obliges us to loſe no time. The young wood of the preceding ſummer's ſhoot is proper for the purpoſe; and the operation is to be performed in the uſual way. In the autumn the bands ſhould be looſed, and in the ſpring the ſtocks dreſſed up, and headed two or three inches above the bud; the buds will be as early in ſhooting out as any of the ſhoots of the growing trees, and will ſoon become good plants for any place.

2. The DOHOON HOLLY is an American plant, particularly of Carolina, where it grows to be nearly as large a tree as our Holly does with us. It naturally riſes with an upright ſtem, which is covered with a brown bark, and this affords plenty of younger branches, whoſe bark is green and very ſmooth. The leaves are pretty large, and of an oval lanceolated figure; they are of a thickiſh compoſition, of a fine green, and grow alternately on the branches. Their edges are ſerrated, though altogether different from the Common Sawed Holly, their ſerratures towards the upper end of the leaf being ſmall and ſharp. The leaf, on the whole, is of a fine compoſition, and grows on ſhort footſtalks on the branches. The flowers are ſmall and white, and a little reſemble thoſe of the Common Holly. They are produced from the ſides of the branches, in ſhort thick cluſters; and are in their native climate ſucceeded by red berries, equalling thoſe of our Common ſort in beauty.

The Dohoon Holly may be PROPAGATED by ſeeds, which we receive from the countries where it grows naturally; for the berries will not ripen, and indeed are very ſeldom produced, in England. The beſt way is to ſow them in pots filled with light ſandy earth, as ſoon as they arrive, and then plunge them up to the rims in the natural mould, where they may remain until the ſpring following; for they rarely ever come up the firſt ſummer. The ſpring after that the plants will appear: and if they have then the aſſiſtance of a hotbed, it will greatly help them forward. They muſt be uſed to the open air ſoon. The pots muſt be taken up and plunged in a ſhady place, and in October they ſhould be removed into the greenhouſe for the winter. In the ſpring the plants in the pots may be thinned by drawing out the ſtrongeſt; and thoſe thus drawn ſhould be planted each in a ſeparate pot, and muſt be ſet forward with a hotbed as before. The others, alſo, may be taken out at two or three years growth, planted in pots, and aſſiſted in the ſame manner. Every October they ſhould be removed into the greenhouſe, ſet out in the ſpring, and treated as greenhouſe plants, until they are at leaſt five or ſix years old; for before then they will be hardly woody enough to venture the planting them out to ſtand. The latter end of March, when the danger of bad weather is chiefly over, is the beſt time for the purpoſe; and if they have a dry ſoil and a warm ſituation, they will bear the cold of our common winters; though if a very ſevere winter ſhould happen before they are got very ſtrong and woody, it is more than probable that all of them will be deſtroyed.

ITEA.

LINNEAN Claſs and Order, *Pentandria Monogynia* Each flower contains five males and one female. There is only one SPECIES:

ITEA *Virginica*: The ITEA; *a deciduous ſhrub*; native of Virginia.

The

The ITEA is a plant of about five or ſix feet in height. The branches are numerous, and are produced irregularly all round. The leaves with which they are ornamented are of a fine green colour, gentle ſerratures poſſeſs their edges, their figure is that of a ſpear, and they grow alternately on the twigs. But the flowers conſtitute the greateſt beauty of theſe ſhrubs; for they are produced in July, at the ends of the young ſhoots, in large erect ſpikes: Their colour is white; and as moſt of the branches will be terminated by them, the tree itſelf appears at a diſtance like one large bunch of white flowers: So delightful is the variety which Nature furniſhes for our contemplation and pleaſure.

The PROPAGATION of this beautiful ſhrub is not very eaſy; though it may be propagated by ſeeds and layers. 1. We receive the ſeeds from abroad. They ſhould be ſown in pots or boxes of fine loamy earth, mixed with driſt or ſea ſand; and theſe ſhould be plunged up to the brim in the moiſteſt part of the garden, where they may remain till the ſpring after; for the ſeeds ſeldom come up the firſt year. In March, therefore, the pots ſhould be taken up, and plunged into a hotbed, which will promote the growth of the ſeeds, and make them become ſtronger by the autumn. After the heat of the bed is over, they may be put in the ſame moiſt places again. The plants ought to be conſtantly weeded and watered; and in the autumn ſhould be removed into the greenhouſe, or placed under a hotbed frame, to be protected in ſevere weather. This care ſhould be continued through the next winter alſo In the ſpring, a damp day being made choice of, and a moiſt part of the nurſery being well prepared, they ſhould be taken out of the pots or boxes, and planted at about a foot aſunder, which will be diſtance enough for their ſtanding two or three years, when they will be of a ſufficient ſize to be finally planted out. 2. Theſe trees are alſo propagated by layers; for which purpoſe, ſome of them ſhould be planted for ſtools in a moiſt rich ſoil. The young ſhoots of the preceding ſummer ſhould be laid in the ground in the autumn; and in order to make them ſtrike root, a little wire ſhould be twiſted pretty cloſe round the bud, where the root is deſired to be: This wire impeding the motion of of the ſap the ſucceeding ſummer, will occaſion them to ſwell in thoſe parts, and ſtrike root. There are other methods by which the operation may be performed; but this has been found the moſt expeditious and ſureſt.

JUGLANS.

LINNEAN Claſs and Order, *Monoecia Polyandria;* Male flowers containing many ſtamina, and female flowers containing two piſtils upon the ſame plant; the males being collected in oblong catkins, the females ſitting in cluſters cloſe to the branches. There are five SPECIES; four of which are ſufficiently hardy for this climate:

1. JU'GLANS *Re'gia:* The COMMON WALNUT; *a deciduous tree*; whoſe native country is uncertain.

2. JU'GLANS *Ni'gra:* The BLACK VIRGINIA WALNUT; *a deciduous tree*; native of Virginia, Carolina, and Maryland.

3. JU'GLANS *A'lba:* The HICKERY or WHITE VIRGINIA WALNUT; *a low deciduous tree*; native of Virginia.

4. JU'GLANS *Cine'rea:* The PENNSYLVANIA WALNUT; *a low deciduous tree*; native of Pennſylvania and other parts of North America.

1. The WALNUT TREE. This as a fruit tree is univerſally known. We ſometimes ſee it riſe to a conſiderable height, and grow to great ſize; in general, however, it does not riſe higher than forty feet; ſpreading out into a globular inelegant head; this added to the late ſeaſon at which it puts forth its leaves, and the ſtiff uncouth appearance it takes after their fall, renders it of low value as an *Ornamental*; unleſs indeed its general appearance be heightened, by the imagination, with the idea of *Utility*. In this point of view, whether we conſider its fruit, or the timber it produces, the Walnut ranks high, and is no doubt an object of notice to the planter. We are far, however, from being ſuch enthuſiaſts

enthusiasts to the Walnut as EVELYN was: indeed its uses as a timber are greatly lessened since his day: Mahogany has superseded it in the more elegant kinds of furniture; and the Beech, being raised at less expence, and, from the cleanness of its texture, being worked with less trouble, has been found more eligible for the commoner sorts; chairs more especially. Nevertheless, the Walnut is still a useful wood: it takes a fine polish, and is in good esteem among the cabinet makers, turners, and gunsmiths. Were the importation of Mahogany to be obstructed, the Walnut it is probable would become a very valuable wood.

The method of PROPAGATING the Walnut is from seeds. EVELYN recommends the Black Virginia sort for timber; and says the nuts should be kept in the husks, or shucks, until March; when they should be planted in the husks; for, he says, "the extreme bitterness thereof is most exitial and deadly to the worm; or it were good to strew some furzes, broken or chopt small, under the ground among them, to preserve them from mice and rats when their shells begin to wax tender." He recommends their being planted in the place where they are to abide; the Walnut being very impatient of transplanting. If, however, it be necessary to remove the plants, he cautions us not to touch the head with the knife; nor even the tap-root, except when very young Speaking of experienced husbandmen, he says, "what they hint of putting a tile shard under the nut, when first set, to divaricate, and spread the roots (which are otherwise apt to penetrate very deep), I like well enough." And from the same source he was informed, "that if they be transplanted as big as one's middle, it may be done safer than when young:" he adds, however, "I do only report it." MILLER also recommends the Black Virginia sort for timber, which he says "is much more inclinable to grow upright than the Common sort; and the wood being generally of a more beautiful grain, renders it preferable to that, and better worth cultivating." "I have seen some of this wood," continues he, "which hath been beautifully veined with black and white, which when polished has appeared at a distance like veined marble." The nuts, he says, should be kept in the husks till February,

February, the proper time of ſowing. If the trees be intended for timber, the nuts ſhould be planted where they are to remain; but if for fruit, in a ſeed bed; becauſe tranſplantation checks their upward growth, and renders them more fruitful. For timber, "they ſhould be planted in lines at a diſtance you intend them to remain; but in the rows they may be placed pretty cloſe, for fear the nuts ſhould miſcarry; and the young trees where they are too thick may be removed, after they have grown two or three years, leaving the remainder at the diſtance they are to ſtand." He alſo cautions againſt too free a uſe of the pruning knife, either to the roots or the branches; but when there is a neceſſity, he ſays, of cutting any of their branches, "it ſhould be done early in September." He adds, "The beſt ſeaſon for tranſplanting theſe trees is as ſoon as the leaves begin to decay, at which time, if they are carefully taken up, and their branches preſerved entire, there will be little danger of their ſucceeding, although they are eight or ten years old; though theſe trees will not grow ſo large or continue ſo long as thoſe which are removed young." HANBURY likewiſe follows EVELYN in recommending the Black Virginia Walnut in preference to the Common ſort for timber. His farther directions concerning the choice of the ſeed and the method of propagation are as follow: "If the fruit of theſe trees are greatly coveted, the utmoſt care ſhould be taken to gather the nuts from thoſe trees which produce the beſt ſorts; and although the Varieties of Walnuts are only ſeminal variations, yet there is the greater chance of having a ſucceſſion of good nuts, if they are gathered from trees that produce good fruit. This maxim holds good in animals: The fineſt breed would degenerate, if attention was not paid to the ſorts for breeding; and the like care muſt be extended throughout the whole ſyſtem of planting, whether for fruit or timber. If for timber, we ſhould be ſolicitous to gather the ſeeds from the healthieſt, the moſt luxuriant and thriving young trees: If for fruit, from thoſe which produce the richeſt and beſt kinds.' Having marked the trees that produce the fineſt nuts either for thinneſs of ſhell or goodneſs of taſte, when they have begun to fall they will be ripe enough for gathering. But

But as collecting them by the hand would be tedious, they may be beat down by long poles prepared for that purpose. Having procured the quantity wanted, let them be preserved, with their husks on, in sand till the beginning of February, which is the time for planting them. This is to be done in the following manner: Let drills be made across the seminary, at one foot asunder, and about two inches and a half deep, and let the nuts be put in these at the distance of about one foot. In the spring the young plants will come up; and here they should continue for two years, being constantly kept clear of weeds; when they will be of a proper size to plant out in the nursery. The ground should be prepared, as has been always directed, by double digging; and the trees being taken out of the seminary, and having their tap-roots shortened, should be planted therein, in rows two feet and a half asunder, and the plants at a foot and a half distance. Here they may remain, with the same culture as has been all along directed for the management of timber trees, till they are of a proper size for planting out for good. If they are designed for standards to be planted in fields, &c. before they are taken out of the nursery they should be above the reach of cattle, which may otherwise wantonly break their leading shoots, though they do not care to eat them on account of their extraordinary bitterness. They ought likewise to be removed with the greatest caution, and the knife should be very sparingly applied to the roots. They must also be planted as soon as possible after taking up; and this work should be always done soon after the fall of the leaf."

Evelyn tells us, that the Walnut Tree may be propagated "by a branch slipped off with some of the old wood, and set in February;" and in another place, "it is certain they will receive their own cions being grafted, and that it does improve their fruit."

It is agreed on all hands, that the Walnut requires a dry, sound, good soil, and will make but little progress, as a timber tree, in a cold barren situation.

2. The Black Virginia Walnut. This is the sort recommended for timber, and will grow to a large tree. The young shoots are smooth, and of a greenish brown,

brown. The leaves are produced irregularly: They are large and finely pinnated, being composed of about eight, ten, twelve, and sometimes fourteen pair of spear-shaped, sharp-pointed folioles, which are terminated by an odd one, sawed at the edges, and the bottom pair are always the least. The flowers give pleasure only to the curious botanist. They blow early in the spring; and the females are succeeded by nuts of different sizes and shapes. The nuts of the Common sort have a very thick shell, inclosing a sweet kernel. They are furrowed, and of a rounder figure than those of the Common Walnut.

There are many *Varieties* of this species, and nuts of different sizes, like those of the Common Walnut, will always be the effect of seed: Some will be small and round; others oblong, large, and deeply furrowed. You must expect also to find a Variety in the leaves; some will have no scent, others will be finely perfumed. Hence the names, *Common Virginian Walnut*, *Aromatic Walnut*, *Deeply-furrowed-fruited Walnut*, &c. have been used to express the different Varieties of this species.

3. The WHITE VIRGINIAN WA NUT, called the Hickery Nut, is a tree of lower stature, seldom rising more than thirty or thirty-five feet high; though the sort called the Shag-Bark is the strongest shooter. The young shoots of all are smooth. The leaves are also pinnated, though some of them are small, the number of folioles being from two or three pair to six or seven, besides the odd one with which they are terminated. The folioles are of a pleasant green colour, narrowest at their base, and serrated at their edges. The flowers are no ornament; and the nuts are small, hard, and of a white colour.

The *Varieties* of this species go by the various names of *Common Hickery Nut*, *Small-fruited Hickery Nut*, *Shag-Bark Hickery Nut*, *&c.*

4. PENNSYLVANIA WALNUT. This species grows to about the height of the former. The leaves are very long, being composed of about eleven pair of folioles, besides the odd one with which they are terminated. The flowers are yellowish, come out at the usual time with the oth rs, and are succeeded for the most part by small,

ſmall, roundiſh, hard-ſhelled fruit; though the nuts will be of different ſizes in the different Varieties.

The method of PROPAGATING theſe trees is from the nuts, which we receive from America, where they grow naturally. Theſe muſt be ſown as ſoon as they arrive, in the manner directed for raiſing the Common Walnut. Their after management muſt alſo be the ſame.

JUNIPERUS.

LINNEAN Claſs and Order, *Dioecia Monadelphia*: Male flowers containing three ſtamina, and female flowers containing three piſtils, ſituated on diſtinct plants. There are ten SPECIES; nine of them as follow:

1. JUNI'PERUS *Commu'nis*: The COMMON JUNIPER; *an evergreen ſhrub*; native of England and Scotland, and of many of the northern parts of Europe.

2. JUNI'PERUS *Oxyce'drus*: The SPANISH JUNIPER; *an evergreen ſhrub or tree*; native of Spain and South of France.

3. JUNI'PERUS *Virginia'na*: The VIRGINIA CEDAR, or the RED CEDAR; *an evergreen tree or ſhrub*; native of Virginia and Carolina.

4. JUNI'PERUS *Bermudia'na*: The BERMUDIAN CEDAR; *an evergreen tree or ſhrub*, native of Bermudas and America.

5. JUNI'PERUS *Barbade'nſis*: The JAMAICA CEDAR; *an evergreen tree or ſhrub*; native of Jamaica and other Weſt India Iſlands.

6. JUNI'PERUS *Thuri'fera*: The SPANISH CEDAR; *an evergreen tree or ſhrub*; native of Spain.

7. JUNI'PERUS *Ly'cia*: The LYCIAN CEDAR; *an evergreen tree or ſhrub*; native of Spain, Italy, and France.

8. JUNI'PERUS *Phœni'cia*: The PHENICIAN CEDAR; *an evergreen ſhrub or tree*; native of Portugal, South of France, and the Eaſt.

9. JUNI'PERUS

9. JUNI'PERUS *Sabi'na:* The SAVIN; *a low evergreen shrub*; native of Italy, Siberia, Mounts Olympus and Ararat, and of Lusitania.

1. The COMMON JUNIPER. This species is divided into two *Varieties:*

The English Juniper.

The Swedish Juniper.

The *English Juniper* grows common upon the chalky hills about Banstead and Leatherhead in Surrey, where it appears in a low weak state, as if cropped and kept down by the browsing of sheep, seldom rising higher than two or three feet; but when planted in a good soil, it will grow to the height of fifteen or sixteen feet, and will produce numerous branches from the bottom to the top, forming a large well looking bushy plant. These branches are tough, and covered with a smooth bark of a reddish colour, with a gentle tinge of purple. The leaves are narrow, and sharp-pointed: They grow by threes on the branches; their upper surface has a grayish streak down the middle, but their under is of a fine green colour, and they garnish the shrub in great plenty. This tree flowers in April and May. The flowers are small, of a yellowish colour, and make no figure. They are succeeded by the berries, which are of a kind of a blueish purple when ripe, which will not be before the autumn twelvemonth following.

The *Swedish Juniper* has a natural tendency to grow to a greater height, and consequently has more the appearance of a tree than the former sort; sixteen or eighteen feet, however, is the highest it commonly grows to; and the plants raised from its seeds have, for the most part, a tendency to grow higher, and become more woody and ramose. The leaves, flowers, and fruit grow in the same manner, and are of the same nature, which shews it to be a Variety only Old Botanists mention it as a distinct species: CASPAR BAUHINE asserts this, and calls one the Shrubby Juniper, and the other Tree Juniper; and he also mentions another sort, which he calls the Lesser Mountain Juniper, with a broader leaf and a larger fruit. This is still a Variety of the Common Juniper. The leaves, flowers, and fruit, however, are much the same; though there may be some difference in the size of their growth. From what

what has been ſaid, the gardener will know, when he meets with them by thoſe different names, where to plant them in ſuitable ſituations. It is obſervable of both theſe ſorts, that in the beginning or middle of May, when they will be in full blow, the farina of the male flowers is diſcharged in ſuch plenty, that upon ſtriking the ſhrub with a ſtick, it will riſe up, in a ſtill air, like a column of white ſmoke, and like that will be wafted with the gentleſt wind, until it is loſt or out of ſight.

2. SPANISH JUNIPER will grow to be rather a higher tree than the Swediſh, in ſome ſoils. It will be feathered from the bottom to the top, if left untouched from the firſt planting, or if not crowded with other trees. The leaves are awl-ſhaped, and finely ſpread open. They are very ſhort, ſharp-pointed, and give the tree a fine look. The flowers are ſucceeded by large reddiſh berries, which are very beautiful when ripe.

3. VIRGINIA CEDAR. This tree is held out by EVELYN and HANBURY as being proper to be planted as a foreſt or timber tree. It grows to near forty feet high, the branches forming a beautiful cone, and, if left unpruned, the tree will be feathered to the very baſe; conſequently it is highly *ornamental.* And HANBURY ſays, it "is valuable for many excellent and rare *uſes*: It will, he ſays, continue ſound and uncorrupt for many ages, being poſſeſſed of a bitter reſin, which prevents the worms from attacking it. The wood may be converted into utenſils of moſt ſorts, as well as applied to great part of the uſes to which the cypreſs is adapted. It is remarkable, however, for being of a very brittle nature, and is therefore not proper to be introduced into buildings where any great weight is to be lodged. Nevertheleſs, in Virginia and Carolina, where they abound, theſe trees are uſed in ſtructures of all kinds, with this precaution; and the inhabitants prefer the timber to moſt other wood for wainſcoting their rooms, and building of veſſels."

4 BERMUDIAN CEDAR. In the iſland from which this tree takes its name, it grows to a timber ſize; but in this country it is a very tender plant, and requires not only a dry warm ſoil, but open mild winters, to

 make

make it continue through them; so that when a persor is desirous of having an extensive collection, then and then only is this sort to be sought after; for, when planted abroad, even in the warmest quarters, the plants require sheds to preserve them from the winter's frost. It is the wood of the Bermudian Cedar of which pencils are made; also drawers of cabinets; and formerly wainscoting was made of this wood. In the island of Bermudas (which is in a manner covered with this tree), as well as upon the Continent of America, ships are built of it: its scent is peculiarly aromatic; but though agreeable to most people, it is offensive to some; and this may be one reason why less of this wood has of late years been imported into this country than was formerly.

5. Jamaica Cedar. This, in Jamaica, is a very large timber tree; but in our climate it is still more delicate than the Cedar of Bermudas; both of which ought to be considered as greenhouse rather than as shrubery plants.

6. Spanish Cedar grows plentifully in the country by whose name it is distinguished: it is a handsome, regular growing tree, rising in a conical form, if the branches are untouched, to the height of thirty or more feet. The leaves are imbricated, and lie over each other four ways; they are acute, and of a fine green colour: From these properties only, an idea of a fine tree may be had. The flowers are insignificant to a common observer; but they are succeeded by berries which make a good show when ripe; for they are very large, and of a fine black colour, and adorn the young branches in great plenty.

7. Lycian Cedar, also common in Spain, will rise to the height of about twenty-five feet; the branches have naturally an upright position, and their bark is of a reddish hue. The leaves are everywhere imbricated, and each is obtuse and of an oval figure. They resemble those of the Cypress, and are very beautiful. The flowers are succeeded by large oval berries, of a brown colour, and will be produced in plenty from the sides of the younger branches all over the tree.

8. Phenician Cedar seldom grows higher than twenty feet, and is a beautiful upright sort, forming a kind

kind of pyramid, if untouched, from the bottom. It has both ternate and imbricated leaves; the under ones grow by threes, and ſpread open; and the upper ones are obtuſe, and lie over each other like the Cypreſs. The flowers are produced from the ends of the branches; and the fruit that ſucceeds them is rather ſmall, and of a yellow colour. It is commonly called the Phenician Cedar, though it is found growing naturally in moſt of the ſouthern parts of Europe.

9. SAVIN. Of this ſpecies there are three ſorts:

Spreading Savin,
Upright Savin, and
Striped Savin.

Spreading Savin is a low ſpreading ſhrub; the branches have a natural tendency to grow horizontally, or nearly ſo; ſo that it muſt be ranked amongſt the loweſt growing ſhrubs; inſomuch that unleſs it is planted againſt a wall, or ſupported in an upright poſition, we ſeldom have it higher than two feet. When it is to be planted and left to Nature, room muſt be firſt allowed for its ſpreading; for it will occupy a circle of more than two or three yards diameter, and will choke any other leſs powerful ſhrub that is placed too near it. The bark on the older ſhoots is of a light brown colour; but the younger, which are covered with leaves running into each other, are of as fine a green as any ſhrub whatever. Theſe leaves are erect, and acute-pointed. They are placed oppoſite, and grow a little like thoſe of the French Tamariſk. This ſhrub ſeldom produces flowers or berries; but when any berries do appear, they are ſmall and of a blueiſh colour. It deſerves a place amongſt low-growing evergreens, on account of the fine ſtrong green of its leaves both in winter and ſummer; but it is valuable for nothing elſe; for it produces neither flowers nor fruit ornamental, and is poſſeſſed of a very ſtrong ſmell; inſomuch that, being ſtirred by whatever runs amongſt it, the whole air is filled with a fetid ſcent, which is emitted from its branches and leaves, and which to moſt people is diſagreeable. It is in great requeſt with horſedoctors and cowleeches, by which they much benefit thoſe creatures in many diſorders. The juice of it, mixed with milk and honey, is ſaid to be good to expel worms from

 children;

children; as well as, without that mixture, to destroy those in horses, for which purpose it is strongly recommended.

Upright Savin is a delightful tree; it will grow to be twelve or fourteen feet high. The branches are numerous and slender, and give the tree a genteel air. The leaves are nearly of the same nature with the other, though they are of a darker green. The flowers, though produced in plenty, make no show; but they are succeeded by berries in such plenty as to cause a good effect. The upright tendency of growth of this tree, together with the very dark green of the leaves, which causes a good contrast with others that are lighter, together with its not being possessed of that strong disagreeable scent of the other sort, makes it valuable for evergreen plantations.

Variegated Savin is a Variety of the former; it has not that tendency to spread like the Common, neither does it grow quite so upright as the Berry-bearing Savin. It is a fine plant, and at present rather scarce. The ends of several of the young shoots are of a fine cream colour; nay, all the smaller branches appear often of that colour, and at a distance will have the appearance of flowers growing on the tree. In short, to those who are fond of variegated plants, this shrub has both beauty and scarcity to recommend itself.

The method of PROPAGATING this genus of plants varies in some degree with the respective species.

The Common Juniper,
The Spanish Juniper,
The Virginia Cedar,
The Spanish Cedar,
The Lycian Cedar, and
The Phenician Cedar,

are raised from seeds procured from the respective places of their growth, and sown the latter end of February or the beginning of March, in beds of light sandy earth, about half an inch deep. In about twelve months after sowing the plants will appear. Having stood two years in the seed bed, they may be removed to the nursery, planting them from one to two feet distance in the rows, with two feet intervals; and here they may remain until wanted for use. They may be transplanted either

either in autumn or in ſpring, care being had to perform the removal in moiſt weather, preſerving as much mould as poſſible amongſt the fibres. HANBURY recommends that the ſeeds of the COMMON JUNIPER be ſown as ſoon as poſſible after they are ripe; for if this precaution, he ſays, is obſerved, they will come up the ſpring following; whereas if they are neglected till the ſpring, they will not appear till the ſpring after that; and ſometimes a great part of them will remain till the ſecond and even third ſeaſon before they come up. The COMMON JUNIPER may alſo be increaſed by layers.

The Bermudian Cedar, and
The Jamaica Cedar,

require that the ſeeds be ſown in pots; that the young ſeedlings be planted out into ſeparate pots; and that theſe be always houſed in the winter.

The COMMON SAVIN is to be increaſed by ſlips, which if planted almoſt at any time, or any how, will grow. The Upright Savin alſo is to be encreaſed by ſlips planted in moiſt weather, in Auguſt, and kept ſhaded and watered in dry weather afterwards. This is the beſt way of treating cuttings of the Upright Savin, though they will often grow if planted at any time, either in winter or ſummer. The Striped Savin alſo is to be encreaſed this way; though care muſt be always uſed to take off thoſe branches that are moſt beautifully variegated, and ſuch alſo as are entirely of a cream colour; for this will be the moſt probable method of continuing it in its variegated beauties. This plant is alſo to be raiſed by berries; and if theſe have the ſame treatment as the other ſorts, it will be very proper; and by theſe the moſt upright and beſt plants are raiſed.

KALMIA.

LINNEAN Claſs and Order, *Decandria Monogynia*: Each flower contains ten males and one female. There are two SPECIES:

 1. KALMIA

1. KA'LMIA *Latifo'lia:* The BROAD-LEAVED KALMIA; *an evergreen shrub*; nativ of Maryland, Virginia, and Pennsylvania.

2. KA'LMIA *Angustifo'lia:* The NARROW-LEAVED KALMIA; *an evergre.n shrub*; native of Pennsylvania and Carolina.

1. The BROAD-LEAVED KALMIA seldom rises to more than four or five feet high; and the branches, which by no means are regularly produced, are hard, and of a grayish colour. The leaves are of an oval, spear-shaped figure, and of a fine shining green colour. Their consistence is rather thick in proportion to their footstalks, which are but slender, and grow irregularly on the branches. The flowers are produced at the ends of the branches, in roundish bunches They are first of a fine deep red, but die away to a paler colour. Each is composed of a single petal, which is tubular at the bottom, spreading open at the top, and has ten permanent corniculæ surrounding them on their outside. They generally flower with us in July; and are succeeded by roundish capsules, full of seeds, which seldom ripen in England. In some places this is a fine evergreen; and in others, again, it often loses its leaves, and that sometimes before the winter is far advanced.

2. The NARROW-LEAVED KALMIA is rather of lower growth than the other, and the branches are more weak and tough. The leaves are very beautiful, being of a fine shining green; they are of a lanceolate figure, and in all respects are smaller than those of the former sort, and stand upon very short footstalks. They are produced in no certain regular manner, being sometimes by pairs, at other times in bunches, growing opposite at the joints. The flowers are produced from the sides of the branches in roundish bunches; they are of a fine red colour, and each is composed of one petal, that has the property of spreading open like the former. They flower in July, and are very beautiful; but are not succeeded by ripe seeds with us.

Both these sorts are to be PROPAGATED three ways; by seeds, layers, and suckers. 1. By seeds. These we receive from abroad; and for their reception we should prepare a compost, consisting of half fresh soil from a rich

rich paſture, taken from thence a year before, and half drift or ſea ſand: theſe being well mixed, will be proper for the reception of the ſeeds, which ſhould be ſown in pots or boxes, half an inch deep. As ſoon as they are ſown, they ſhould be removed into a ſhady place, to remain until the ſpring following, and all this time nothing but weeding will be wanted; for they ſeldom if ever come up the firſt ſummer. About the beginning of March it will be proper to plunge theſe pots into a hotbed, and this will fetch the plants up, and make them grow ſtrong. They muſt be hardened by degrees to the air, and then ſet in a ſhady place. Watering muſt be now and then given them, if the ſeaſon proves dry; and at the approach of winter they may be removed into the greenhouſe, or ſet under a hotbed frame, but ſhould always have the free air in open weather. In theſe pots or boxes they ſhould remain until they are two-years-old ſeedlings; when they ſhould be ſhaken out, and planted in a ſeparate pot. They ſhould then be forced, by plunging the pots into a hotbed. Afterwards, they may be removed into the ſhade; and if they are kept growing in the pots, and removed under ſhelter in hard weather for a year or two, they may be afterwards planted out finally. 2. Theſe ſhrubs are propagated by layering. It ſhould be done in the autumn; and the young wood of the preceding ſummer's ſhoot is proper for the purpoſe. If the ſoil is free and light, they will ſtrike root pretty readily; though we muſt ſometimes wait two years before we find any: But by this way the ſtrongeſt plants are obtained in the leaſt time. 3. They are alſo encreaſed by ſuckers; for if the ſoil be light and fine, and is what agrees with them, after ſtanding a few years, they naturally ſend out ſuckers in plenty. Theſe ſhould be taken off in the ſpring: and thoſe with bad roots ſhould be ſet in pots, and plunged into a hotbed, to make them grow.

LAVATERA.

LINNEAN Clafs and Order, *Monadelphia Polyandria*. Each flower contains many males and many females; the males being joined together at the bafe. There are nine SPECIES; four of which are cultivated in our open grounds:

1. LAVATE'RA *Arbo'rea*: The COMMON LAVATERA, or MALLOW TREE; *a deciduous fhrub*; native of Italy.

2. LAVATE'RA *Trilo'ba*: The THREE-LOBED LAVATERA, or MALLOW TREE; *a deciduous fhrub*; native of Spain

3. LAVATE'RA *O'lbia*: The FIVE-LOBED LAVATERA, or MALLOW TREE; *a deciduous fhrub*; native of the South of France.

4. LAVATE'RA *Mi'cans*: or the GLITTERING LAVATERA; or the SHINING-LEAVED MALLOW TREE; *a deciduous fhrub*; native of Spain and Portugal.

1 The COMMON LAVATERA is a well known plant: It ufually grows to eight or ten feet high, and in a rich foil will grow to twelve, or more. The ftem is thick and ftrong, and divides near the top into feveral branches, which ar clofely ornamented with large downy leaves; they are foft to the touch, plaited, and their edges are cut into many angles. The flowers are produced in clufters, from the wings of the leaves, in June, and there will be a fucceffion of them until late in the autumn. Each flower has its feparate footftalk: Their colour is purple; their fhape like that of the Common Mallow; and they would make a great fhow, were they not much obfcured by the largenefs of the leaves. The whole tree has a noble look; and its continuing for about three months in flower makes it very valuable. But though its fhort-lived continuance is much to be regretted, yet Nature feems to have made fome amends for this, by furnifhing it with good feeds in very great plenty; for by thefe thoufands of plants may be foon raifed; nay, they will fometimes fhed themfelves, and come up without any art. But when they

they are to be regularly ſown, let it be done in April, in the places where they are deſigned to remain, and they will flower the ſummer after. Though this plant is called a biennial, in ſome warm dry ſituations the ſtalks become hard and woody, and the plants will continue to produce flowers and ſeeds for many years.

There are ſeveral *Varieties* of this ſpecies; the leaves of ſome being round and indented, others acutely cut, others waved: Theſe among old gardeners go by the names of the *Round-leaved, Waved-leaved Mallow Tree, &c.*

2. The THREE-LOBED LAVATERA. This ſpecies is very ornamental in the front, or among the low ſhrubs in the wilderneſs quarters, or when ſtationed in large borders in pleaſure grounds, as it is naturally of low growth, ſeldom riſing to above four or five feet high. It has rather a large ſpreading root in proportion to the ſize of the ſhrub. The branches are numerous, and of a paliſh green colour; and the leaves are of different figures, though chiefly trilobate, or compoſed of three lobes, that are indented on their edges. They vary much in their ſize, ſome being larger, ſome ſmaller, and ſome more divided than others. Their colour, when the plant is in perfect health, is a very pleaſant green; but they will often ſhew themſelves a little variegated; at which time you may be aſſured the ſhrub is in a ſickly ſtate. This often does not continue long, and the plant will aſſume its former verdure, and as frequently and very ſpeedily relapſe into its weak ſtate; which ſhews that, though hardy with reſpect to cold, it is rather of a ſickly nature in this country. The flowers are produced ſingly, on ſhort footſtalks: they grow from the joints, at the boſoms of the leaves; three or four of them will appear at each joint; and being large, they make a fine ſhow in Auguſt, the time of flowering.

There are *Varieties* of this ſpecies, differing in the ſhape of the leaves and ſize of the flowers, which ſtill have names among old Botaniſts.

3. The FIVE LOBED LAVATERA is a diſtinct ſpecies from the preceding, though it differs little from it, except in the nature of the leaves, each of which is compoſed

posed of five lobes, that are hastated, or pointed like a spear, and in the flowers of this shrub being smaller. They will be in full blow in August, and there will often be a succession of them till the early frosts advance. The leaves of this species vary: Some are shaped like Briony; others are nearly round; and the lobes of others are very acute. Hence the names Briony-leaved, Round leaved, Acute-leaved Mallow Tree, &c. have been used to express them.

4. GLITTERING LAVATERA grows to about the height of the former. The leaves are large, septangular, plaited, downy, white, and glisten towards the sun. The flowers are produced in bunches, from the ends of the branches: They are shaped like those of the Common Mallow, come out in July, and continue in succession until the end of autumn.

The *Varieties* of this species go by the names of *Waved leaved*, the *Common Spanish*, the *Sulphur-leaved Mallow Tree*, &c. These three sorts are easily PROPAGATED by cuttings, which should be planted, early in the spring, in a shady border of light rich earth. Many of them will grow; and the plants may stand two or three years before they are removed to the places where they are designed to remain.

The best method of PROPAGATING all the sorts is by seeds; and by this way fresh Varieties may be obtained. The seeds should be procured from Spain, where the plants naturally grow; for none, except the first sort, ripen well here. Having got a sufficient quantity, sow them in a border of light, fine, rich earth, about the middle of March. They will easily come up, and nothing but weeding and watering in dry weather will be required until the spring after; when they should be planted in nursery lines, there to remain until they are set out to stand. The leaves of all the sorts continue until the frosts come on; so that if an open winter happens, they will continue in verdure the greatest part of the season.

The Lavatera tribe affect a warm sandy situation and soil, in which they will sometimes continue to exhibit their beauties for many years; but in general they are short-lived, continuing only two or three years: this re-

renders them peculiarly eligible to be ſcattered plentifully in a newly made ſhrubery; they will add warmth to young plants, and will die away themſelves before the ſpaces they occupy will be required by the ſurrounding ſhrubs.

LAURUS.

LINNEAN Claſs and Order, *Enneandria Monogynia*: Each flower contains nine males and one female. There are eleven SPECIES, four of which are adapted to our purpoſe:

1. LAU'RUS *No'bilis*: The EVERGREEN BAY, or the COMMON BAY; *an evergreen ſhrub or tree*; native of Italy, Greece, and Aſia.

2. LAU'RUS *Æſtiva'lis*: The DECIDUOUS BAY; *a tall deciduous ſhrub*; native of ſwampy places, and the ſides of brooks and rivers in Virginia.

3. LAU'RUS *Be'nzoin*: The BENZOIN TREE, or BENJAMIN TREE, *a deciduous tree or ſhrub*; native of Virginia and Pennſylvania.

4. LAU'RUS *Sa'ſſafras*: The SASSAFRAS TREE; *a deciduous tree or ſhrub*; native of Virginia, Carolina, and Florida.

1. The EVERGREEN BAY. EVELYN ſays, he has ſeen Bay Trees near thirty feet high, and almoſt two feet in diameter, and enumerates the Bay amongſt *uſeful* trees. HANBURY catches at this idea, and tells us in general terms, that "it will grow to thirty feet in height, with a trunk of two feet in diameter," and, accordingly, he arranges it amongſt his Foreſt Trees: he acknowledges however at the ſame time that the wood is of little value. The Bay is nevertheleſs a fine aromatic and a beautiful evergreen: It is ſaid to be the true *Laurus* or Laurel of the Antients, with which they adorned the brows of their ſucceſsful Generals. The leaves ſtand cloſe, are about three inches long and two broad; are hard, rigid, and of a deep green colour. The Bay too, like the Holly, Box, and Laurel, will bear

bear the ſhade and drip of taller trees, and it is upon the whole a very deſirable, as being a very *ornamental* evergreen. There are ſeveral *Varieties*; as the Broad-leaved Bay, the Narrow-leaved Bay, and the Wave-leaved Bay.

This tree is PROPAGATED by layers, or by the berries. In order to raiſe a quantity of theſe trees by layers, ſome ſtools ſhould be planted for the purpoſe; and after theſe are ſhot about a yard high, the branches muſt be brought down to the ground in the winter, all the preceding ſummer's ſhoots laid on it, and pegged down (being firſt ſlit in the joint), and the leaves taken off, which would otherwiſe be under ground. In one year's time theſe layers will have taken root; and in the ſpring they ſhould be taken up, and planted in the nurſery a foot aſunder, in rows two feet diſtance. After they are planted out, if the weather ſhould prove dry, they muſt be conſtantly watered; for without ſuch care, it is difficult to make this tree grow. After they have taken well to the ground, they will require no farther trouble than keeping them clean from weeds, and digging between the rows each winter, till they are finally planted out. 2. In order to raiſe this tree from the berries, they ought to hang on the trees till about January before they are gathered. A well ſheltered ſpot of ground for the ſeminary muſt be made choice of; and having the mould ſmooth and fine, they ſhould be ſown ſoon after they are gathered in beds, or drills, rather more than half an inch deep. Towards the cloſe of the ſpring the plants will come up, and during ſummer muſt be duly attended, by watering and weeding. In the winter following, their ſheltered ſituation muſt not be truſted to, to defend them from the froſt: Furze buſhes, or ſome ſuch things, ought to be ſtuck in rows, between the beds or drills, to guard them from the black froſts. Indeed, without this precaution, if the winter ſhould prove very froſty, few of the young ſeedlings will be alive in ſpring. During the following ſummer, weeding and watering muſt be obſerved, and the winter after that they ſhould be defended with covering as before; for they will be ſtill in danger of being deſtroyed by ſevere froſts. In the enſuing ſpring, the ſtrongeſt may be taken out of the ſeed beds, and planted

planted in the nurſery way; though if they have not by that time made good ſhoots, it will be adviſeable to let them remain in their beds till the third ſpring; for a ſmall plant of this kind is with more difficulty made to grow than one which is larger. When they are planted in the nurſery, the diſtance which ſhould be allowed them is the ſame as the layers, a foot aſunder and two feet diſtance in the rows; and this will not be found too cloſe; for notwithſtanding the greateſt care is exerted in planting them in the nurſery, even making choice of rainy and cloudy weather, which muſt always be obſerved in ſetting them out, many of them will be loſt by being tranſplanted. After they are thus planted out in the nurſery, whether layers or ſeedlings, they muſt be ſtill watered in dry weather, kept free from weeds, and the rows dug between every winter. You will even find, that thoſe plants which ſuffer leaſt by being tranſplanted will have met with a check, which they will not recover in two or three years; and till they have acquired new ſtrength they ſhould not be taken from the nurſery; but when they appear to be good ſtiff plants, having the year before made a vigorous ſhoot, they will be then proper plants for planting out where they are to remain. Holes ſhould be got ready for their reception; and as ſoon as the firſt autumnal rains fall, the work ſhould be ſet about, eſpecially if the land be gravelly or dry; but if it be moiſt, the ſpring will do as well. Being now planted at one yard diſtance, they will make a poor progreſs for two or three years more; but after this, when they have overcome all theſe difficulties, they will grow very faſt, and arrive to be good trees in a few years.

Although this tree flouriſhes beſt in old gardens, where the ſoil has been made rich and deep, and loves the ſhade, HANBURY tells us, "it thrives nevertheleſs exceedingly well in our hotteſt gravels and ſands; and, after it has ſurmounted the hardſhips of tranſplanting, will grow in ſuch ſituations extremely faſt, and arrive to a larger bulk."

2. The DECIDUOUS BAY, in a moiſt rich ſoil, in which it principally delights, will grow to be about ſixteen feet high; but in ſome ſoils, that are poſſeſſed of the oppoſite qualities, it will hardly arrive at half that

that height. The branches are not very numerous, but they are ſmooth, and of a purpliſh colour, look well in winter, and in ſummer exhibit their leaves of an oval ſpear-ſhaped figure: They are about two or three inches in length, are proportionally broad, and placed oppoſite to each other on the branches. Their upper ſurface is ſmooth, and of a pleaſant green colour, whilſt their under is rough and veined. The flowers are ſmall and white, make no figure, come out from the ſides of the branches in May, and are ſucceeded by large red berries, which never ripen in England: So that, notwithſtanding the leaves in ſummer are very pretty, and the colour of the bark makes a variety in winter, it is principally the ſcarcity of this plant which makes it valuable.

3. The BENZOIN TREE will grow to a much larger ſize than the other, and its branches are rather numerous: They are ſmooth, and of a fine light green colour. The leaves are oval, acute, near four inches long, and two broad; their upper ſurface is ſmooth, and of a fine light green colour, but their under ſurface is venoſe, and of a whitiſh caſt: When bruiſed, they emit a fine fragrance. The flowers make no figure: They are ſmall and yellowiſh, come out from the ſides of the branches in little cluſters, and are ſucceeded by large blackiſh berries, which never ripen in England.

4. The SASSAFRAS TREE. The wood of the Saſſafras is well known in the ſhops, where it is ſold to be made into tea, being eſteemed an excellent antiſcorbutic and purger of the blood. A decoction of the leaves and bark is alſo ſaid to poſſeſs the ſame virtues, and is drank by many perſons for thoſe purpoſes. This tree will grow to nearly the height of the others, though the branches are not ſo numerous. Its bark is ſmooth, and of a red colour, which beautifully diſtinguiſhes it in winter; whilſt the fine ſhining green of its leaves conſtitutes its greateſt beauty in ſummer. In theſe, indeed, there is a variety, and a very extraordinary one. Some are large, and of an oval figure; others are ſmaller, and of the ſame ſhape; whilſt others, again, are ſo divided into three lobes, as to reſemble the leaves of ſome ſorts of the Fig tree. Their edges are entire; their under ſurface is of a whitiſh caſt; their

foot-

footſtalks are pretty long, placed alternately on the branches, and die to a red colour in the autumn. The flowers are ſmall and yellowiſh: They are produced in cluſters on longiſh pedicles, and are ſucceeded by blackiſh berries, which never ripen in England.

The PROPAGATION of theſe three ſorts of trees may be performed two or three ways. 1. By the ſeeds. Theſe we receive, from the places where the trees grow naturally, in the ſpring. They ſhould be preſerved in ſand; and, as ſoon as they arrive, ſhould be ſown in largiſh pots, an inch deep. The ſoil for their reception ſhould be taken from a rich paſture at leaſt a year before, with the ſward. It ſhould alſo be laid on a heap, and frequently turned, until the ſward is grown rotten, and the whole appears well mixed and fine. If the paſture from whence it was taken near the ſurface is a ſandy loam, this is the beſt compoſt for theſe ſeeds; if not, a ſmall addition of drift or ſea ſand ſhould be added, and well mixed with the other mould. After filling the pots with this ſoil, the ſeeds ſhould be ſown an inch deep; and then they ſhould be plunged into common mould up to the rim. If the ſoil be naturally moiſt, it will keep them cooler, and be better; and if the place be well ſheltered and ſhaded, it will be better ſtill. Nothing more than weeding, which muſt be conſtantly obſerved during the ſummer, will be neceſſary; and in this ſtation they may remain until the March following; about the middle of which month, having prepared a good hotbed, the pots ſhould be taken up and plunged therein. Soon after the ſeeds will come up; and when the young plants have ſufficiently received the benefit of this bed, they ſhould be enured by degrees to the open air. Weeding and watering muſt be obſerved during the ſummer; and, at the approach of the cold weather in the autumn, they ſhould be removed under a hotbed frame, or ſome cover, to be protected from the froſts during the winter. In the ſpring, when this danger is over, they ſhould reſume their firſt ſtation; namely, the pots ſhould be plunged up to the rim, as when the ſeeds were firſt ſown; and if this place be well ſheltered, they may remain there all winter; if not, and ſevere froſts threaten, they ſhould be taken up and placed under

under cover as before. After they have been thus managed three years from the feeds, they ſhould be taken out of the pots with care, and planted in the nurſery ground, at ſmall diſtances; where they may remain until they are ſtrong enough to be finally ſet out. By ſowing the feeds in pots, and aſſiſting them by an hotbed, a year at leaſt is ſaved; for they hardly ever come up, when ſown in a natural border, under two years from the feeds; nay, they have been known to remain three, and even ſome plants to come up the fourth year after ſowing; which at once ſhews the preference of the former practice, and ſhould caution all who have not ſuch convenience, not to be too haſty in diſturbing the beds when the feeds are ſown in the natural ground; as, eſpecially if they are not well preſerved in mould or ſand, theſe may be ſome years before they appear. Indeed, it is the long time we are in obtaining theſe plants, either by feeds, layers, &c. that makes them at preſent ſo very ſcarce amongſt us. 2. Theſe plants may alſo be encreaſed by layers; but very ſlowly, for they will be two, and ſometimes three, or even four years, before they have ſtruck out good roots; though the Benjamin tree is propagated the faſteſt by this method. The young twigs ſhould be laid in the ground in the autumn; and it will be found that twiſting the wire round the bud, ſo as in ſome degree to ſtop the progreſs of the ſap, and taking away with a knife a little of the bark, is a more effectual method of obtaining good roots ſoon than by the ſlit or twiſting, eſpecially when practiſed on the Saſſafras Tree. 3. Plants of theſe ſorts are likewiſe ſometimes obtained by ſuckers, which they will at all times throw out, and which may be often taken off with pretty good roots; but when they are weak, and with bad roots, they ſhould be planted in pots, and aſſiſted by a moderate heat in a bed: With ſuch management they will be good plants by the autumn, and in the ſpring may be planted out anywhere. 4. Cuttings of theſe trees, when planted in a good bark bed, and duly watered, will alſo oftentimes grow. When this method is practiſed, and plants obtained, they muſt be enured by degrees to the open air, till they are hardy enough to be finally planted out.

LIGUS-

LIGUSTRUM.

LINNEAN Clafs and Order, *Diandria Monogynia* Each flower contains two males and one female. There is only one SPECIES:

LIGU'STRUM *Vulga're:* The PRIVET; a well known *deciduous or evergreen ſhrub*; common in the woods and hedges of many parts of England, and almoſt all Europe.

The PRIVET is diviſible into two *Varieties:*

The Deciduous or Common Privet, and
The Evergreen Privet.

The *Deciduous Privet* will grow to the height of about ten or twelve feet. The branches are very numerous, ſlender, and tough; covered with a ſmooth gray bark; and, when broken, emit a ſtrong ſcent. The young twigs are generally produced oppoſite, and alternately of contrary directions on the older branches. The leaves alſo are placed oppoſite by pairs in the ſame manner. They are of an oblong figure, ſmall, ſmooth, of a dark green colour, have a nauſeous diſagreeable taſte, and continue on the trees very late. The flowers are produced in cloſe ſpikes, at the ends of the branches, in May, June, and often in July: They are white, very beautiful, and ſucceeded by black berries, which in the autumn will conſtitute the greateſt beauty of this plant; for they will be all over the tree, at the ends of the branches, in thick cluſters. They are of a jet black; and will thus continue to ornament it in this ſingular manner during the greateſt part of the winter.

The PROPAGATION of the Privet is eaſy; for it may be encreaſed, 1. By the ſeeds; and by this way the ſtrongeſt plants may be obtained. The ſeeds, ſoon after they are ripe, ſhould be ſown in any bed of common garden mould made fine. They ought to be covered about an inch deep; and all the ſucceeding ſummer ſhould be kept clean from weeds; for the plants never, at leaſt not many of them, come up until the ſpring after. After they are come up, they will require no other care

 than

than weeding; and in the ſpring following may be planted in the nurſery ground, where they will require very little care beſides keeping the weeds down, until they are taken up to plant. 2. Theſe plants may be encreaſed by layers; for the young ſhoots being laid in the ground in the autumn, will by that time twelvemonth have taken good root; the largeſt of which may be planted out to ſtand, and the ſmalleſt ſet in the nurſery, to gain ſtrength. 3. Cuttings alſo, planted in October, will ſtrike root freely; and if the ſoil is inclined to be moiſt, and is ſhaded, it will be the better for them, eſpecially if the ſucceeding ſummer ſhould prove a dry one. If theſe cuttings are thinly planted, they will require no other removing till they are finally ſet out. If a large quantity is deſired, they may be placed cloſe, within about two or three inches of each other, and then taken up and planted in the nurſery the autumn following, to remain there until they are wanted for the above purpoſe. It propagates itſelf by *ſuckers* and by *layers*, whenever the twigs or branches touch the ground.

The Privet, of all others, will thrive beſt in the ſmoke of great cities; ſo that whoever has a little garden in ſuch places, and is deſirous of having a few plants that look green and healthy, may be gratified in the Privet, becauſe it will flouriſh and look well there. It will alſo grow very well under the ſhade and drip of trees.

The *Evergreen Privet.* This Variety has a tendency to grow to be a taller and a ſtronger tree than the common deciduous ſort. The leaves are rather larger, more pointed, of a thicker conſiſtence, of a dark green colour, and they continue on the ſame plant ſo long as to entitle it to the appellation of Evergreen; though it may often be obſerved to be almoſt deſtitute of leaves early in winter, eſpecially thoſe that were on the ends of the higheſt branches, which are often taken off by the firſt cutting winterly winds. In order to have this tree keep up the credit of an evergreen, it ſhould have a well ſheltered ſituation; for although it be hardy enough to bear with impunity the ſevereſt cuts of the northern blaſts, on the tops of hills, craggy rocks, &c. yet without ſome ſhelter the leaves are ſeldom preſerved all winter, and with protection it is generally allowed

allowed to be a handſome evergreen. As it is a Variety of the deciduous ſort, the ſame flowers and fruit may be expected.

It is to be raiſed in the ſame manner, by layers or cuttings; and the ſeeds of this ſort ſometimes produce plants of the like ſort, that retain their leaves. It will bear a very moiſt ſituation.

LIQUIDAMBER.

LINNEAN Claſs and Order, *Monoecia Polyandria*: Male flowers containing many ſtamina, and female flowers containing two piſtils, ſituated upon the ſame plant; the males being collected into long conical looſe catkins, and the females forming a globe ſituated at the baſe of the male ſpike. There are only two SPECIES:

1. LIQUIDA'MBER *Styraci'flua*: The VIRGINIA LIQUIDAMBER, or the MAPLE-LEAVED LIQUIDAMBER; *a deciduous tree*; native of the rich moiſt parts of Virginia and Mexico.

2. LIQUIDA'MBER *Peregri'num*: The CANADA LIQUIDAMBER, or the SPLEENWORT-LEAVED GALE; *a deciduous tree*; native of Canada and Pennſylvania.

1. The VIRGINIA LIQUIDAMBER will ſhoot in a regular manner to thirty or forty feet high, having its young twigs covered with a ſmooth light brown bark, while thoſe of the older are of a darker colour. The leaves grow irregularly on the young branches, on long footſtalks: They reſemble thoſe of the Common Maple in figure; the lobes are all ſerrated; and from the baſe of the leaf a ſtrong midrib runs to the extremity of each lobe that belongs to it. They are of a lucid green, and emit their odoriferous particles in ſuch plenty as to perfume the circumambient air; nay, the whole tree exſudes ſuch a fragrant tranſparent reſin, as to have given occaſion to its being taken for the Sweet Storax. Theſe trees, therefore, are very

proper to be planted singly in large opens, that they may amply display their fine pyramidal growth, or to be set in places near seats, pavilions, &c. The flowers are of a kind of saffron colour: They are produced at the ends of the branches the beginning of April, and sometimes sooner; and are succeeded by large round brown fruit, which looks singular, but is thought by many to be no ornament to the tree.

2. CANADA LIQUIDAMBER. The young branches of this species are slender, tough, and hardy. The leaves are oblong, of a deep green colour, hairy underneath, and have indentures on their edges alternately, very deep. The flowers come out from the sides of the branches, like the former; and they are succeeded by small roundish fruit, which seldom ripens in England.

The PROPAGATION of both these species is the same, and may be performed by seeds or layers; but the first method is the best. 1. We receive the seeds from America in the spring. Against their arrival a fine bed, in a warm well sheltered place, should be prepared. If the soil is not naturally good, and inclined to be sandy, it should be wholly taken out near a foot deep, and the vacancy filled up with earth taken up a year before, from a fresh pasture, with the sward and all well rotted and mixed by being often turned, and afterwards mixed with a sixth part of drift or sea sand. A dry day being made choice of, early in March let the seeds be sown, and the finest of this compost riddled over them a quarter of an inch deep. When the hot weather in the spring comes on, the beds should be shaded, and waterings given often, but in very small quantities, only affording them a gentle, nay, a very small sprinkling at a time. MILLER says, the seeds of these plants never come up under two years. But, continues HANBURY, with this easy management, I hardly ever knew it longer than the end of May before the young plants made their appearance. The plants being come up, shading should still be afforded them in the parching summer, and a watering every other night; and this will promote their growth, and cause them to become stronger plants by the autumn. In the autumn, the beds should be hooped to be covered with mats in the severe fevere frofts. Thefe mats, however, fhould always be taken off in open weather; and this is all the management they will require during the firft winter. The fucceeding fummer they will require no other trouble than weeding; though, if it fhould prove a very dry one, they will find benefit from a little water now and then. By the autumn they will be grown ftrong enough to refift the cold of the following winter, without demanding the trouble of matting, if the fituation is well fheltered; if not, it will be proper to have the hoops prepared, and the mats ready, againft the black northern frofts, which would endanger at leaft their lofing their tops. After this, nothing except weeding will be wanted; and in the fpring following, that is, three years from their firft appearance, they fhould be taken up (for they fhould not be removed before, unlefs fome of the ftrongeft plants be drawn out of the bed) and planted in the nurfery, a foot afunder, and two feet diftant in the rows. Hoeing the weeds in the rows in the fummer, and digging them in the winter, is all the trouble they will afterwards occafion until they are finally planted out. 2. Thefe plants are eafily encreafed by layers. The operation muft be performed in the autumn, on the young fummer's fhoots; and the beft way is by flitting them at a joint, as is practifed for carnations. In a ftrong dry foil, they will be often two years or more before they ftrike root; though, in a fine light foil, they will be found to take freely enough. By this method good plants may be obtained, though it is not fo eligible as the other, if we have the conveniency of procuring the feeds.

LIRIODENDRON.

Linnean Clafs and Order, *Polyandria Polygynia*: Each flower contains many males and many females. There are two Species; one of them bearing a tulip-like, the other a lily-like flower; the former is not uncommon in our open grounds:

LIRIODE'NDRON *Tulipi'fera:* The TULIP TREE, or the VIRGINIA TULIP TREE; *a deciduous tree*; native of most parts of America.

The TULIP TREE. In those parts of America where it grows common, it will arrive to a prodigious bulk, and affords excellent timber for many uses; particularly, the trunk is frequently hollowed, and made into a canoe sufficient to carry many people; and for this purpose no tree is thought more proper by the inhabitants of those parts. With us, it may be stationed among trees of forty feet growth. The trunk is covered with a gray bark. The branches, which are not very numerous, of the two years old wood, are smooth and brown; whilst the bark of the summer's shoots is smoother and shining, and of a blueish colour. They are very pithy. Their young wood is green, and when broken emits a strong scent. The leaves grow irregularly on the branches, on long footstalks. They are of a particular structure, being composed of three lobes, the middlemost of which is shortened in such a manner, that it appears as if it had been cut off and hollowed at the middle: The two others are rounded off. They are about four or five inches long, and as many broad. They are of two colours; their upper surface is smooth, and of a stronger green than the lower. They fall off pretty early in autumn; and the buds for the next year's shoots soon after begin to swell and become dilated, insomuch that, by the end of December, those at the ends of the branches will become near an inch long, and half an inch broad. The outward lamina of these leaf buds are of an oval figure, have several longitudinal veins, and are of a blueish colour. The flowers are produced with us in July, at the ends of the branches: They somewhat resemble the Tulip, which occasions its being called the Tulip tree. The number of petals of which each is composed, like those of the Tulip, is six; and these are spotted with green, red, white, and yellow, thereby making a beautiful mixture. The flowers are succeeded by large cones, which never ripen in England.

The PROPAGATION of the Tulip tree is very easy, if the seeds are good, for by these, which we receive from abroad, they are to be propagated. No particular compost

post need be sought for; neither is the trouble of pots, boxes, hotbeds, &c. required: They will grow exceedingly well in beds of common garden mould, and the plants will be hardier and better than those raised with more tenderness and care. Therefore, as soon as you receive the seeds, which is generally in February, and a few dry days have happened, that the mould will work freely, sow the seeds, covering them three quarters of an inch deep; and in doing of this, observe to lay them lengthways, otherwise, by being very long, one part, perhaps that of the embryo plant, may be out of the ground soon, and the seed be lost. This being done, let the beds be hooped; and as soon as the hot weather and drying winds come on in the spring, let them be covered from ten o'clock in the morning until sunset. If little rain happens, they must be duly watered every other day; and by the end of May the plants will come up. Shade and watering in the hottest summer must be afforded them, and they will afterwards give very little trouble. The next winter they will want no other care than, at the approach of it, sticking some furze bushes round the bed, to break the keen edge of the black frosts; for it is found that the seedlings of this sort are very hardy, and seldom suffer by any weather. After they have been two years in the seed bed, they should be taken up and planted in the nursery, a foot asunder, and two feet distant in the rows. After this, the usual nursery care of hoeing the weeds, and digging between the rows in the winter, will suffice till they are taken up for planting out.

LONICERA.

Linnean Class and Order, *Pentandria Monogynia*: Each flower contains five males and one female. There are fourteen Species; eleven of which will bear the open air of this country.

1. Lonice'ra *Caprifolium*: The Italian Honey-

SUCKLE; *a deciduous or evergreen climber*; native of Italy and the South of Europe.

2. LONICE'RA *Pericly'menum*: The ENGLISH HONEYSUCKLE, or WOODBINE; *a deciduous or evergreen climber*; native of England, Germany, and the midland parts of Europe.

3. LONICE'RA *Sem'ervi'rens*: The TRUMPET HONEYSUCKLE; *a deciduous or evergreen climber*; native of America.

4. LONICE'RA *Diervi'lla*: The DIERVILLA, or ACADIAN HONEYSUCKLE; *a deciduous shrub*; native of Acadia and Nova Scotiá.

5. LONICE'RA *Symphorica'rpus*: ST. PETER'S WORT, or the VIRGINIA HONEYSUCKLE; *a deciduous shrub*; native of Virginia and Carolina.

6. LONICE'RA *Cœru'lea*: The BLUE-BERRIED HONEYSUCKLE, or the UPRIGHT BLUE-BERRIED HONEYSUCKLE; *a deciduous shrub*; native of Switzerland.

7. LONICE'RA *Alpigéna*: The RED-BERRIED HONEYSUCKLE; or the UPRIGHT RED-BERRIED HONEYSUCKLE; or the ALPINE HONEYSUCKLE; *a deciduous shrub*; native of Savoy and the Helvetian and Pyrenean Mountains.

8. LONICE'RA *Ni'gra*: The BLACK-BERRIED HONEYSUCKLE; or the BLACK-BERRIED UPRIGHT HONEYSUCKLE; *a deciduous shrub*; native of Switzerland and the Alps.

9. LONICE'RA *Xylosteum*: The FLY HONEYSUCKLE; *a deciduous shrub*; native of most of the coldest parts of Europe.

10. LONICE'RA *Pyrena'ica*: The PYRENEAN HONEYSUCKLE, or DWARF CHERRY; *a deciduous shrub*; native of the PYRENEAN MOUNTAINS.

11. LONICE'RA *Tartarica*: The TARTARIAN HONEYSUCKLE, or DWARF CHERRY, or the DWARF CHERRY WITH HEART-SHAPED LEAVES; *a deciduous shrub*; native of Tartary.

1. The ITALIAN HONEYSUCKLE. The *Varieties* of this species are, Early White Italian Honeysuckle, Early Red Italian Honeysuckle, Yellow Italian Honeysuckle, Late Red-flowered Italian Honeysuckle, Evergreen Italian Honeysuckle.

The *Early White Italian Honeysuckle* is that which first makes

makes its appearance in May. The leaves of this ſort are oval, and placed oppoſite by pairs, cloſe to the branches, at the extremity of which the leaves quite ſurround it. The flowers grow in bunches round the ends of the branches, and have a very fine ſcent. Their blow will be ſoon over; and they are ſucceeded by red pulpy berries, which will be ripe in the autumn.

The *Early Red* differs from the preceding in that the leaves are narrower, the fibres of the flowers are more ſlender, and it blows a little later in the ſpring.

The *Yellow Italian Honeyſuckle* does not blow quite ſo early as the other, and the flowers are yellow: In other reſpects it is very much like the former.

Late Red-flowered Italian Honeyſuckle is one of the beſt we have. The ſtem is tolerably firm; the branches are few, and the leaves large; the flowers are alſo large, of a deep red colour, though leſs ſcented than the earlier ſorts.

Evergreen Italian Honeyſuckle. This is a ſtronger ſhooter than any of the ſorts. The joints are more diſtant from each other. The leaves are large, of a thick conſiſtence, unite, and ſurround the ſtalk with their baſe, and continue all winter. The flowers are large, of a good red colour, with ſome paler ſtripes, and often continue to blow to the end of autumn.

2. ENGLISH HONEYSUCKLE. The *Varieties* of this ſpecies are, The Common Woodbine of our Hedges, The Oak leaved Honeyſuckle, Red Dutch Honeyſuckle, Midſummer Honeyſuckle, Late German Honeyſuckle, Long-blowing Honeyſuckle, Evergreen Honeyſuckle.

The *Common Woodbine* is known all over England, in our woods and hedges. There are ſtill Varieties of this ſort, in its wild ſtate; ſome having prodigious weak trailing branches; others again with tolerably woody ſtems. Some of the flowers are whitiſh, others are of a greeniſh caſt; whilſt others are poſſeſſed of a reddiſh tinge. As the flowers of none of theſe are nearly ſo beautiful as thoſe of the cultivated ſorts, only a plant or two of them ſhould be introduced; which will cauſe ſome variety, and ſerve as a foil to ſet the others off. There is a *Sub-Variety* of this ſort, with ſtriped leaves.

Oak-

Oak-leaved Honeyſuckle is an accidental variety of our Common Woodbine. It differs in no reſpect from it, only that ſome of the leaves are ſhaped like thoſe of the Oak tree, on which account it is valuable, and makes a pretty variety in collections.

There is alſo a *Sub-Variety* of this ſort, with leaves beautifully variegated, called *Striped Oak-leaved Honeyſuckle.*

Red Dutch Honeyſuckle is a very good ſort. It flowers in June, and will often continue in blow a month or two. The branches have a ſmooth purpliſh bark, and may be known from the others even in winter, when they will appear with their ſwelled buds alſo of that colour. The leaves are of an oblong oval figure, and ſtand oppoſite by pairs on the branches, on ſhort footſtalks. The flowers are produced in bunches at the ends of the branches: Their outſide is red, but within they are of a yellowiſh colour, and poſſeſſed of a delightful odour.

The *Midſummer Honeyſuckle* is very much like the former, only the ſtalks are more ſlender, of a lighter brown colour, and the tubes of the flowers are ſmaller, neither are they ſo red. It will be in blow about Midſummer; and the plant, whether ſet againſt a wall, pales, a hedge, or in the ground, will be all over covered with bloom, making an enchanting appearance to the eye, and perfuming the air all around to a conſiderable diſtance.

Late German Honeyſuckle is very much like the Red Dutch, only it blows later. It will flower in July and Auguſt; and has all the properties of the other ſorts, as to fragrance and beauty.

The *Long-blowing Honeyſuckle* is ſtill another Variety of the Dutch. It will often exhibit flowers in June, July, and Auguſt, though the profuſion will not be ſo great as that of the other ſorts.

Evergreen Honeyſuckle is another Variety which retains its leaves all winter. It often flowers late in the autumn; and ſometimes, in mild ſeaſons, retains its bloom until Chriſtmas, which makes it ſtill more valuable.

3. TRUMPET HONEYSUCKLES. Of theſe are the following *Varieties:* Virginian Trumpet Honeyſuckle, Caro-

Carolina Trumpet Honeyfuckle, Evergreen Trumpet Honeyfuckle.

Virginian Trumpet Honeyfuckle is the moft beautiful of all the forts, though Nature has denied it fmell. The branches are flender, fmooth, and of a reddifh colour. The leaves fit clofe to the branches by pairs. They are of an oblong oval figure, and their lower furface is not of fo fhining a green as the upper. Thofe at the extremity of the branches near the flowers furround the ftalk, through which it comes. The flowers grow in bunches, at the ends of the fhoot, and are of a bright fcarlet colour. They will often be in blow from June to October; but the flowers have no fcent.

Carolina Trumpet Honeyfuckle differs in no refpect from the former, only that the branches are more flender, and the leaves and flowers alfo are proportionally fmaller, thereby making a pretty variety. This fort was introduced into our gardens from Carolina, as was the preceding from Virginia.

Evergreen Trumpet Honeyfuckle. The leaves are of a thicker fubftance, and continue on the plants all winter; but the flowers are of a deep fcarlet, like the other, and are poffeffed of little or no fragrance.

The PROPAGATION of thefe forts is very eafy. 1. The young branches being laid in the ground any time in the winter, with no other art, will become good plants by the autumn following, and may be then taken off for ufe. 2. But our common method of propagating thefe forts is by cuttings. The beft month for this work is October. By this way prodigious quantities of plants may be raifed, and hardly any of them will fail growing. So eafily may thefe delightful plants be multiplied, when a plant of each fort is once obtained.

The Evergreen Italian Honeyfuckle (the beft of the evergreen forts) does not readily take by *cuttings*; fo that in order to make fure of this plant, the young branches muft be *layered*, any time in the autumn or winter, and by the autumn following they will have plenty of roots, and be good plants fit for removing to any place.

The Evergreen Honeyfuckles, though climbing plants, fhould occafionally be ftationed in the evergreen quarters, as fhould all the other forts among the deciduous

deciduous trees and ſhrubs; being ſo managed that their appearance may agree with thoſe of upright growth. This is done by nipping off the young ſhoots (which will ſoon get rambling and out of reach), that the plants may be kept within bounds, and made to join in the collection with great beauty. Neither may they only be kept low, to almoſt what height is required; but they may, by fixing a ſtake for their ſupport, be trained up to a ſtem, which will every year grow more and more woody and firm: ſo that in this caſe the eye muſt frequently overlook the tree, to take off the young ſhoots as they grow out, and not permit the head to grow too large and ſpreading for the ſtem, which it ſoon would do without this care; and with it, the head may be ſo kept in order as to bear good proportion to the ſtem, thereby cauſing the tree to have the appearance of an upright ſhrub.

4. DIERVILLA is a ſhrub of about the height of three or four feet. The branches are few, and larger in proportion than the height of the ſhrub; they are very full of pith, and when broken emit a ſtrong ſcent. The leaves are placed oppoſite by pairs, on ſhort footſtalks: They are near three inches long, and about half as broad; and of an oblong heart-ſhaped figure, finely ſerrated, and end in acute points: Their upper ſurface is ſmooth, and of a fine green colour; their under is lighter, and has five or ſix pair of ſtrong nerves running irregularly from the midrib to the borders. The flowers are produced in looſe bunches, both at the ends and at the ſides of the branches: Each is formed of one leaf; the tube is long, and the top is divided into five parts, which turn backward. They are of a yellow colour, and will be in blow in May, and ſometimes moſt of the ſummer months. Theſe flowers are ſucceeded, in the countries where they grow naturally, by black oval berries, each containing four cells. Diervilla forms an agreeable variety amongst other shrubs of its own growth, though the flowers make no great figure. It is very hardy with respect to cold; and may be planted in any part of the nursery where it is wanted.

No art is required to PROPAGATE this plant; it spawns, and thus propagates itself in great plenty.

Theſe

These ſuckers ſhould be taken up in autumn, and planted out in the nurſery: After remaining there a year or two, they may be finally taken up. This tree may be alſo encreaſed by cuttings. They ſhould be planted in October, very cloſe, if a quantity are wanted. By the autumn following, they will have good roots. They may be taken up and planted in the nurſery, like the ſpawn, for a year or two, and then ſet out to ſtand. Plants raiſed this way will not be quite ſo ſubject to throw out ſuckers as the others.

5. St. PETER's WORT. St. Peter's Wort will ariſe to the height of about four or five feet. The main ſtems are ragged, and of a dirty dark brown. The branches are numerous and ſhort, though oftentimes it ſends out ſome trailing ſlender branches, which will grow to a great length. The leaves of this ſhrub conſtitute, its greateſt beauty: They are very numerous, ſmall, about half an inch long, and of an oval figure. Their foot-ſtalks are exceedingly ſhort, and they ſtand oppoſite by pairs on the ſlender branches: Theſe die in the autumn to a dark brown. The time of this plant's flowering is Auguſt. The flowers grow round the ſtalks: They are ſmall, of an herbaceous colour, and make no figure.

The PROPAGATION is very eaſy. 1. If a ſpadefull of mould be thrown over each of the trailing branches, any time in the winter, they will by the autumn following have ſtruck root; and theſe may be planted out in the nurſery, to ſtand until they are of a proper ſize to be planted out for good. 2. This ſhrub may be alſo propagated by cuttings; and in order to obtain good cuttings for the purpoſe, the year before the plants ſhould be headed near the ground, which will make them ſhoot vigorouſly the ſummer following. Theſe young ſhoots muſt be the cuttings to be planted. October is the beſt month for the work; and if they are planted in a moiſtiſh ſoil, and have a ſhady ſituation, they will have taken good root by the autumn. If they are planted very thick, as cuttings commonly are, they ſhould be all taken up and planted in the nurſery a foot aſunder, and two feet diſtant in the rows; but if the living cuttings are no nearer than about a foot, they may remain without removing until they are planted out.

6. The

6. The Blue-berried Honeysuckle is a ſhrub of about four feet in growth. The branches are round, ſmooth, and of a reddiſh purpliſh colour. The leaves are oblong, ſpear-ſhaped, of a fine green, and ſtand oppoſite by pairs on the branches. The flowers, which are white, are produced in May from the ſides of the branches, and are ſucceeded by blue berries, that will be ripe in Auguſt.

7. The Red-berried Honeysuckle will grow to the height of about five feet. The branches are very upright; the young ſhoots are angular, and covered with a brown bark. The leaves are tolerably large, ſpear-ſhaped, a little reſembling thoſe of the mock Orange, and grow oppoſite to each other. The flowers are produced from the ſides of the branches, on long footſtalks: They are of a red colour, come out in April, and are each ſucceeded by a pair of red berries, which will be ripe the end of July or early in Auguſt.

8. Black-berried Honeysuckle differs from the Blue-berried only in that the ſeeds of this are black, and grow two together; whereas thoſe of the Blue-berried are ſingle and diſtinct. Except this, there is hardly any difference to be perceived.

9. Fly Honeysuckle will grow to the height of about ſeven or eight feet. The bark on the branches is of a whitiſh colour, which cauſes a variety, and makes it diſtinguiſhed in the winter ſeaſon. The leaves, which are placed oppoſite by pairs, are downy, and of an oblong oval figure. The flowers are white and erect: They are produced from the ſides of the branches in June, and are ſucceeded by two red berries, which will be ripe in September.

10. The Pyrenean Honeysuckle, or Dwarf Cherry, is but a low ſhrub: It ſeldom arrives to more than a yard in height. The branches are produced irregularly. The leaves are ſmooth, oblong, and placed oppoſite by pairs. The flowers are white, produced from the ſides of the branches, on ſlender footſtalks, in April; and are ſucceeded by roundiſh berries, which will be ripe in September.

11. Tartarian Honeysuckle, or Dwarf Cherry with Heart-shaped Leaves, is a ſhrub of about three or four feet high. Its branches are erect,

erect, like the upright sorts; and it differs in few respects from them, except that the leaves are heart-shaped. It exhibits its flowers in April; and these are succeeded by twin red berries, which will be ripe in August.

These are the *Upright* sorts of the Lonicera; to which one method of PROPAGATION is common; and that may be performed two ways. 1. By seeds. Common garden mould, dug fine, and cleared of the roots of all weeds, will serve for their reception. In this the seeds should be sown soon after they are ripe, about half an inch deep. After the beds are neated up, they will require no other care until the spring; when the weeds should be picked off as fast as they appear. Some of the plants by this time will have come up; but the far greater part will remain until the second spring before they shew themselves; so that the beds must be entirely untouched until at least two years after sowing. They will require no care all this time, except being kept clear of weeds; though if watering be afforded them in dry weather, it will be the better. After they are all up, and have stood a year or two in the seed bed, they may be taken up and planted in the nursery, at small distances; and in two or three years they will be of a proper size to plant out to stand. 2. All these sorts may be also propagated by cuttings. These should be planted in October, in any sort of garden mould that is tolerably good. If a quantity is wanted, they may be placed very close; and a small spot of ground will hold thousands. If the place be shaded, it will be a great advantage, as most cuttings are in danger of suffering by the violence of the sun's rays before they have struck, or whilst they are striking root. The winter following, they may be all taken up and planted out in the nursery, a foot asunder, and two feet distant in the rows, where they may stand until they are finally taken up for planting.

LYCIUM.

LINNEAN Clafs and Order, *Pentandria Monogynia*: Each flower contains five males and one female. There are eight SPECIES; one only of which is hardy enough to ftand a fevere winter in our climate.

LY'CIUM *Ba'rbarum*: The BOXTHORN; *a deciduous creeper*; native of Afia, Africa, and Europe.

The BOXTHORN. This fpecies affords two *Varieties*:

The Broad-leaved Boxthorn.

The Narrow-leaved Boxthorn.

The Broad-leaved Boxthorn is a rambling plant, and will, if let alone, in a few years overfpread every thing that is near it. The branches are very many, and fpread about in all directions. They will lie upon the ground, if unfupported, and will fhoot, in a good foil, fixteen feet in length in one fummer. Thofe branches that lie upon the ground will ftrike root; fo that from every part frefh fhoots will be fet forth the next fpring; and thus in a few years they will occupy a large compafs of ground; fo that whenever this plant is defired, they fhould be conftantly kept within bounds. Indeed, from its exceedingly rambling nature, not above a plant or two for variety or obfervation fhould be admitted in hardly any place. The branches of this plant are covered with a gray or whitifh bark. The leaves are of a light whitifh green, and of a thick confiftence. They grow on the branches, on all fides, by threes. This plant, of all the forts, is poffeffed of the longeft fpines (fome of which are a foot or more in length). Thefe fpines are garnifhed with leaves; and on thefe they for the moft part ftand fingly in an alternate manner. On the branches where they grow by threes, the middle one is always the largeft. They are all of an oval, fpear-fhaped figure, are very fmooth, a little gloffy, and often continue till the middle of winter before they fall off. Befides the long leafy thorns before mentioned, it produces many fhort fharp fpines, of a white colour, near the ends of the fhoots. The flowers are produced in Auguft,

August, and there will be often a succession of blow until the frosts come on. They grow singly at the joints, on short footstalks. They are of a purplish colour, small, and are succeeded by no fruit with us, as I could observe.

The *Long Narrow-leaved Boxthorn* is also a very great rambler. The branches are many, and are produced irregularly on all sides. It is possessed of spines, but these are very short, and the bark with which they are all covered is pretty white. The leaves are of a lanceolate figure, and are narrow and long. Their colour is that of a whitish green, and they grow alternately on the branches. The flowers are small, and appear in July; and are succeeded by red berries, which ripen in September, and at that time are very beautiful.

The PROPAGATION of these sorts is by cuttings: for they will grow, if planted at any time in any manner, and in almost any soil or situation, except a white clay. In a black rich earth, they will be the most healthful and most vigorous shooters; and though the cuttings will grow at all times, yet the winter months are to be preferred for the purpose.

M A G N O L I A.

LINNEAN Class and Order, *Polyandria Polygynia*: Each flower contains many males and many females. There are four SPECIES:

1. MAGNO'LIA *Glau'ca*: The SEA-GREEN MAGNOLIA, or the BAY-LEAVED TULIP TREE, or the SMALL MAGNOLIA; *a tall sub-evergreen shrub*; native of Virginia and Pennsylvania.

2. MAGNO'LIA *Acumina'ta*: The LONG-LEAVED MAGNOLIA; *a sub-evergreen shrub or tree*; native of Pennsylvania.

3. MAGNO'LIA *Tripe'tala*: The UMBRELLA TREE; *a sub-evergreen shrub or tree*; native of Carolina and Virginia.

4. MAGNO'LIA *Grandiflóra*: The EVERGREEN MAGNOLIA, or LAUREL-LEAVED TULIP TREE; *an evergreen tree*; native of Florida and Carolina.

1. The SEA-GREEN or SMALL MAGNOLIA grows with us to about the height of ten or twelve feet. The wood is white; and the branches, which are not very numerous, are covered with a ſmooth whitiſh bark. The leaves are tolerably large, and of two colours; their upper ſurface being ſmooth, and of a fine green, whilſt their under is hoary. They are of an oval figure, have their edges entire, and often continue the greateſt part of the winter before they fall off the trees. The flowers are produced at the ends of the branches, in May: Their colour is white; and the petals of which they are compoſed are concave and large; ſo that, together with the numerous ſtamina in the center, they preſent a beautiful appearance. They are alſo remarkable for their ſweet ſcent; and are ſucceeded by conical fruit, which never ripens in England; but in the places where they grow naturally, a ſingular beauty and oddity is added to theſe trees by the fruit; for the ſeeds are large, and lodged in cells all around the cone. When quite ripe, theſe are diſcharged from their cells; and hang each by a long narrow thread, cauſing thereby an uncommon and pleaſing effect.

2. LONG-LEAVED MAGNOLIA will grow to be near twenty feet high. The wood of this ſort is yellow, and the branches are covered with a ſmooth light bark. The leaves are very large, being near ten inches long; their figure is oval, ſpear-ſhaped, and all end in points. The flowers, which are produced in May, are white, and compoſed of twelve obtuſe petals, which, together with the number of ſtamina, make a good ſhow. Theſe alſo are ſucceeded by conical fruit, which never ripens in England.

3. The wood of the UMBRELLA TREE, which grows to about twenty feet in height, is more ſpongy than any of the other ſpecies of Magnolia. It is called the *Umbrella Tree*, from its manner of producing the leaves; for theſe are exceedingly large, and ſo produced as to form the appearance of an umbrella. The flowers of this ſort alſo are white, and the number of petals of which each is compoſed is about ten: They are ſucceeded

ceeded by fruit of a conical figure, with many cells all round for the seeds, which never ripen in England.

All these sorts may be PROPAGATED by seeds, layers, and cuttings. By the first of these methods the best plants are raised, though it is a very tedious way, and must be followed with great patience and trouble. We receive the seeds from those parts of America where they grow naturally. These are always preserved in sand, but, nevertheless, will not always prove good. As soon as possible after they arrive, which is generally in February, they should be sown in pots about half an inch deep. The best compost for them is a fresh loamy earth, mixed with a fourth part of drift sand; and the seeds should be thinly sown in each pot After this is done, the pots should be plunged up to the rims in the natural mould, under a warm hedge, where they may reap the benefit of the sun during the month of March and part of April; but when the rays of the sun begin to be strong and powerful, drying the mould in the pots very fast, they should be taken up and plunged again up to the rims in a shady border. By the end of May, if the seeds were good, the plants will come up; and all the summer they must be constantly attended with weeding and watering. At the approach of winter, they should be removed into the greenhouse, or placed under some cover; but in mild weather should always have the benefit of the open air and gentle showers In March, the pots with their seedlings should be plunged into a hotbed to set them forwards. Tanners bark is what the hotbed should be composed of; and as much air as the nature of the bed will allow, should always be afforded them. Water also must be given pretty often, though in small quantities, and the glasses must be shaded in the heat of the day. After this, about June, they should be inured to the open air; watering must still be afforded them; and this is what they require during the second summer. It has been a practice to plunge the pots into a hotbed soon after the seeds are sown; but this is a very bad method, for the young plants being thereby forced, grow thin and slender, and are seldom made to live longer than the first year. The second summer's management also has usually been, to plant the seed-

 lings

lings in March, in little pots, and then plunge them into a hotbed: but this is alſo a very bad way; for theſe ſeedlings, whether raiſed on hotbeds or the common ground, will be ſmall, and not of conſiſtence ſufficient to draw the juices, though the powers of vegetation are aſſiſted by a hotbed: Thus, hardly any of them ſurvive this early tranſplanting. This having been the general practice, theſe plants have been always thought very difficult to preſerve the ſecond year; whereas all thoſe difficulties vaniſh, by obſerving the above-directed method; for by letting the ſeeds have only the natural ſoil, they will the firſt ſummer be formed into young plants, which, though ſmall, will nevertheleſs be plants, and healthy. Thus being in the ſpring in their natural ſtate, with their pores open to receive the nutritious juices, and not having ſuffered by being tranſplanted, the hotbed will ſo help them, that they will be pretty plants by the autumn. At the approach of winter, they muſt be removed again under cover, and the former aſſiſtance of a hotbed ſhould be afforded them; and this ſhould be repeated until the plants are grown to be a foot or more in length. The ſpring following, the mould ſhould be turned out of the pots and ſhaken from the roots, and each plant put into a ſeparate pot. For theſe, a hotbed of tanners bark ſhould be ready, which will promote their growth, and make them healthy and fine. During the time they are in the bed, they ſhould be ſhaded; and about Midſummer the pots may be taken out and placed in a ſhady border. The winter following, it will be proper to houſe them in ſevere froſty weather; but always obſerve to place them abroad in mild ſeaſons. In March they may be turned out of the pots, the mould hanging to the roots, and planted with that in the places where they are to remain. 2. Theſe plants may be alſo propagated by layers. The young ſhoots in the autumn are moſt proper for the purpoſe; and it is found that a gentle twiſt, ſo as juſt to break the bark about the joint, is a better method than any other in practice. Theſe will ſometimes ſtrike root in one year, and ſometimes you muſt wait more than two before you find them with any. After they have ſtruck root, and are taken up, the

the beſt time for which is March, it is moſt eligible to plant each ſeparately in a pot, and plunge them into a hotbed, as directed for the ſeedlings; and by the ſpring following they will be ſtrong good plants for any place. 3. Theſe plants may likewiſe be increaſed by cuttings; by which they may be procured in plenty, if a perſon has the conveniency of a good ſtove; and without one this method ſhould not be attempted. Theſe cuttings ſhould be planted in pots; and after they are ſet in the ſtove, muſt be duly watered and ſhaded: By obſerving theſe directions many of them will grow. After this, they ſhould be brought by degrees to the open air; the winter following they ſhould be placed under a hotbed frame, or ſome ſhelter; and in the ſpring planted out to remain.

Theſe plants often retain their leaves, eſpecially when young, all winter, or the greateſt part of it, in ſome ſituations; and in ſuch they paſs for evergreens.

4. The Evergreen Laurel-leaved Magnolia. In the countries where it grows naturally, it arrives to the height and bulk of a timber tree. Thoſe countries are adorned with woods that are chiefly compoſed of this plant; and indeed, a wood of ſo noble a tree, luxuriantly ſhooting, flowering, and ſeeding, healthy and ſtrong, in ſoil and ſituation wholly adapted to its nature, muſt be a ſight of which we can hardly form an adequate idea, or have a juſt conception of its beauty or grandeur; for the tree naturally aſpires with an upright ſtem, and forms itſelf into a regular head. Many other trees do the ſame; but its moſt excellent properties conſiſt of the ſuperlative beauties of the leaves, flowers, and ſeeds. The leaves much reſemble thoſe noble leaves of the Laurel, from which it is ſo called, only they are larger, and of a thicker conſiſtence: Many of them will be ten inches or more in length, and four broad, and all are firm and ſtrong. Their upper ſurface is of a ſhining green, but their under is lighter, and often of a browniſh colour. This tinge, which is not always found in all trees, is by ſome thought a great beauty, and by others an imperfection; ſo various is the taſte of different people. Theſe leaves are produced without any order on the

tree, and ſit cloſe to the branches, having no ſeparate footſtalks. The idea we can form of a tree, of ſeventy or eighty feet high, plentifully ornamented with ſuch large and noble leaves, muſt be very great, and will induce us on their account only to endeavour to naturalize ſo noble a plant to our country. But let us conſider their flowers. These we find large, though ſingle, and of a pure white. They are produced at the ends of the branches, in July, and each is compoſed of about nine or ten large ſpreading petals. They have the uſual properties of thoſe that are broad and rounded at their extremity, of being narrow at the baſe, and their edges are a little undulated or waved. In the center of theſe petals are ſituated the numerous ſtamina, which the Botaniſt will be more curious in obſerving than the Gardener. But what affects all equally alike that have the ſenſe of ſmelling is, their remarkable fragrance, which indeed is of ſo great a degree, as to perfume the air to ſome diſtance; and if one tree, when in blow, is ſufficient to effect this, what conception ſhould we form of the odours diffuſed in the countries where there are whole woods of this tree in full vigour and blow! The fruit is nearly of the ſhape and ſize of a large egg; but what make it moſt ſingular and beautiful are the pendulous ſeeds, of a fine ſcarlet, which being diſcharged from their cells, hang by long threads, and have an effect both ſtriking and uncommon.

Rules have been given above for PROPAGATING deciduous Magnolias: the ſame rules obſerved, whether for ſeeds, layers, or cuttings, will raiſe plenty of this ſort; neither need any thing be added, except hinting to the Gardener, that this is more tender than the other ſorts, and that from thence he ſhould learn not to be over-haſty in committing theſe plants to the winter's cold, and planting them finally out. Snow is peculiarly injurious to them while young; ſo that, at the approach of ſuch weather, they muſt be particularly covered; and if ſnow ſhould happen to fall unawares, it ſhould be carefully cleared off the leaves and ſtems. When theſe plants are ſet abroad to remain, if the place is not exceedingly well ſheltered, it will be proper to have a ſhed at hand, which the

Gardener may put together, to ſcreen them from the ſevere northern froſts, and the black eaſterly winds, from which this ſhrub is moſt likely to ſuffer damage; and theſe froſty winds are the moſt deſtructive to it when they come early in the winter, while the ſhoots are rather tender, for then they are often deſtroyed, and the tree rendered unſightly for ſome time, though it will ſhoot out again. When this ſhrub is to be encreaſed by layers, it will be neceſſary, after the operation is performed, to make a hedge of reeds, or ſomething, at a little diſtance round it, to keep off the ſtrong winds, and prevent them from blowing the layers out of the ground; for without ſome guard this will be in danger of being done; ſince the leaves being very large and ſtrong, the wind muſt have great power over them.

MEDICAGO.

LINNEAN Claſs and Order, *Diadelphia Decandria*: Each flower contains ten males and one female; the males being connected at the baſe in two diviſions. There are twenty-four SPECIES; one of which, only, is admiſſible into our collection; the reſt being herbaceous plants.

MEDICA'GO *Arbórea*: The TREE LUCERNE, or TREE MEDICK, or MOON TREFOIL; *an evergreen ſhrub*; native of Italy, Crete, and the iſlands of the Archipelago.

The TREE LUCERNE will grow to be ſix or ſeven feet high, and divides without any order into many branches, which are covered with a gray bark. There is a delicacy in the young ſhoots beyond what is found in moſt trees; for they are white and ſilvery, and at the ſame time covered with the fineſt down. Theſe young ſhoots are plentifully ornamented with leaves, many of which come out from a bud. They are trifoliate, and grow on long ſlender footſtalks. One of the folioles is cuneiform, or ſhaped like a wedge; the

 others

others grow out more into a lanceolate figure, have alſo a whitiſh look, and are downy, though not to ſo great a degree as the young twigs on which they grow. They have a large midrib, which contracts the borders in the evening, and this alters their poſition of ſides on the alteration of weather. The flowers are produced from the ſides of the branches, in cluſters on long footſtalks. Each of theſe cluſters will be compoſed of ten or twelve flowers, which are of a beautiful yellow. They are of the butterfly kind; and re ſucceeded by moon-ſhaped pods, that ripen their ſeeds very well. One or other of theſe trees is to be found in blow almoſt at all times The beginning of the blow is generally ſaid to be in April or May; and indeed then we may expect to ſee the flowers largeſt and in the greateſt perfection; but the flowers of theſe trees may be ſeen in July, Auguſt, and September; and in greenhouſes have been known to blow all winter; which makes the tree more valuable to thoſe who are deſirous of ſeeing flowers in unuſual months.

This ſhrub is by many ſuppoſed to be the true *Cytiſus* of Virgil. it grows plentifully in Italy, in the iſlands of the Archipelago, and many other parts, where it is eſteemed excellent fodder for cattle. "For this purpoſe, continues HANBURY, the raiſing of it has been recommended in England; but there ſeems no probability of ſuch a ſcheme being brought to bear here; neither is it any way neceſſary to give ourſelves the trouble to try experiments of this kind, as, ſhould it even ſucceed to our utmoſt wiſhes, we have many ſorts of fodder that will exceed it in quantity and quality, without any proportion to the extraordinary expence which muſt attend the raiſing any quantity of theſe ſhrubs, to cut for that uſe. The flowers, leaves, and top ſhoots have, however, a fine peaſe like taſte, which is what, I make no doubt, moſt cattle would be fond of, and of which the inhabi ants of ſome countries where it grows naturally reap the advantage; for the goats that feed on it yield a greater quantity as well as a more excellent kind of milk, from which good cheeſe is at length obtained, where theſe creatures have plenty of theſe ſhrubs to brouze upon.

"In our wilderneſs quarters we muſt give this tree a very

very dry ſoil and a well ſheltered ſituation; for with us it is rather a tender ſhrub, and has been frequently treated as a greenhouſe plant; and this is another argument againſt any attempt to raiſe theſe ſhrubs for fodder in England: They are too tender to bear our ſevere winters without ſhelter; and ſhould we proceed in raiſing ſixty or ſeventy acres, a thorough froſty winter would deſtroy the greateſt part of them; or, if the winter ſhould not be ſo ſevere as totally to kill them, yet their end ſhoots would be ſo nipped and damaged, that it would be late in the ſummer before they would ſhoot out and recover this injury, and conſequently ſmall crops muſt be expected."

This plant is eaſily PROPAGATED by ſeeds or cuttings. 1. The ſeeds ſhould be ſown in the ſpring, a quarter or half an inch deep, in beds of fine light garden mould. After they are come up, the uſual care of weeding muſt be afforded them; and if they are ſhaded and now and then watered in hot weather, it will be ſo much the better. The beds muſt be hooped againſt winter, and plenty of mats muſt be ready to cover the plants when the froſt comes on; and if this ſhould be very ſevere, their covering ſhould be encreaſed, or there will be danger of loſing them all. In the ſpring the ſtrongeſt may be drawn out, and planted in pots, to be houſed for a winter or two, until they are got ſtrong; but where a quantity is wanted, and there is no ſuch conveniency, it may be proper to let them remain in the ſeed bed another winter, for the conveniency of being covered in bad weather; and then in the ſpring they may be planted out in the nurſery, in lines two feet aſunder, and at one foot diſtance. This nurſery ſhould be in a well ſheltered warm place, and they will be ready for tranſplanting whenever wanted. 2. Theſe plants may be raiſed by cuttings. If a few only are wanted for ornamenting a ſhrubery, the beſt way will be to plant theſe in pots, and ſet them up to the rims in a ſhady place, that they may have the conveniency of being houſed in winter. When a quantity is wanted, they muſt take the chance of wind and weather, and the moſt we can then do is to plant them in fine light ſoil in a well ſheltered place. The latter end of March is the beſt time for the purpoſe; they will

ſtrike

ſtrike root freely, eſpecially if they are ſhaded and watered in dry weather; and from this place they need not be removed until they be finally ſet out.

MELIA.

LINNEAN Claſs and Order, *Decandria Monogynia*: Each flower contains ten males and one female. There are two SPECIES; one of which will bear the open air; the other is a hothouſe plant.

MELIA *Azedarach*: The BEAD TREE; *a deciduous tree*; native of Syria.

The BEAD TREE is a large plant: in its native country it will grow to the ſize of one of our pear trees; and there is no doubt, if our ſoil and ſituation ſuited it, that it would arrive to near that magnitude with us. The trunk is covered with a gray bark; and the young branches, which are not very numerous, are quite ſmooth and green. The leaves are a very great ornament to this tree: They are compound, and very large, the whole leaf being a foot and a half, and ſometimes near two feet long. Each is compoſed of a great number of folioles, which are all terminated by an odd one. Theſe little leaves have their upper ſurface of a ſtrong ſhining green; their under is paler; and their edges are indented. The flowers are produced in July, from the ſides of the branches, in long cluſters: They are, ſeparately, ſmall, of a blueiſh colour, very fragrant, and each ſtands on a long footſtalk. The flowers are ſucceeded by a yellow fruit, tolerably large, in which ſome nuts are encloſed, uſed in the Catholic countries to compoſe ſome ſorts of roſaries; on which account this tree is called the Bead Tree.

"It is generally preſerved in winter as a greenhouſe plant; and indeed a few plants of this fine ſhrub ought always to be introduced in ſuch places deſigned for trees as are proper for them. The reaſon of its being treated as a greenhouſe plant is, becauſe it is rather of a tender nature; and as the plants are not yet very plentiful in England,

England, to this may be added, the desire of preserving those few a person has obtained. But notwithstanding the Bead Tree's being looked upon as a greenhouse plant, some gardeners have ventured to set them abroad against warm walls, where they have stood the winter, and flourished exceedingly well; others have planted them out in well sheltered places only, where they have flourished and stood the brunt of many winters. What inclines me to introduce the Melia amongst our hardy trees is, that I have planted it in an open cold expanse, in a naturally damp and moist soil, where it has flourished for more than seven years, and displayed its beautiful foliage, every summer, to the great pleasure of all beholders. This treatment and practice, however, must be used with caution; and whoever ventures to plant them abroad must have a dry soil as well as a warm and well sheltered situation, and then nothing but our hardest frosts will deprive the owner of these treasures. But, were they more tender, and if a person has no greenhouse, it will be worth while to venture the planting a few abroad, though there should be little chance of his keeping them longer than two or three winters, as they are scarce plants with us, and the leaves, the only beauties the tree can afford in that time, are compounded in such a manner as to afford admiration and pleasure." HANBURY.

PROPAGATION. Care and trouble must be used before we can raise these plants to be of sufficient strength and hardiness to defend themselves, when planted finally out. They are all to be raised from seeds; and these are to be procured from the places where they commonly grow, which is in most of the Catholic countries. These seeds must be sown in pots, filled with light sandy earth, half an inch deep, the end of March. This done, the pots should be plunged into a bark bed, which will cause them to come up. When the plants appear, they must have plenty of air and water; and the open air must be afforded them pretty soon in the summer, that they may be hardened before winter. After they are taken out of the beds, they should be set in a shady place, and every other day watered till the autumn; and at the approach of winter, they should be removed into the greenhouse, with the hardiest of those

thoſe plants. In April following, the plants ſhould be taken out of the pots, and each planted in a ſeparate ſmall pot; and after this is done, they ſhould have the benefit of the bark bed as before, to ſet them a-growing. Care muſt be taken to give them ſufficient air, and not to draw them too much; and after they are well entered upon a growing ſtate, they muſt be hardened to the open air as ſoon as poſſible, and the pots taken out, and plunged up to the rims in a ſhady border, which will prevent the mould in the pots drying too much. They will require little watering, if this method be uſed, during the ſummer; and at the approach of winter, they muſt be removed into the greenhouſe as before, or placed under a hotbed frame, or ſome ſhelter. The next ſpring they muſt be ſet out with other greenhouſe plants, and managed accordingly, and removed into the houſe again with them. Every other year, they ſhould be ſhifted out of their pots, with the earth to their roots, and planted in larger; and by thus treating them as greenhouſe plants, and letting them have larger pots as they encreaſe in ſize till they are ſix or eight years old, they will arrive to be good ſtrong trees. Then in April, having made choice of the drieſt, warmeſt, and beſt ſheltered ſituation, there they may be planted, taking them out of the pots with all their mould; which if done with care, they will never droop on being removed.

MENISPERMUM.

LINNEAN Claſs and Order, *Dioecia Dodecandria*: Male flowers containing twelve ſtamina, and female flowers containing two piſtils, are ſituated upon diſtinct plants. There are eight SPECIES; three of them as follow:

1. MENISPE'RMUM *Canade'nſe*: The CANADA MOONSEED; *a ligneous climber*; native of Canada and Virginia.

2. MENISPE'RMUM *Virgi'nicum*: The VIRGINIA MOONSEED;

MOONSEED; *a ligneous climber*; native of the ſea ſhore of Virginia and Carolina.

3. MENISPE'RMUM *Caroli'num*: The CAROLINA MOONSEED; *an herbaceous climber*; native of Carolina.

1. The CANADA MOONSEED will twine round trees to the height of fifteen or ſixteen feet; and if there be no trees near for it to aſpire by, its almoſt numberleſs branches will twiſt and run one among another, ſo as to form a thick cloſe-ſet buſh. Theſe twining ſtalks are covered with a ſmooth green bark, though in ſome places they are often reddiſh, and in winter often of a brown colour. The leaves are very large, and ſtand ſingly upon long green footſtalks, which alſo have a twining property, and aſſiſt the plant to climb. Theſe leaves have their upper ſurface ſmooth, and of a ſtrong green colour, but are hoary underneath. They are what are called peltated leaves: The footſtalk is not near the middle of the leaves, but within about a quarter of an inch of the baſe, and from thence it branches into ſeveral veins unto the extremity. Theſe peltated leaves are of a roundiſh figure in the whole, though they are angular, and being large, and of a good green, make it a valuable climber. The flowers are produced in July, from the ſides of the ſtalks. They grow in bunches, and are of a greeniſh colour. They are ſucceeded by ſeeds, which often ripen well here.

2. The VIRGINIA MOONSEED differs very little from the other, except in the ſhape of the leaves; for it has the ſame kind of twining ſtalks, produced in great plenty, and the flowers and fructification are the ſame; ſo that nothing more need be obſerved of this, only that the leaves are often heart-ſhaped, and many of them have lobes like thoſe of the common Ivy.

3. The CAROLINA MOONSEED is an herbaceous climber, and will, by the aſſiſtance of trees, riſe to be ten or twelve feet high. The twining ſtalks are garniſhed with heart-ſhaped leaves, which do not divide into lobes like the others. Theſe leaves, which are of a good ſtrong green colour, have their under ſurface hairy, and are much ſmaller than either of the other ſorts; the ſpecies itſelf being of all the leaſt valuable, as it is ſcarcely ever known to produce flowers here.

All theſe ſorts PROPAGATE themſelves very faſt.

1. If

1. If they are planted in a light ſoil, their roots will ſo ſpread and multiply the ſhoots, that in a few years after planting, each of them being wholly taken up, they may be parted, often into ſome ſcores of plants, which will be fit to ſet out, the weakeſt in the nurſery to gain ſtrength, and the ſtrongeſt where they are to remain. Any time from October to March will do for taking off the ſuckers or parting the roots. 2. The young ſhoots alſo, being covered with mould, will grow, and be good plants in one year. 3. They may be likewiſe raiſed by ſeeds; for if theſe are ſown in the ſpring, in a bed of light earth, half an inch deep, they will come up, and require no other trouble than weeding until they are finally planted out, which may be two years after their appearance, and which may be done very well from the ſeed bed, without previous planting in the nurſery.

MESPILUS.

LINNEAN Claſs and Order, *Icoſandria Pentagynia*: Each flower contains about twenty males and five females. There are nine SPECIES; ſeven of which are here treated of:

1. ME'SPILUS *Germa'nica*: The GERMAN MEDLAR, or DUTCH MEDLAR; *a deciduous tree*; native of the South of Europe.

2. ME'SPILUS *Arbutifo'lia*: The ARBUTUS-LEAVED MEDLAR, or the VIRGINIA WILD SERVICE TREE; *a deciduous ſhrub*; native of Virginia.

3. ME'SPILUS *Amela'nchier*: The AMELANCHIER; *a deciduous ſhrub*; native of Auſtria, France, and Italy.

4. ME'SPILUS *Canade'nſis*: The CANADA MEDLAR, or SNOWY MESPILUS; *a deciduous ſhrub*; native of Canada and Virginia.

5. ME'SPILUS *Cotonea'ſter*: The DWARF QUINCE; *a deciduous ſhrub*; native of the Pyrenees, Ararat, and many of the cold parts of Europe.

6. ME'SPILUS *Chamæ-Me'ſpilus*: The BASTARD QUINCE,

QUINCE; *a deciduous ſhrub*; native of the Auſtrian and Pyrenean Mountains.

7. ME'SPILUS *Pyraca'ntha* The PYRACANTHA, or EVERGREEN THORN; *an evergreen ſhrub or climber*; native of Italy and the South of France.

1. The GERMAN MEDLAR in ſome ſituations grows to be a moderately large tree. It grows irregularly, and the branches are frequently crooked. The leaves are ſpear-ſhaped, large, entire, downy underneath, and grow on very ſhort channelled footſtalks. The flowers, which grow ſingly from the ſides of the branches, are very large, and of a white colour. They come out the end of May, and are ſucceeded by that well known fruit called *The Medlar.*

The *Varieties* of this ſpecies are, *The Pear-fruited Medlar*, and *The Nottingham Medlar*. Theſe are plants of more upright growth than the Dutch Medlar. Their leaves are narrower, and their flowers and fruit ſmaller.

2. ARBUTUS-LEAVED MEDLAR. This is frequently called Virginia Wild Service Tree with an Arbutus Leaf. It is a ſhrub about ſix feet high, frequently ſending forth many ſuckers from the root, and branches from the ſides of the plant. The leaves are ſpear-ſhaped, downy underneath, and indented. They grow alternately on very ſhort footſtalks. Their upper ſurface is a fine green colour, though white below; and they die to a purple colour in the autumn. The flowers are produced in bunches from the ends and ſides of the branches: They are ſmall, white, come out in May, and are ſucceeded by a dark brown fruit, like the common Haw, which will ſometimes be ripe in the autumn.

3. AMELANCHIER. The ſtalks of this ſpecies are ſlender, branching a little, and grow to about four feet high. The young branches are of a reddiſh purple colour, and the whole plant is altogether deſtitute of thorns. The leaves are oval and ſerrated, about three quarters of an inch long, half an inch broad, green on their upper ſurface, and woolly underneath. The flowers are produced in bunches from the ends of the branches: Their colour is white; and they are ſucceeded by ſmall black fruit, of a ſweetiſh taſte, which will be often ripe in the autumn. This is a beautiful ſhrub,

ſhrub, and in different parts goes by the various names of *The Dwarf Black-fruited Medlar*, *The New England Quince*, *Vitis Idæa*, *&c.* The young ſhoots which ſupport the flowers are woolly underneath; but this by degrees wears off, and they ſoon become of a purple colour, which remains all winter.

4. CANADA MEDLAR. This ſhrub, which riſes to about five feet high, is free from thorns, and divides into a few branches, which are ſmooth, and of a purpliſh colour. The leaves are oval, oblong, ſmooth, ſlightly ſerrated, and grow on long ſlender footſtalks. The flowers are white, and terminate the branches in ſmall bunches: They come out in May; and are ſucceeded by a purpliſh fruit, hardly ſo large as the common Haw.

5. DWARF QUINCE grows to about four or five feet high. The branches are few, ſmooth, and of a reddiſh purple colour. The leaves are oval, entire, and grow on very ſhort footſtalks. The flowers are produced, two or three together, from the ſides of the branches, without any footſtalks. They are ſmall, of a purpliſh colour, come out in May, and are ſucceeded by round fruit, of a bright red colour when ripe, in the autumn.

6. BASTARD QUINCE. This ſpecies grows to about four or five feet high. The branches are few, ſmooth, ſlender, and covered with a purpliſh bark. The leaves are oval, ſmooth, ſerrated, of a yellowiſh green, and grow on pretty long footſtalks. The flowers are produced in ſmall heads, from the wings of the ſtalks; and between them are long narrow bracteæ, which fall off before the flowers decay. Both flowers and bracteæ are of a purpliſh colour: The fruit is ſmall, and of a red colour when ripe.

All theſe ſorts are to be PROPAGATED from the ſeeds, from layers, and by budding them upon Hawthorn ſtocks. 1. The ſeeds ſhould be ſown in the autumn, ſoon after they are ripe, in a bed of good earth, in a moiſt part of the garden. They uſually lie two years before they make their appearance; during which time the bed muſt be kept clean from weeds. When the plants come up, they muſt be frequently watered, if dry weather ſhould happen; and this ſhould occaſionally be repeated all the ſummer. Weeds muſt be eradicated

eradicated as they arise; and in the autumn, winter, or spring, the strongest plants may be drawn out, and set in the nursery ground, a foot asunder, in rows two feet distant from each other; whilst the others may remain in the seed beds a year longer, to gain strength. In the nursery the Medlars should be trained for standards, if designed for fruit; or they may be headed to any height if for other purposes, while the lower kinds will require no other management than keeping them clean from weeds, and digging the ground between the rows in winter. 2. These plants may be also raised by layers, especially the five last sorts. The young branches should be laid early in the autumn; and by the autumn following many of them will have struck root, when they should be taken up, and planted in the nursery ground, like the seedlings, to remain there for a year or two, before they are finally set out. 3. But the most expeditious, and by far the best way of raising these sorts is, by budding them upon stocks of the White Thorn. The Haws to raise the stocks should be gathered from such trees as are largest, shoot freest, and have the largest leaves and fewest thorns. When the stocks are one year old, they should be set in the nursery at the before-mentioned distance. By the end of July, many of them will be ready for working; when they should be budded in the usual way, and they will easily take. Seldom any other method than this is practised for raising Medlars; and the other sorts, when growing on so firm a basis as the White Thorn, will be larger, have a better look, and be more fertile in flowers and fruit.

7. The PYRACANTHA, or EVERGREEN THORN, has been chiefly used to ornament or hide the ends of houses, barns, stables, or other buildings that break in upon the view; and for this purpose no plant is better adapted, as by its evergreen leaves, closely set, it will not only keep from sight whatever cannot regale that sense, but will be to the highest degree entertaining by the profusion of berries it will produce, and which will be in full glow all winter. But though the hiding as well as ornamenting of walls, &c. has been the chief use for this tree, it is with very good reason planted as an evergreen in shrubery quarters, where, notwithstanding

its branches againſt walls, &c. are very flexible, it will become ſtronger and more woody, and will diffuſe its leafy branches in an agreeable manner. The branches will be terminated with its fine fruit, which will glow in the quarters all winter, if they are not eaten by the birds; ſo that the tree before us is proper for any place. A farther account of this ſhrub is almoſt needleſs, as it is well known; there being few towns which have not a houſe or two whoſe front is ornamented with them, being trained up to a great height; but when planted ſingly in quarters, though their ſtems naturally become ſtronger, they ſeldom grow higher than twelve or fourteen feet; and they will ſpread abroad their ſlender branches, and will often have a buſhy, though not unpleaſing form. Theſe branches are covered with a ſmooth bark, which is of a dark greeniſh brown colour, and often ſpotted with grayiſh ſpots; and they are often poſſeſſed of thorns, which, though not numerous, are ſharp and ſtrong. The leaves are ſpear-ſhaped, oval, and their edges are crenated. Their upper ſurface is ſmooth, and of a fine ſhining green; their under is paler; and they are produced in much plenty all over the ſhrub. The flowers are produced in bunches, like thoſe of the common Hawthorn; though they are ſmall, and not of ſo pure a white. They are often later before they are produced; and are ſucceeded by thoſe large delightful bunches of berries, which are of a fiery red, and which are as ornamental in the winter as any that are produced on trees of the berry-bearing tribe.

This plant is eaſily PROPAGATED by the berries, or from layers. 1. The berries ſhould be ſown in any common garden mould made fine, an inch deep; and theſe will remain two years before they appear: though if the berries are old ones (for they will often remain on the tree two years) they will frequently come up the ſucceeding ſpring. After the plants have ſtood one or two years in the ſeed bed, in the ſpring they ſhould be planted out in the nurſery, at ſmall diſtances; and in about two years more they will be good plants, fit for any place. 2. They are eaſily propagated by layers; and this buſineſs ſhould be performed in the autumn, on the young ſhoots. A gentle twiſt may be given them;

them; though, if they are only laid down, and covered with earth, they will ſtrike root by the next autumn; nay, continues HANBURY, "I have known that, by ſome mould being accidentally thrown on a branch which was near the ground, roots have ſhot from almoſt every joint." Theſe layers ſhould be taken off any time in the winter; the ſtrongeſt will be fit for immediate uſe, while the weaker may be ſet in the nurſery, like the ſeedlings, and in a very little time they will grow to be good plants. It diſlikes a very moiſt ſituation.

M O R U S.

LINNEAN Claſs and Order, *Monoecia Tetrandria*: Male flowers containing four ſtamina, and female flowers containing two piſtils, upon the ſame plant; the male flowers being collected in a catkin. There are ſeven SPECIES; four of which are proper for our collection:

1. MO'RUS *A'lba*: The WHITE MULBERRY, or the SILKWORM MULBERRY; *a deciduous tree*; native of China, and cultivated almoſt univerſally for the feeding of ſilkworms.

2. MO'RUS *Ni'gra*: The BLACK MULBERRY, or the COMMON GARDEN MULBERRY; *a deciduous tree*; native of Perſia, and the maritime parts of Italy.

3. MO'RUS *Papyri'fera*: The PAPER MULBERRY; *a low deciduous tree*; native of Japan.

4. MO'RUS *Ru'bra*: The VIRGINIA MULBERRY; *a low deciduous tree*; native of Virginia.

1. The WHITE or SILKWORM MULBERRY will grow to a large ſize: Its leaves are of a clear light green; and open conſiderably earlier in the ſpring than thoſe of the other ſpecies of Mulberry: Its fruit is alſo paler coloured than that of the other ſorts, which makes this take the name of the White Mulberry. "This tree (ſays HANBURY) poſſeſſes the peculiar property of breeding no vermin either growing or cut down;

Q 2 neither

neither does it harbour any ſort of caterpillar, the Silkworm only excepted, whoſe food is its leaves. The Mulberry tree was very earneſtly recommended by King James to be planted in great quantities to feed theſe worms, in order to have ſilk of our own working: and, indeed, if we conſider what vaſt ſums the produce of ſilk brings in to other States, we might find an undertaking of this nature worthy of a princely care and aſſiſtance." The Mulberry delights moſt in a light dry ſoil; but there is very little land in this kingdom, generally ſpeaking, which might not be planted with theſe trees, and probably to great national advantage. Be this as it may, it is ſufficiently *ornamental* to be admitted into a large collection: And, beſides the *uſes* of its leaves to the Silkworm, EVELYN and HANBURY recommend it very ſtrongly as a foreſt or timber tree, and enumerate ſome of the uſes of its wood; none of them, however, ſufficiently ſtriking to induce us to recommend it to the planter's notice merely as a timber tree.

2. The BLACK OR GARDEN MULBERRY is principally cultivated for the fruit; and in ornamental plantations a few of them will be ſufficient, to make the collection general, as well as to be ready at all ſeaſons for the notice and obſervation of the Botaniſt.

There is a *Variety* of it, with jagged leaves, which makes it eſteemed on that account; but the fruit is ſmaller than that of the common ſort.

3. The PAPER MULBERRY is ſo called, becauſe the inhabitants where the trees grow naturally make paper of the bark. It will grow to the height of about thirty feet; and exhibits its fine large leaves of different ſhapes, many of them being divided into ſeveral lobes, whilſt others again are entire. They are of a fine ſtrong green colour, though the under ſurface is paler than the upper. The flowers, as has been obſerved, are male and female; and the females are ſucceeded by ſmall black fruit. It is the bark of the young ſhoots of which the paper is made; and for this uſe it is cultivated much in China, as well as Japan, where large plantations are raiſed. The plants are headed to within about a foot of the ground; and every year the crop of the ſummer's ſhoots is taken.

4. The

4. The VIRGINIA MULBERRY TREE will grow to be thirty or more feet high. It ſends forth many large branches; and the bark of the young ſhoots is of a blackiſh colour. The leaves are larger than the Common Mulberry, and rougher; though in other reſpects they ſomewhat reſemble them. It produces plenty of catkins, in ſhape like thoſe of the Birch tree; and the female flowers are ſucceeded by a dark reddiſh fruit. This is a very ſcarce plant at preſent; and is coveted by none but thoſe who are deſirous of making their collection general.

Theſe ſeveral ſpecies of Mulberry may be PROPAGATED from ſeeds, by layers, and from cuttings. 1. Where the ſeeds can be procured, it is the moſt expeditious way of raiſing great quantities; and whoever has a correſpondence in the South of France, or in Italy, may through that channel obtain them. Having the ſeeds ready, let a fine warm border of rich mellow earth be prepared, and let this border be hooped, in order to ſupport mats to defend the young plants, when they appear, from froſts. If no ſuch border can be eaſily had, it will be proper to make a gentle hotbed, and cover it with fat mould: This alſo muſt be hooped, as the border. Then ſow the ſeeds in little drills, about a quarter of an inch deep. The middle of March is the beſt time for this work; and when the young plants appear, which will be in about ſix weeks, they muſt be conſtantly covered with the mats in the night, if any appearance of froſts preſents itſelf, as there often is at that ſeaſon. During the ſummer they ſhould be kept clear from weeds, and covered from the extreme heat of the ſun while the hot months continue. Whenever any cloudy or rainy weather approaches, the mats ſhould be always taken off, that the plants may enjoy the benefit of it. By thus carefully nurſing the beds, keeping them clear from weeds, watering the plants in dry ſeaſons, covering them from the parching ſun, and uncovering them again in the night, cloudy or rainy weather, the plants by autumn will be got pretty ſtrong; though not ſo ſtrong as to be left to themſelves. The following winter they will require ſome care. When the froſts approach, they muſt be carefully covered with the mats, as in the ſpring; for without this protection, many of them would be destroyed, and the greatest part killed, at least down to the ground. In this bed they may stand two years, when they will be strong enough to plant out in the nursery. The ground for this purpose being double dug, the young plants should be set in rows, at two feet and a half distance, and one foot and a half asunder in the rows. Here they may remain till they are of a sufficient size to be finally planted out. 2. Another method of propagating this tree is by layers. Whoever has not the conveniency of obtaining the seeds, must procure a number of plants to be planted for stools. The ground on which these stools are to stand should be double dug, and the trees may be planted for this purpose two yards asunder. The size of the ground, and the quantity of trees for the stools, must be proportioned according to the number of plants wanted; though the reader should observe, that a few stools will soon produce many layers, as they throw out plenty of young branches, when the head is taken off. Having a sufficient quantity of stools that have shot forth young wood for layering, in the beginning of winter perform this business as follows: Let the earth be excavated around each stool, and let the preceding summer shoot be slit at a joint, and laid therein; a peg would be proper, to keep them from being torn up, and the fine mould should fill the interstices; the ground must be levelled, and the young twigs cut down to one eye above the surface, that it may just appear above the ground. Such is the method of layering this tree; and whoever performs the operation in this manner, will find in the autumn following, that the plants will have all taken good root, and made a considerable shoot in the stem. These plants will be now ready for the nursery ground, in which they should be planted and managed in the same way as the seedlings. The stools, the second year after, will have exhibited a fresh crop of young wood for layering: And thus may this operation be performed every second year, till the desired quantity is raised. 3. By *cuttings* also all the sorts may be propagated, and this may be done two ways: By *cuttings planted in autumn*. These should be strong shoots of the last year's wood; and if the tree to be encreased is not in so

flourishing

flourishing a state as to make such shoots, it should be headed the year before, and you will have cuttings proper for your purpose. The strongest shoots are the best; and October is the best month for the business. They should be a foot and a half long, and must be planted a foot deep, in a shady well sheltered place, and a moist soil well worked and fine: By this method many good plants may be raised. These trees may also be encreased by *cuttings planted in the summer*. The latter end of June, or the beginning of July, is a proper time for the work, and the management must be as follows: Having a sufficient number of pots ready, the cuttings, or rather slips, from the trees should be gathered, and planted in these pots, in any sort of common garden mould made fine. After this, they should have a good watering, and the pots be plunged up to their rims in the stove. Here, if water and shade be constantly afforded them, they will strike root and become good plants. It may be proper to observe farther in this place, that cuttings planted in pots in March, and managed this way, will readily grow. After they have struck root, they may be hardened by degrees to the open air. They should remain under cover in the pots all winter; for they will be rather tender at first, by being so nicely nursed; but in the spring, when all danger of frost is over, they may be turned out, with the mould, either in nursery lines at a foot distance and two feet asunder in the rows, or else in the places where they are designed to remain; for they will be hardy enough, after growing openly this summer, to be in little danger of suffering by almost any weather.

M Y R I C A.

LINNEAN Class and Order, *Dioecia Tetrandria*: Male flowers containing four stamina, and female flowers containing two pistils upon distinct plants. There are six SPECIES; two of which are of a fragrant quality, and may be admitted into ſhrubery quarters.

1. MYRI'CA *Ceri'fera:* The CANDLEBERRY MYRTLE, or WAX-BEARING MYRICK; *a deciduous ſhrub*; native of Carolina, Virginia, and Pennſylvania.

2. MYRI'CA *Ga'le:* The GALE, or DUTCH MYRTLE; *a low deciduous ſhrub*; native of heathy bogs in many parts of England, and alſo of moſt of the northern parts of Europe.

1. CANDLEBERRY MYRTLE is a ſhrub about five feet in growth. Many ſlender branches are produced from the ſtalk: They are tough, ſmooth, and of a yellowiſh brown, having the older ſpotted with gray ſpots. The leaves grow irregularly on them all round; ſometimes by pairs, ſometimes alternately, but generally at unequal diſtances. They are of a lanceolated figure; and ſome are ſerrated at the top, whilſt others have their edges wholly entire. They ſtand on very ſhort foot-ſtalks, having their upper ſurface ſmooth, and of a ſhining green colour, whilſt their under is of a more duſky hue. The branches of the old plants ſhed their leaves in the autumn; but the young plants, raiſed from ſeeds, retain them the greateſt part of the winter; ſo as during that ſeaſon to have the appearance of an ever-green. But this beauty will not be laſting; for they ſhed their leaves proportionally earlier as the plants get older. There are both male and female trees of this ſort. The flowers are ſmall, of a whitiſh colour, and make no figure; neither does the fruit that ſucceeds the female, which is a ſmall, dry, blue berry, though produced in cluſters, make any ſhow: So that it is from the leaves this tree receives its beauty and value; for theſe being bruiſed, as well as the bark of the young ſhoots, emit the moſt refreſhing and delightful fragrance, that is exceeded by no Myrtle, or any other aromatic ſhrub.

There is a *Variety* of this ſpecies, of lower growth, with ſhorter but broader leaves, and of equal fragrance. This grows commonly in Carolina; where the inhabitants collect, from its berries, a wax, of which they make candles, and which occaſions its being called the *Candleberry Tree*. It delights in a moiſtiſh ſoil,—will grow in a very moiſt one.

2. The

2. The GALE, or SWEET GALE, is a ſhrub of about the ſame growth with the other. The branches are tough and ſlender, and covered with a ſmooth yellowiſh brown bark. The leaves are of the ſame figure with the other, though not ſo large: They are placed in the ſame irregular manner on the branches; and when bruiſed, like them, emit a delightful and refreſhing ſcent. The flowers will appear in June, and the berries, which ſucceed them in cluſters make no figure to any except a Botaniſt; ſo that where that ſcience has no ſhare in view, it is on account of its fragrance that it is propagated. This ſort grows wild upon bogs, in many parts, particularly the northern parts of England; ſo that when it is deſigned to be in the ſhrubery, the moiſteſt parts muſt be aſſigned it.

Both theſe ſorts may be PROPAGATED by ſeeds or layers. 1. The ſeeds of the Candleberry Myrtle, and the Spleenwort-leaved Gale, we receive from abroad; thoſe of the Sweet Gale, from the bogs where they grow in England. The beſt way is to ſow them in boxes of earth from a rich paſture, well broken and fine. They ſhould be ſown about half an inch deep; and when the hot weather comes on, ſhould be ſet in the ſhade. They will often remain until the ſecond year before they come up, eſpecially thoſe ſeeds that come from abroad. If the boxes are ſet in the ſhade, and the plants come up, they will require no other trouble the firſt ſummer than keeping clean from weeds; in winter they ſhould be removed to a warm hedge or wall, where they may enjoy the benefit of the ſun. In the following ſpring they will come up in plenty. In the beginning of May they ſhould reſume their ſhady ſituation; and this ſummer they will require no other trouble than weeding and watering in dry weather. In the winter they ſhould be removed into a well ſheltered place; and this may be repeated two years; when, in the ſpring, they ſhould be taken out of the boxes, and planted in the nurſery, at about a foot aſunder. 2. Theſe ſorts may be alſo eaſily propagated by layers; for this operation being performed on the young wood in the autumn, will occaſion them to ſhoot good roots by the autumn following;

many

many of which will be good plants, fit for any place, 3. These plants may likewise be encreased by suckers; for many of them often throw them out in vast plenty; so that these being taken out, the strongest and best rooted may be finally set out; whilst the weaker, and those with less root, may be planted in the nursery.

NYSSA.

LINNEAN Class and Order, *Polygamia Dioecia:* Male flowers containing ten stamina, and hermaphrodite flowers containing five males and one female each, upon distinct plants. There is only one SPECIES:

NY'SSA *Aqua'tica:* The TU'PELO TREE; *a deciduous tree or shrub*; native of watery places in America.

The TUPELO comprehends two *Varieties:*

The Entire-leaved Tupelo.

The Serrated-leaved Tupelo.

The *Entire-leaved Tupelo Tree*, in its native country, will grow to be near twenty feet high; with us, its size will vary according to the nature of the soil or situation. In a moist rich earth, well sheltered, it will bid fair for twenty feet; in others, that are less so, it will make slower progress, and will in the end be proportionally lower. The branches are not very numerous; and it rises with a regular trunk, at the top of which they chiefly grow. The leaves are of a lanceolated figure, and of a fine light green colour. They end in acute points, and are very ornamental, of a thickish consistence, soft, grow alternately on pretty long footstalks, and often retain their verdure late in the autumn. The flowers, which are not very ornamental, are produced from the sides of the branches, growing sometimes singly, sometimes many together, on a footstalk. They are of a greenish colour; and, in the countries where they naturally grow, are succeeded by oval drupes, inclosing oval, acute,

furrowed

furrowed nuts. In England, they feldom produce fruit.

The Serrated-leaved Tupelo Tree grows ufually to be near thirty feet high, and divides into branches near the top like the other. The leaves are oblong, pointed, of a light green colour, and come out without order on long footftalks. The flowers come out from the wings of the leaves, on long footftalks. They are fmall, of a greenifh colour; and are fucceeded by oval drupes, containing fharp-pointed nuts, about the fize of a French Olive.

The PROPAGATION of thefe forts is from feeds, which we receive from America. As foon as they arrive, they fhould be fown in large pots of light fandy earth, one inch deep. The Gardener (who muft not expect to fee any plants come up the firft fpring), after this work is done, fhould plunge his pots up to their rims in the natural ground; and if it be a moiftifh place, it will be the better. Weeding muft be obferved all fummer; and a few furze bufhes ought to be pricked round the pots in November, which will prevent the ground from freezing, and forward the coming up of the feeds. In the next fpring the pots fhould be plunged into a hotbed, and after that the feeds will foon come up. As much air as poffible, and watering, fhould be afforded them; and they muft be hardened foon, to be fet out. The pots fhould be then plunged to their rims again in the natural mould; where they may remain until October. Watering muft be given them, and they fhould alfo be fhaded in the heat of the day. In October they fhould be houfed, with other greenhoufe plants, or elfe fet under a hotbed frame, or fome other cover, all winter. The third fpring they fhould be taken out of the larger pots, and each planted in a fmaller, in which their growth may be affifted by a gentle heat in a bed; but if they are planted up to the rims in a moiftifh place, and fhaded in dry weather, they will grow very well. Though by this time they may have become hardy, yet it will be proper to fhelter them the winter following in bad weather. They will require little more care during their ftay in the pots, which may be either two, three, or more years, if they are large enough; when

when in some spring they may be turned out, with the mould, into the places where they are to remain, which ought always to be moist and well sheltered.

ONONIS.

LINNEAN Class and Order, *Diadelphia Decandria*: Each flower contains ten males and one female; the males being divided at the base into two divisions. There are twenty-nine SPECIES; one of which, being of a shrubby nature, is proper for our purpose:

ONO'NIS *Fruticósa*: The SHRUBBY ONONIS, or PURPLE SHRUBBY RESTHARROW; *a low deciduous shrub*; native of the Alps and other mountainous parts of Europe.

The SHRUBBY ONONIS, or RESTHARROW, is a flowering shrub of about a yard in growth. The branches are numerous, slender, and covered with a purplish brown bark, having no spines. The leaves are trifoliate, grow irregularly on the branches, sit close, are narrow, spear-shaped, and their edges are serrated. The flowers come out in panicles from the ends of the branches: They are of the papilionaceous kind, and their general characters will indicate their structure. They stand on long footstalks, usually three on one. They are large, red, appear in May, and are succeeded by short turgid pods, which will have ripe seeds by July or August.

This sort may be PROPAGATED by the seeds. Common garden mould of almost any soil, made fine, will do for the purpose. The beds should be made and the seeds sown in March, and covered about half an inch deep. In May the plants will appear; and all the summer they must be weeded, and duly watered in dry weather. In the spring they should be taken out of the seed bed, and planted in the nursery, a foot asunder, where they may stand a year or two, and then be planted out for good.

As

As the ſeeds of this ſort ripen exceedingly well with us, a few may be ſown in different parts of the garden, and ſticks placed for a direction. Where there are too many come up to grow together, they may be drawn, and tranſplanted for other places, or thrown away, if plenty of ſeeds can always be had; and thus may theſe plants be raiſed in their proper places, without the trouble of removing.

PASSIFLORA.

LINNEAN Claſs and Order, *Gynandria Pentandria:* Each flower contains five males and three females; the males and females growing together at the baſe. There are twenty-eight SPECIES; one of which is proper for our collection:

PASSIFLO'RA *Cæru'lea:* The PALMATED PASSION FLOWER, or the COMMON PASSION FLOWER; *a ligneous climber*; native of the Brazils.

The PASSION FLOWER will climb to a prodigious height; MILLER ſays, "to forty feet, with ſtalks almoſt as large as a man's arm;" and adds, that it will make ſhoots of twelve or fifteen feet long in one ſummer. The leaves are palmated, being ſhaped like the hand; each is compoſed of five folioles, the middle one of which is, like the fingers of the hand, longer, and the reſt are ſhorter in the ſame proportion. Theſe folioles are ſmooth, and have their edges free from ſerratures, and all together form a fine leaf. The leaves grow from the joints, on ſhort footſtalks, from whence alſo the claſpers come out. From the joints, alſo, the flowers are produced, in July, Auguſt, and September. They are well known; and in ſome countries ſerve as monitors to the religious, as ſhewing the inſtruments of our Bleſſed Saviour's Paſſion; for they bring in the leaves of ſome of the ſorts to repreſent ſome part of it, and the contorted cirrhi for the *flagella* with which he was ſcourged.

This

This extraordinary plant is very easily PROPAGATED; for it takes freely either by cuttings, layers, or seeds. 1. By cuttings. These should be planted in a moist rich soil, at the beginning of March. The beds should be immediately hooped, and every day, during the drying March winds and sun, should be covered with mats; and all that time they should have frequent waterings in the evening. In moist, hazy, or cloudy weather, they should be constantly uncovered; and with this management many of them will strike root. If, through the heat of summer, the mats be applied, and evening waterings continued, the plants being thus kept cool and moist, will shoot to be good ones by the autumn. During the winter, the mats must be applied in frosty weather; and in the spring they may be set out to stand. 2. Good plants are obtained by layers; for these being laid in the ground in the spring, will have struck root, and be good plants for removing the spring following. 3. By seeds. These should be sown in pots filled with fine sandy soil, from a rich meadow; and these plunged up to the rims in a shady border. In these pots they will readily come up; and at the approach of winter should be removed into the greenhouse, or set under a hotbed frame. In the spring following they may resume their old place; and the spring after that may be set out for good.

The after management will be, if planted to climb up trees in warm well sheltered places, to take away the dead shoots in the spring that have been killed by the frosts; for these will not only appear unsightly, but by shortening the branches it will cause them to shoot stronger and flower better. If planted against high walls, they must be constantly nailed up as they shoot, and in the spring following the branches must be shortened, and the others taken away. If they be reduced to about a yard or four feet in length, and all weak shoots cut out, you will be pretty sure of having plenty of good bloom the summer after. This sort is succeeded by a large, oval, yellow fruit, which also looks well. As this plant is rather tender, and requires mats to be nailed before it in very hard frost, these mats must be always taken off immediately on the alteration

alteration of weather; for otherwise the stems will grow mouldy, and be destroyed that way. And as it is usual to lay straw, dung, &c. about the stems to prevent the frost penetrating the ground, this dung, &c. must not be laid up to the stem so as to touch it, but all round it; for if it is laid up to the stem, the bark will be destroyed, and the tree killed, and also very little chance remain of the root's throwing out fresh shoots, as it often does when the plant is killed down to the ground.

PERIPLOCA.

LINNEAN Class and Order, *Pentandria Digynia:* Each flower contains five males and two females. There are five SPECIES; one of which is sufficiently hardy for this climate.

PERIPLOCA *Græca:* The PERIPLOCA, or VIRGINIA SILK, or CLIMBING DOG'S BANE; *a ligneous climber*; native of Syria.

The PERIPLOCA is a fine climbing plant, that will wind itself with its ligneous branches about whatever tree, hedge, pale, or pole is near it; and will arise, by the assistance of such support, to the height of above thirty feet; and where no tree or support is at hand to wind about, it will knit or entangle itself together, in a most complicated manner. The stalks of the older branches, which are most woody, are covered with a dark brown bark, whilst the younger shoots are more mottled with the different colours of brown and gray, and the ends of the youngest shoots are often of a light green. The stalks are round, and the bark is smooth. The leaves are the greatest ornament to this plant; for they are tolerably large, and of a good shining green colour on their upper surface, and cause a variety by exhibiting their under surface of a hoary cast. Their figure is oblong, or rather more inclined to the shape of a spear, as their ends are pointed, and they stand opposite by pairs, on short footstalks. Their flowers

flowers afford pleasure to the curious examiner of nature. Each of them singly has a star like appearance; for though it is composed of one petal only, yet the rim is divided into segments, which expand in such a manner as to form that figure. Their inside is hairy, as is also the nectarium, which surrounds the petal. Four or five of the flowers grow together, forming a kind of umbel. They are of a chocolate colour, are large, and will be in blow in July and August, and sometimes in September. In the country where this genus grows naturally, they are succeeded by a long taper pod, with compressed seeds, having down to their tops.

The PROPAGATION of this climber is very easy; for if the cuttings are planted in a light, moist soil, in the autumn or in the spring, they will readily strike root. Three joints at least should be allowed to each cutting: They should be the bottom of the preceding summer's shoot; and two of the joints should be planted deep in the soil.

Another, and a never-failing method is by layers; for if they are laid down in the ground, or a little soil only loosely thrown over the young preceding summer's shoots, they will strike root at the joints, and be good plants for removing the winter following.

PHILADELPHUS.

LINNEAN Class and Order, *Icosandria Monogynia*: Each flower contains about twenty males and one female. There are only two SPECIES:

1. PHILADE'LPHUS *Corona'rius*: The COMMON SYRINGA, or the CUCUMBER TREE, or the MOCK ORANGE, or the PHILADELPHUS; *a deciduous shrub*; native place uncertain.

2. PHILADE'LPHUS *Inodo'rus*: The CAROLINA SYRINGA, or PHILADELPHUS, or the SCENTLESS SYRINGA; *a tall deciduous shrub*; native of Carolina.

1. The

1. The MOCK ORANGE or PHILADELPHUS admits of three remarkable *Varieties*: Common Syringa, Double Syringa, and Dwarf Syringa.

The *Common Philadelphus*, or *Mock Orange*, is a very beautiful ſhrub, about ſix feet in growth. It ſends forth numerous branches from the root, which are brittle and full of pith. Theſe alſo ſend out others from their ſides that are ſhorter, ſtand generally oppoſite by pairs, and are alternately of contrary directions These younger ſhoots are ſlender, jointed, and covered, ſome with a ſmooth pale brown bark, others with a ſmooth bark of a darker colour. The leaves are large, and placed oppoſite, by pairs, on ſhort footſtalks. They are of an oval, ſpear-ſhaped figure, of a ſtrong green colour, and have the flavour of a cucumber. Their edges are irregularly indented, their ſurface is rough, and they fall off early in the autumn. This ſhrub, by its flowers, makes a fine figure in May and June; for they are produced in cluſters both at the ends and from the ſides of the branches. They are of a fine white colour, and exceedingly fragrant. The petals of which each is compoſed are large, and ſpread open like thoſe of the Orange; and then forming branches, which ſtand each on its own ſeparate ſhort footſtalk, and being produced in plenty all over the ſhrub, both at once feaſt the eye and the ſmell: The eye, by the pleaſing appearance it will then have; the ſmell, as the air at ſome diſtance will be replete with the odoriferous particles conſtantly emitted from thoſe fragrant flowers. Theſe flowers, however, are very improper for chimneys, water-glaſſes, &c. in rooms; for in thoſe places their ſcent will be too ſtrong; and for the ladies in particular, often too powerful.

The *Double-flowering Philadelphus* or *Mock Orange* is a low variety of this ſpecies, ſeldom riſing to more than a yard high. The deſcription of the other belongs to this ſort except that the leaves and branches are proportionally ſmaller and more numerous, and the bark of the ſhoots of a lighter brown. It is called the Double flowering Syringa, becauſe it ſometimes produces a flower or two with three or four rows of petals; whereas in general, the flowers, which are very few, and ſeldom produced, are ſingle. They are much ſmaller than thoſe of the other; and you will not ſee a flower of any kind on this ſhrub oftener perhaps than once in five years. It is hardly worth propagating on this account; ſo that a few plants only ought to be admitted into a collection, to be ready for obſervation.

The *Dwarf Philadelphus* or *Mock Orange* is ſtill of lower growth than the other, ſeldom ariſing to more than two feet in height. The deſcription of the firſt ſort ſtill agrees with this; only that the branches and leaves are ſtill proportionally ſmaller and more numerous, and the bark is ſtill of a lighter brown. It never produces flowers.

2. The CAROLINA MOCK ORANGE OR PHILADELPHUS is the talleſt grower by far of any ſort of the Syringa, and makes the grandeſt ſhow when in blow; though the flowers are deſtitute of ſmell. It will grow to about fourteen feet in height, the branches are numerous and ſlender, and the bark on the young ſhoots is ſmooth and brown. The leaves alſo are ſmooth and entire, and placed oppoſite by pairs on longiſh footſtalks. The flowers, which are produced at the ends of the branches, are of a fine white colour, and, being larger than thoſe of the firſt ſort, have a noble look.

The PROPAGATION of all the ſorts is very eaſy. They are encreaſed by layers, cuttings, or ſuckers. 1. The moſt certain method is by layers; for the young twigs being laid in the earth in the winter, will be good rooted plants by the autumn following. 2. Theſe plants may be encreaſed by cuttings, which, being planted in October, in a ſhady moiſt border, many of them will grow; though it will be proper to let thoſe of the Carolina ſort remain until ſpring, and then to plant them in pots, and help them by a little heat in the bed. By this aſſiſtance, hardly one cutting will fail. 3. They may be alſo encreaſed by ſuckers; for all the ſorts throw out ſuckers, though the Carolina Syringa the leaſt of any. Theſe will all ſtrike root, and be fit for the nurſery ground: Nay, the Double-flowering and the Dwarf ſorts are always encreaſed this way; for theſe plants having ſtood five or ſix years, may be taken up and divided into ſeveral ſcores. All the plants however, whether raiſed from layers, cuttings, or ſuckers, ſhould be planted in the nurſery ground to get ſtrength, before they are let out for good. They ſhould be planted a foot aſunder, and the diſtance in the rows ſhould be two feet. After this, they will require no other care than hoeing the weeds, until they have ſtood about two years, which will be long enough for them to ſtand there. The Mock Orange diſlikes a wet ſituation.

PHILLYREA.

LINNEAN Claſs and Order, *Diandria Monogynia:* Each flower contains two males and one female. There are three SPECIES:

1. PHILLYRE'A *Média:* The OVAL-LEAVED PHILLYREA, or MOCK PRIVET, or the MEDIAL-LEAVED PHILLYREA; *a tall evergreen ſhrub*; native of the South of Europe.

2. PHILLYRE'A *Latifólia:* The BROAD-LEAVED PHILLYREA, or MOCK PRIVET; *a tall evergreen ſhrub*; native of the South of Europe.

3. PHILLYRE'A *Anguſtifólia:* The NARROW-LEAVED PHILLYREA, or MOCK PRIVET; *a deciduous ſhrub*; native of Spain and Italy.

1. The OVAL-LEAVED or MIDDLE PHILLYREA has the following *Varieties:* Common Smooth-leaved *Phillyrea*, Privet-leaved *Phillyrea*, Olive-leaved *Phillyrea*.

The *Common Smooth-leaved Phillyrea* will grow to be twelve or fourteen feet high, and the branches are many; the older of which are covered with a dark brown bark, but the bark on the young ſhoots is of a fine green colour. They are oval, ſpear-ſhaped, and grow oppoſite, by pairs, on ſtrong ſhort footſtalks. The flowers are produced in cluſters, from the wings of the young branches. They are ſmall, and of a kind of greeniſh white colour; they appear in March, and are ſucceeded by berries, which are firſt green, then red, and black in the autumn when ripe.

Privet leaved Phillyrea will grow to be ten or twelve

 feet

feet high, and the branches are covered with a brown bark. The leaves a little refemble the Privet; they are of a fine green colour, and grow by pairs on the branches. They are of a lanceolate figure, and their edges are entire, or nearly fo, for fome figns of ferratures fometimes appear. The flowers grow like others, in clufters, in March. They are whitifh, and are fucceeded by fmall black berries.

The *Olive-leaved Phillyrea* is the moft beautiful of all the forts. It will grow to be about ten or twelve feet high; and the branches, which are not numerous, fpread abroad in a free eafy manner, which may not improperly be faid to give the tree a fine air. They are long and flender, and are covered with a light brown bark; and on thefe the leaves ftand oppofite by pairs, at proper intervals, on fhort footftalks. They refemble thofe of the Olive-tree, and are of fo delightful a green as to force efteem. Their furface is exceedingly fmooth, their edges are entire, and the membrane of a thickifh confiftence. The flowers are fmall and white, and like the other forts make no fhow. They are fucceeded by fingle roundifh berries.

2. The BROAD-LEAVED PHILLYREA will grow to be about twelve feet high. The branches feem to be produced ftronger and more upright than thofe of the former fpecies. The bark is of a gray colour, fpotted with white, which has a pretty effect; and the leaves grow oppofite by pairs. They are of a heart fhaped oval figure, of a thick confiftence, and a ftrong dark green colour. Their edges are fharply ferrated, and they ftand on fhort ftrong footftalks. The flowers grow from the wings of the leaves in clufters, in March. They are of a kind of greenifh white colour, make no fhow, and are fucceeded by fmall round black berries.

The *Varieties* of this fpecies are, the *Ilex*-leaved *Phillyrea*, the Prickly *Phillyrea*, the Olive *Phillyrea* with flightly-ferrated edges.

3. The NARROW-LEAVED PHILLYREA is of lower growth, feldom rifing higher than eight or ten feet. The branches are few and flender, and they alfo are beautifully fpotted with gray fpots. The leaves, like the others, ftand oppofite by pairs. They are long and narrow, fpear-fhaped and undivided, of a deep green colour,

colour, and of a thick confiftence. Their edges are entire, and they alfo ftand on fhort footftalks. The. flowers, like the others, make no fhow. They are whitifh, and grow in clufters from the wings of the branches, in March; and are fucceeded by fmall round black berries.

The *Varieties* of this fpecies are, the Rofemary *Phillyrea*, Lavender *Phillyrea*, Striped *Phillyrea*, &c.

The *Phillyreas* are to be PROPAGATED by feeds or layers. 1. By feeds. Thefe ripen in the autumn, and fhould be fown foon after. The mould muft be made fine, and if it is not naturally fandy, if fome drift fand be added, it will be fo much the better. The feeds for the moft part remain until the fecond fpring before they come up; and if they are not fown foon after they are ripe, fome will come up even the third fpring after. They muft be fown about an inch deep; and during the following fummer fhould be kept clean from weeds. After they are come up, the fame care muft be obferved, and alfo watering in dry weather; and if the beds are hooped, and the plants fhaded in the hotteft feafon, they will be fo much the better for it. However, at the approach of winter they muft be hooped, and the beds covered with mats in the hardeft frofts, otherwife there will be danger of lofing the whole crop; for thefe trees, though they are very hardy when grown tolerably large, are rather tender whilft feedlings. It will be proper to let them remain in the feed beds, with this management, for two fummers, and then, waiting for the firft autumnal rains, whether in September or October (and having prepared a fpot of ground), they fhould at that juncture be planted out, and this will occafion them immediately to ftrike root. The diftance they fhould be planted from each other need not be more than a foot, if they are not defigned to remain long in the nurfery: If there is a probability of their not being wanted for fome years, they fhould be allowed near double that diftance; and every winter the ground in the rows fhould be well dug, to break their roots, and caufe them to put out frefh fibres, otherwife they will be in danger of being loft, when brought into the fhrubery quarters. 2. By layers they will eafily grow. The autumn is the beft time for this operation, and the young ſhoots are fit for the purpoſe. The beſt way of layering them is by making a ſlit at the joint; though they will often grow well by a twiſt being only made. When the gardener chooſes the method of twiſting a young branch for the layers, he muſt be careful to twiſt it about a joint ſo as only to break the bark; for if it is too much twiſted, it will die from that time, and his expectations wholly vaniſh. But if it be gently twiſted with art and care, it will at the twiſted parts be preparing to ſtrike root, and by the autumn following, as well as thoſe layers that had been ſlit, will have good roots; the ſtrongeſt of which will be fit for planting where they are wanted to remain, whilſt the weaker and worſt rooted layers may be planted in the nurſery ground like the ſeedlings, and treated accordingly.

PHLOMIS.

LINNEAN Claſs and Order, *Didynamia Gymnoſpermia*: Each flower contains four males and one female; two of the males being ſomewhat longer than the other two; and the ſeeds being naked. There are fourteen SPECIES; two of which are adapted to the ſhrubery.

1. PHLO'MIS *Frutico'ſa*: The YELLOW PHLOMIS, or JERUSALEM SAGE; *a non deciduous hoary ſhrub*; native of Spain and Sicily.

2. PHLO'MIS *Purpu'rea*: The PURPLE PHLOMIS, or PORTUGAL SAGE, *a non-deciduous hoary ſhrub*; native of Portugal and Italy.

1. The YELLOW PHLOMIS, or JERUSALEM SAGE. The *Varieties* of this ſpecies are, The Broad-leaved Sage Tree of Jeruſalem, The Narrow-leaved Jeruſalem Sage Tree, The Cretan Sage Tree.

The *Broad-leaved Jeruſalem Sage Tree* is now become very common in our gardens, which indeed is no wonder, as its beauty is great, and its culture eaſy. It will grow to be about five feet high, and ſpreads its branches without order all around. The older branches are

are covered with a dirty, greenifh, dead, falling, ill-looking bark; and this is the worft property of this fhrub: But the younger fhoots are white and beautiful; they are four-cornered, woolly, and foft to the touch. The leaves are roundifh and oblong, and moderately large; and thefe grow oppofite at the joints of the fhrub on long footftalks. They are hoary to a degree of whitenefs, and their footftalks alfo are woolly, white, tough, and ftrong. The flowers are produced in June, July, and Auguft, at the top joints of the young fhoots, in large whorled bunches. They are of the labiated kind, each confifting of two lips, the upper end of which is forked, and bends over the other. A finer yellow can hardly be conceived than the colour of which they are poffeffed; and being large, they exhibit their golden flowers at a great diftance, caufing thereby a handfome fhow.

The *Narrow-leaved Jerufalem Sage Tree* is of lower growth than the other, feldom rifing higher than a yard or four feet. This fhrub is in every refpect like the other; only the fhoots feem to have a more upright tendency of growth. The leaves alfo, which are narrower, are more inclined to a lanceolate form: They are numerous in both the forts, and hide the deformity of the bark on the older ftems, which renders them lefs exceptionable on that account. In fhort, thefe forts are qualified for fhruberies of all kinds, or to be fet in borders of flower gardens, where they will flower, and be exceeded even in that refpect by very few fhrubs.

Cretan Sage Tree is ftill of lower growth than either of the former, feldom arriving to a yard in height. The leaves are of the fame white hoary nature; they are very broad, and ftand on long footftalks. The flowers are alfo of a delightful yellow colour, very large, and grow in large whorls, which give the plant great beauty.

2. Purple Phlomis, or Portugal Sage. The ftalks of this fpecies are woody, four feet high, and fend forth feveral angular branches, which are covered with a white bark. The leaves are fpear-fhaped, oblong, woolly underneath, crenated, and grow on fhort footftalks. The flowers are produced in whorls, from the joints of the branches. They are of a deep purple

 colour,

colour, and have narrow involucra. They appear in June and July, but are not succeeded by ripe seeds in England.

There is a *Variety* of this species, with iron-coloured flowers; and another with flowers of a bright purple.

There are some other shrubby sorts of *Phlomis*, of great beauty; but these not only often lose their leaves, and even branches, from the first frost, but are frequently wholly destroyed, if it happens to be severe. They are low shrubs, very beautiful, and look well among perennial flowers, where they will not only class as to size with many of that sort, but, being rather tender, may with them have such extraordinary care as the owner may think proper to allow them.

The PROPAGATION of the above sorts is very easy, either by layers or cuttings. 1. If a little earth be thrown upon the branches, any time in the winter, they will strike root, and be good plants by the autumn following, fit for any place. Thus easy is the culture by that method. 2. The cuttings will also grow, if planted any time of the year. Those planted in winter should be the woody shoots of the former summer: These may be set close in a shady border; and being watered in dry weather, will often grow. This shrub may be propagated by young slips, also, in any of the summer months. These should be planted in a shady border, like Sage, and well watered. If the border is not naturally shady, the beds must be hooped, and covered with matting in hot weather. Watering must be constantly afforded them; and with this care and management many of them will grow.

P I N U S.

LINNEAN Class and Order, *Monoecia Monadelphia*. Male flowers containing many stamina joined at the base, and female flowers containing one pistil upon the same plant; the males being disposed in scaly bunches, the

the females in imbricated cones. There are twelve SPECIES:

1. PI'NUS *La'rix:* The LARCH, or DECIDUOUS PINE; *a tall deciduous tree*; native of Switzerland, the Alps, and some parts of Italy.

2. PI'NUS *Sylve'stris:* The WILD PINE; *a tall evergreen tree*; native of Scotland and the northern parts of the continent of Europe.

3. PINUS *Stro'bus:* The WEYMOUTH PINE, or the WHITE PINE; *a tall evergreen tree*; native of New England, Virginia, Canada, and Carolina.

4. PI'NUS *Pine'a:* The STONE PINE; *an evergreen tree*; native of Spain and Italy.

5. PI'NUS *Ce'mbra:* The CEMBRO, or the CEMBRO PINE; *an evergreen tree*; native of Switzerland, the Alps, Siberia, and Tartary.

6. PI'NUS *Tæ'da:* The SWAMP PINE; *an evergreen tree*; native of the Swamps of Virginia and Canada.

7. PI'NUS *Ce'drus:* The CEDAR OF LEBANON; *an evergreen tree*; native of Mount Lebanon.

8. PI'NUS *Pice'a:* The YEW-LEAVED FIR; *a tall evergreen tree*; native of Scotland, Sweden, and Germany.

9. PI'NUS *A'bies:* The EUROPEAN SPRUCE FIR; *a tall evergreen tree*; native of the northern parts of Europe and of Asia.

10. PI'NUS *Canade'nsis:* The AMERICAN SPRUCE FIR, or the NEWFOUNDLAND SPRUCE FIR; *a tall evergreen tree*; native of Canada, Pennsylvania, and other parts of North America.

11. PI'NUS *Balsa'mea:* The HEMLOCK FIR; *a low evergreen tree*; native of Virginia and Canada.

12. PI'NUS *Orienta'lis:* The ORIENTAL FIR; *a low evergreen tree*; native of the East.

1. The LARCH. This is a lofty tree: its branches are slender, and incline downward: the leaves are of a light green; and, as those of the Cedar of Lebanon, are bunched together like the pencils or little brushes of the painter. In spring, when the leaves and flowers are breaking out, the Larch has a particularly elegant appearance; and in winter, it gives variety to a wooded scene by the bright colour of its naked branches: It is in good esteem as an *Ornamental*; and its timber is of the

the more *useful* kind: it is superior to that of most of the *Pinus* tribe. HANBURY says, "Many encomiums have been bestowed on the timber of the Larch: and we find such a favourable account of it in antient authors, as should induce us to think it would be proper for almost any use. Evelyn recites a story of Witsen, a Dutch writer, that a ship built of this timber and Cypress, had been found in the Numidian sea, twelve fathoms under water, sound and entire, and reduced to such a hardness as to resist the sharpest tool, after it had lain submerged above a thousand four hundred years. Certain it is, this is an excellent wood for ship and house building. At Venice this wood is frequently used in building their houses, as well as in Switzerland, where these trees abound: So that, without all doubt, the Larch excels for masts for ships, or beams for houses, doors, windows, &c. particularly as it is said to resist the worm.

"In Switzerland, their houses are covered with boards of this wood, cut out a foot square; and as it emits a resinous substance, it so diffuses itself into every joint and crevice, and becomes so compact and close, as well as so hardened by the air, as to render the covering proof against all weather. But as such covering for houses would cause great devastation in case of fire, the buildings are confined to a limited distance, by an order of police from the magistrates. The wood, when first laid on the houses, is said to be very white; but this colour, in two or three years, is changed, by means of the sun and resin, to a black, which appears like a smooth shining varnish."

Of the *Common Larch* there are several *Varieties*. The flowers which the commonest sort exhibits early in the spring are of a delicate *red* colour; another sort produces *white* flowers at the same season, and these have a delightful effect among those of the Red sort; whilst another, called the *Black Newfoundland Larix*, encreases the variety, though by an aspect little differing from the others. There are also Larches with *greenish* flowers, *pale red*, &c. all of which are accidental Varieties from seeds. These Varieties are easily distinguished, even when out of blow: The young shoots of the White-flowering Larch are of the lightest green, and

and the cones when ripe are nearly white. The Red-flowering Larch has its ſhoots of a reddiſh caſt, and the cones are of a brown colour; whilſt the cones and ſhoots of the Black Newfoundland Larch are in the ſame manner proportionally tinged. The cones, which are a very great ornament to ſeveral ſorts of the Pines, are very little to theſe. Their chief beauty conſiſts in the manner of their growth, the nature and beauty of their pencilled leaves, and fair flowers; for the cones that ſucceed them are ſmall, of a whitiſh, a reddiſh, or a blackiſh brown colour, and make no figure.

The method of PROPAGATION is from ſeed: The cones may be gathered in November, and ſhould be left in a dry place till the ſpring. Juſt before ſowing, let the cones be opened or torn into four quarters by a knife, the point of which muſt be thruſt exactly down the center, ſo that the ſeeds in their reſpective places may not be damaged. Formerly, great pains were beſtowed in getting at the ſeeds, by cutting off the ſcales of the cones ſingly, and letting the ſeeds drop. This occaſioned great expence to thoſe who wanted a quantity of ſeeds; ſo that it is wholly laid aſide now, for the more eaſy method of opening them with knives, and then threſhing them. A certain price is generally allowed per thouſand to the poor for opening them. When a ſufficient quantity is opened, they ſhould be threſhed in a room, which will divide the ſcales, and diſlodge the ſeeds, without injuring many of them. Three thouſand cones will generally produce about a pound of good ſeeds. The cones being ſufficiently broken, and the ſeeds threſhed out, they ſhould be winnowed or ſieved to have clear ſeeds; after which they will be ready for ſowing *. Let the ſeminary conſiſt of a ſpot of fine light earth; and let the ſeeds be ſowed in beds a quarter of an inch deep. In the ſpring, when the plants appear, they ſhould be gently refreſhed with water in dry weather, and carefully kept

* The method now chiefly in practice by thoſe who raiſe Larches on a large ſcale, is to ſcatter the cones upon the prepared ſeed beds, and leave it to the run and air to extract the ſeeds. When a ſufficient number are diſcharged, the cones are raked off; and either removed to another ſeed bed, or laid up for another ſeaſon.

clean

clean from weeds during the whole ſummer. By the autumn they will not have ſhot more than an inch or two; and in ſpring they ſhould be pricked out in beds about three inches aſunder. The ſpring following, they muſt be taken out of theſe beds with care, and planted in the nurſery ground, three feet aſunder in the rows, and two feet diſtance; and here they may remain until they are fit to be planted out finally, which will be about the ſecond or third year after. If they grow well in the nurſery, it is adviſeable to plant them where they are to continue after having attained two years ſtrength in that place, if the ground can poſſibly be prepared for their reception; ſince theſe trees always thrive beſt when removed ſmall from the nurſery, if they are of a ſufficient ſize not to be injured by the weeds; if they are ſmaller, the owner muſt keep them clean. The Larch Tree will grow extremely well on almoſt any ſoil, as well in clays as in other ſorts; it thrives amazingly on the declivities of hills, and ſides of high mountains; it is hardy enough to reſiſt the ſevereſt cold, therefore proper for all expoſed places: And, as the timber is ſo valuable, and its growth ſo quick, it is a tree which may be propagated to the great advantage of the owner.

It is almoſt impoſſible to ſay too much in favour of this Tree. It grows on the barreneſt ſoils, and in the bleakeſt ſituations. In rich genial ſites it luxuriates too much, grows top heavy, and either loſes its head, or is bowed down into an unſightly form, and becomes unprofitable. Its timber, whether in the water, or in contact with the earth, is durable almoſt beyond compariſon.

2. The WILD PINE. This ſpecies includes two *Varieties:* The Scotch Fir and the Pineaſter.

The *Scotch Fir.* This tree is too well known to require any deſcription: and the method of propagating it will be found fully treated of under the Article TIMBER GROVES.

The *Pineaſter.* This is a large timber tree, and naturally throws out very large arms, ſome of which will be nearly horizontal. "Some people think theſe trees are very ornamental on their account; for in the winter eſpecially they appear naked, and are of a yellowiſh colour;

colour; and being spread abroad thus large, and without order, in the mixture of the more regular sorts of growing Firs, they make a good contrast. The Gardener must observe, that the leaves of this sort are very large and long, and of a lighter green than those of the Scotch Fir, which is another circumstance to direct him to its situation; and he must also observe, that those long and large leaves which ornament the younger branches only, give the tree a majestic air: and as the larger arms appear naked to view, so the younger, being thus plentifully furnished, have a noble effect, besides what beauty it receives from its numerous cones." HANBURY.

Its PROPAGATION may be the same as that of the Scotch Fir.

3. The WEYMOUTH PINE. This is a princely tree, majestic and elegant in the highest degree. HANBURY says, "It will grow to more than a hundred feet high, and makes such excellent masts for ships, that the Legislature, in the reign of Queen Anne, enacted a law enforcing the encouragement of the growth of these trees in America, where they abound." As an *Ornamental*, it stands first of all the Pines. The bark is smooth and soft to the touch, and, though of a dusky brown colour, on the whole has a delicate look. The leaves are truly ornamental, though their colour is nothing extraordinary; but they are long and slender, and are formed into tassels, which hang in so easy and elegant a manner, as "to make one in love with the tree."

The PROPAGATION of the Weymouth Pine is not so difficult as has been heretofore understood: It may be raised in common seed beds with ordinary care. HANBURY gives us the following directions: "The seeds of the Weymouth Pine are larger than those of the Scotch Fir; and in order to raise the young plants, it will be proper to sow them in pots or boxes, which may be removed into the shade after the plants are come up, when the sun's rays are violent. If they are sown in beds of fine light earth, they should be hooped and constantly covered with mats from the sun's heat, and as carefully uncovered when he sets. In about six or seven weeks after sowing, the young plants will

appear,

appear, when they ſhould be regularly guarded from birds, otherwiſe all your ſeeds, time, and trouble, will be loſt; for if the birds take to them at their firſt coming up, and are unmoleſted, they will not leave a ſingle plant. The plants being now above ground, the weeds ſhould be conſtantly picked out, as they appear, leſt, the fibres of their roots mixing with thoſe of the Firs, many of the latter may be drawn out with them. In dry weather they ſhould be refreſhed with water: But this muſt be done ſparingly, and with the utmoſt caution; for as the ſtems of the young plants are very ſlender, by over-watering they are frequently thrown aſide, which they hardly ever recover. Thus (continues HANBURY) I have known Gentlemen who, in attempting to raiſe theſe trees, have ſeen the young plants go off without perceiving the cauſe; and the more watering and pains they have taken, have found the plants perſiſt in this way more and more, to their great mortification and aſtoniſhment. In the ſpring following theſe plants ſhould be pricked out in beds half a foot aſunder each way; and here they may ſtand two years, when they may be either finally planted out, or removed into the nurſery, at the diſtance of one foot aſunder, and two feet in the rows. If care has been taken of them in the nurſery, they may be removed at a conſiderable height with great aſſurance of ſucceſs; for it is much eaſier to make this Pine grow than any of the other ſorts: So that where they are wanted for ornament in parks, open places, &c. a ſhow of them may be made in a little time.

"The ſoil the Weymouth Pine delights in moſt is a ſandy loam; but it likes other ſoils of an inferior nature: and although it is not generally to be planted on all lands, like the Scotch Fir, yet I have ſeen it luxuriant and healthy, making ſtrong ſhoots, on blue and red clays, and other ſorts of ſtrong ground. On ſtony and ſlaty ground, likewiſe, I have ſeen ſome very fine trees: So that I believe whoever is deſirous of having plantations of this pine, need not be curious in the choice of his ground."

4. The STONE PINE will not grow to the height of the former; and the bark is rough, and on ſome trees of a reddiſh colour. The leaves are long, very ornamental,

mental, and of a fine sea green colour. The cones give this tree the grandest look; for they are sometimes near six inches long, and are large, thick, and turbinated. The scales are beautifully arranged, and the whole cone is large and curious. "The kernels are eatable, and by many preferred to almonds; in Italy they are served up at table in their desserts; they are exceedingly wholesome, being good for coughs, colds, consumptions, &c. on which account only this tree deserves to be propagated. HANBURY continues, It may be very proper here to take notice of a very great and dangerous mistake Mr. MILLER has committed, by saying, under this article of Stone Pine, that seeds kept in the cones will be good, and grow, if they are sown ten or twelve years after the cones have been gathered from the trees; whereas the seeds of this sort, whether kept in the cones or taken out, are never good after the first year; and though sometimes a few plants will come up from the seeds that are kept in the cones for two years before, yet this is but seldom; neither must a tenth part of a crop be expected. This caution is the more necessary, as several Gentlemen who had cones, upon reading Mr. MILLER's book, and finding the seeds would take no damage when kept there, deferred the work for a season or two, when they thought they should have more conveniency either of men or ground for their purpose; and were afterwards wholly disappointed, no plants appearing, the seeds being by that time spoiled and worth nothing."

The PROPAGATION of the Stone Pine is from the seeds, which may be procured from their large cones by the help of a vise; for this will so effectually break the cones, without hurting the seeds, that they may be taken out with pleasure. The cones should be fresh, not older than a year or two at farthest, or the seeds will not be good; for although it has been asserted, that the seeds of Pines in general will keep in their cones many years, yet the cones of this species of Pine are an exception, as the seeds are rarely found good after the cones are one year old. The season for sowing these seeds is the middle of March. The weather being fine, and the ground fit for working, they should be sown about half an inch deep, in beds of fine light earth.

earth. In about ſeven weeks the plants will appear, which muſt be kept clean from weeds, and now and then watered in dry weather until July, by which time they will have made a tolerable ſhoot. In the month of July they ſhould be taken out of the ſeed beds, and pricked in others four inches aſunder. Rainy and cloudy weather muſt be made choice of for this work; and after they are planted, the beds ought to be hooped, in order to be covered with mats in the heat of the day, which, however, ſhould be always uncovered in the night. When they have taken to the ground, farther covering will be needleſs; and here they may remain, with only now and then watering, and keeping them clear of weeds, till the ſpring twelvemonth following; when, in the beginning of April, they ſhould be planted out in the nurſery, in well prepared ground, a foot aſunder, and at two feet diſtance in the rows. Here they may ſtand two years, and then ſhould be finally planted out. But if the trees are deſired to be larger before they are brought to the ſpot where they are to ſtand, they muſt be kept conſtantly removing every two years in the nurſery; for without this management this is a very difficult tree to be improved.

The Stone Pine delights in a ſandy loam; though, like moſt other Pines, it will grow well in almoſt any land.

5. The CEMBRO PINE is a fine tree, though of lower growth than any of the former, and the leaves are very beautiful; for they are of a lighter green than moſt of the ſorts, and are produced five in a ſheath. They are pretty long and narrow; and as they cloſely ornament the branches all round, they look very beautiful, and render the tree on their account valuable. The cones of theſe trees alſo on their waving heads, have a good effect; for they are larger than thoſe of the Pineaſter, and the ſquamæ are beautifully arranged.

6. The SWAMP PINE. Of this ſpecies there are many *Varieties:* HANBURY gives us the following account of them:

" The *Three-leaved American Swamp Pine* is a very large growing tree, if it has the advantage of a moiſt ſituation. The leaves are of a fine green colour, and are exceedingly long, ſlender, and beautiful; three iſſue out

out of one ſheath, and they cloſely garniſh the younger branches. This is a tree worthy of propagation, whether we regard its timber, or its fine appearance when growing. Its timber is ſaid to be equal in value to that of moſt ſorts of the Pine; and beſides the beauty it receives from its fine long three-ſheathed leaves, its head will be ornamented with very large cones, the good effect of which may be eaſily conceived.

"The *Two-leaved American Pine* will grow to be a large tree, and the leaves are long; two only grow in each ſheath, which occaſions its being ſo diſtinguiſhed. The leaves are of a lighter colour than many of the others. On the whole, it is a fine tree, but will make very little variety, unleſs cloſely examined. The cones of this ſort are much larger, and the ſcales more beautifully arranged, than thoſe of the Scotch Fir, though they are not of the ſize of the former ſort. This Fir alſo likes a moiſt ſoil.

"The *Yellow American Pine*, the *Yellow Tough Pine*, and the *Tough Pine of the Plains*, I received by thoſe names: There is ſome difference in the ſize and ſhape of the cones, though that ſeems inconſiderable. Theſe three ſorts make very little variety among themſelves; for they have nearly the ſame manner of growth; and though I have none that are yet grown to any large ſize, yet they all ſeem to have a tendency to throw out large arms, a little like the Pineaſter. How valuable the timber may be, I cannot tell; but the younger ſhoots of all of them are exceeding tough, and had we plenty, would make excellent bands for fagoting. The leaves are long, and of a yellowiſh green colour; there are three, and ſometimes two only, in a ſheath. If a large quarter of theſe were to be planted, to be ſeen at a diſtance, by any of the darker-coloured ſorts of Pines, their very different ſhade muſt have a delightful effect.

"*Baſtard Pine* is another ſort we receive from America, though it differs very little from ſome of the other American ſorts. The leaves are long and ſlender; ſometimes two and ſometimes three grow in each ſheath. They are generally of a yellowiſh colour towards their baſe, though their ends are green. The cones are rather long and ſlender, and the ends of the ſcales are ſo pointed, as to occaſion its being called by ſome the Prickly-coned Pine.

" *Frankincenſe Pine* is another American ſort, which we receive under that name. The leaves of it are long, and of a fine green colour. They are narrow, and three are contained in each ſheath. They cloſely ornament the younger branches all around. This tree, however, beautiful as it is on their account, makes little variety among the Pines, for many others look like it; but by the cones it makes a ſtriking difference; for theſe are exceeding large, even as large as thoſe of the Stone Pine; but their ſcales are looſer, and their arrangement is not quite ſo beautiful.

" The *Dwarf Pine*, as its name imports, is the leaſt grower of all the ſorts of Pines. It is an American plant, and the leaves grow two in a ſheath; theſe are ſhort, and of a pretty good green colour. This ſort is coveted by ſome, on account of its low growth; but it is the leaſt beautiful of any of the Pines, and has naturally a ſhabby look. The cones are ſmall, and the ſcales are pointed. There is very little in the plant to make it deſirable.

" There are many other ſorts of American Pines, which we receive from thence with the like cant names as thoſe of the above, which I have choſe to retain, as they will probably be continued to be ſent over, and that the gardener receiving them as ſuch may beſt know what to do with them. In many of thoſe ſorts I ſee at preſent no material difference, ſo am induced to think they are the ſame, ſent over with different names. Some of the ſorts above mentioned differ in very few reſpects; but I have choſe to mention them, as a perſon may be ſupplied with the ſeeds from Pennſylvania, Jerſey, Virginia, Carolina, &c. where they all grow naturally: and having once obtained the ſeeds, and from them plants, they will become pleaſing objects of his niceſt obſervations."

Theſe may all be propagated in the ſame manner as the WEYMOUTH.

7. The CEDAR OF LEBANON. This, in its native ſoil, has always been conſidered as the moſt majeſtic tree in nature. The leaves grow in pencils like thoſe of the Larch; and the extremities of its branches are like-

likewise declining, as those of the Deciduous Pine; to which at first sight it bears a strong resemblance; excepting in that it is less lofty and more spreading. There are some very fine Cedars of Lebanon in Stow Gardens. It ranks among the first of the *ornamental* tribe; and the *uses* of its timber are universally acknowledged. HANBURY enumerates the following: "It was greatly used in the building of Solomon's Temple, which at once convinces us of its superlative excellence. It is said to continue sound for two thousand years; and we are told, that in the Temple of Apollo at Utica there was found cedar wood of that age. The magnificent temples of the Pagans, as well as those of the true God, were chiefly built of this famous timber. The statue of the Great Goddess at Ephesus was made of this material; and if this tree abounded with us in great plenty, it might have a principal share in our most superb edifices. The effluvia constantly emitted from its wood are said to purify the air, and make rooms wholesome. Chapels and places set apart for religious duties, being wainscoted with this wood, inspire the worshippers with a more solemn awe. It is not obnoxious to worms; and emits an oil which will preserve cloth or books from worms or corruption. The sawdust will preserve human bodies from putrefaction, and is therefore said to be plentifully used in the rites of embalming, where practised."

The method of PROPAGATION is this: Having procured the cones, whether from the Levant or of our own growth, the seeds, a little before sowing, should be got out in this manner: Let a hole be bored with a gimblet exactly up the center of each cone, from the base to the apex; put them into a tub of water, where they may remain till the next day; then having a wooden peg, rather bigger than the gimblet, let it be thrust down the hole, and it will so divide the cones, that the different scales may be taken away, and the seeds picked out. In doing this, great care must be taken not to bruise and hurt the seeds, which will then be very tender. The soil in which you sow these seeds should be rather of a sandy nature; or, for want of this, some mould taken from a rich pasture, and sieved with a little drift sand, will serve the purpose.

Having the mould and ſeeds ready,.in the beginning of March let the latter be ſown in pots or boxes near half an inch deep: In about ſeven or eight weeks the plants will come up, when they ſhould be removed into the ſhade from the heat of the ſun; where they may ſtand, but not under ſhelter, all the ſummer; during which time they ſhould be kept clean from weeds, and watered now and then. In the winter ſeaſon they muſt be removed into a warmer ſituation; or, if it is likely to prove very ſevere, they ſhould be ſheltered either by mats, or removed into the greenhouſe, or covered with a hotbed frame; for they are ſubject to loſe their young tops at firſt, by the ſeverity of froſts. In the beginning of April following, theſe plants may be pricked out in beds four inches aſunder; and if the weather proves dry, they ſhould be ſhaded and watered till they have taken root; after which, they will want little ſhading and leſs watering. Indeed, nothing more is required than keeping them clean from weeds, and covering the ground ſo as to keep it moiſt, and prevent its chapping by the ſun's rays. In theſe beds they may remain two years; when, in the ſpring, they ſhould be tranſplanted to the nurſery, where they may remain till they are finally planted out. During the time they are in the nurſery, and after planting out, many will frequently have a tendency to droop in their leading ſhoot: As ſoon, therefore, as this is perceived, an upright ſtake muſt be driven into the ground, to which the ſhoots ſhould often be tied with baſs matting to keep them in their upright growth. This, however, will not always effect it; for ſome, after being tied, ſo effectually turn the ſhoot downwards over the bandage, though looſe, as to appear as if they were beat down on purpoſe. The Larch alſo will ſometimes rebel in this way: So that it would not be amiſs, in both caſes, whenever they firſt diſcover any ſigns of ſuch a tendency, to lighten the head, by nipping off the extremities of ſome few of the largeſt branches.

When theſe trees are planted out to remain, they ſhould be left to Nature, after being properly fenced: Not a knife nor a hatchet ſhould come near them; lopping even their loweſt branches is ſo injurious, that it both retards their growth and diminiſhes their beauty.

The

The Cedar of Lebanon wlll grow well in almoſt any ſort of ſoil or ſituation. As a proof of this, we need only obſerve, that in its native ſituation the roots are during part of the year covered with froſt and ſnow.

8. The YEW-LEAVED FIR. This ſpecies includes the Silver Fir and the Balm of Gilead Fir.

The *Silver Fir* is a noble upright tree *. The branches are not very numerous, and the bark is ſmooth and delicate. The leaves grow ſingly on the branches, and their ends are ſlightly indented. Their upper ſurface is of a fine ſtrong green colour, and their under has an ornament of two white lines, running lengthways on each ſide the midrib, on account of which ſilvery look this ſort is called the Silver Fir. The cones are large, and grow erect; and when the warm weather comes on, they ſoon ſhed their ſeeds; which ſhould be a caution to all who wiſh to raiſe this plant, to gather the cones before that happens.

The *Balm of Gilead Fir* has of all the ſorts been moſt coveted, on account of the great fragrance of its leaves; though this is not its only good property: for it is a very beautiful tree, naturally of an upright growth, and the branches are ſo ornamented with their balmy leaves, as to exceed any of the other ſorts in beauty. The leaves, which are very cloſely ſet on the branches, are broad; and their ends are indented. Their upper ſurface, when healthy, is of a fine dark green colour, and their under has white lines on each ſide the midrib lengthways, nearly like thoſe of the Silver Fir. Theſe leaves, when bruiſed, are very finely ſcented; and the buds, which ſwell in the autumn for the next year's ſhoot, are very ornamental all winter, being turgid, and of a fine brown colour: and from theſe alſo exſudes a kind of fine turpentine, of the ſame kind of (though heightened) fragrancy. The tree being wounded in any part, emits plenty of this turpentine; and HANBURY ſays, "it is ſuppoſed by many to be the ſort from whence the Balm of Gilead is taken, which occaſions

* Mr. MARSHAM ſays, "The talleſt trees I have ſeen were Spruce and Silver Firs, in the vallies in Switzerland I ſaw ſeveral Firs in the Dock-yards in Venice 40 yards long; and one of 39 yards was 18 inches diameter at the ſmall end. I was told they came from Switzerland."

this tree being ſo called. But this is a miſtake; for the true Balm of Gilead is taken from a kind of *Terebinthus*; though I am informed, that what has been collected from this tree has been ſent over to England from America (where it grows naturally), and often ſold in the ſhops for the true ſort."

Theſe trees are PROPAGATED by ſowing the ſeeds in a ſhady border, about the middle of March. They will readily come up if the ſeeds are good; but as this is not often the caſe, eſpecially if they are procured from the ſeedſmen, they ſhould be ſown very cloſe, otherwiſe you will be certain of having a very thin crop. The ſucceeding ſummer the plants will require no trouble, except keeping them clean from weeds; and the ſpring after that they ſhould be pricked out in beds at about four inches diſtance from each other. Here they may ſtand for two years, when they ſhould be planted in the nurſery, in rows a foot aſunder every way. The year, or at fartheſt two years, after they have been ſet in the nurſery, they ſhould be finally planted out; for if they are continued longer, many of them will die in the removal, and thoſe which grow frequently loſe their leading ſhoot, and meet with ſo great a check as to be hardly able to get into a good growing ſtate for ſeveral years.

The Silver Fir is exceedingly hardy, and will grow in any ſoil or ſituation, but always makes the greateſt progreſs in a good rich loamy earth.

The latter muſt be planted in a deep, rich, good earth; neither will it live long in any other ſort of ſoil. It matters little whether it be a black mould, or of a ſandy nature, provided it be deep, and there is room for the roots to ſtrike freely. As theſe trees have hitherto been planted without this precaution, and as ſuch a kind of ſoil does not often fall in the ordinary courſe of gardening, very few trees that have been planted many years are in a flouriſhing ſtate; for if they do not like the ſoil, or if the roots begin to meet with obſtructions, they ſoon begin to decline, which will be frequently in leſs than ſeven years; the firſt notice of which is, their leaves, which are naturally of a fine ſtrong green colour, loſe their verdure, and appear with a yellow tinge; and this colour grows upon them daily,

daily, until the appearance of the tree is changed. Another ſign of this tree being at its *ne plus ultra* is, its producing vaſt plenty of cones; this argues a weakneſs, and they generally die away by degrees ſoon after. This is always the caſe where the ſoil does not wholly agree with them; but where it is deep and good, they will be healthy and flouriſhing, and produce cones for ſeeds.

9. The EUROPEAN SPRUCE FIR. This ſpecies includes the Norway Spruce, and the Long-coned Corniſh Fir.

The *Norway Spruce* is a tree of as much beauty while growing, as its timber is valuable when propagated on that account. Its growth is naturally like the Silver, upright; and the height it will aſpire to may be eaſily conceived, when we ſay that the white deal, ſo much coveted by the joiners, &c. is the wood of this tree; and it may perhaps ſatisfy the curious reader to know, that from this Fir pitch is drawn. The leaves are of a dark green colour; they ſtand ſingly on the branches, but the younger ſhoots are very cloſely garniſhed with them. They are very narrow, their ends are pointed, and they are poſſeſſed of ſuch beauties as to excite admiration. The cones are eight or ten inches long, and hang downwards.

The manner of PROPAGATING this tree is nearly the ſame as that of the Scotch Fir, only this will more eaſily grow when of a large ſize, and conſequently will not require removing ſo often in the nurſery. In the middle of March, having got the ſeeds out of the cones, ſow them in a north border; for when they come up, by being conſtantly ſhaded all the ſummer in ſuch a ſituation, they will ſhoot much ſtronger, and be better to prick out the ſpring following in the nurſery. In about ſix or ſeven weeks after ſowing, the young plants will appear, when they ſhould be ſcreened with the uſual care from the birds, which otherwiſe would ſoon deſtroy them. By the autumn, many of theſe young plants, if they are kept clean from weeds, and watered in dry weather, will have ſhot three or four inches: and in ſpring they ſhould be carefully taken out of their ſeed beds, ſo that the fibres may by no means be broken off or injured. Being thus cautiouſly taken up, they ſhould be as carefully planted in the nurſery ground, at the diſtance of one foot aſunder each way. Here they may remain, with keeping them free from weeds, for three years, when they ſhould be ſet out in the places where they are deſigned to remain. But if larger trees are deſired for this purpoſe, they ſhould be taken up and planted in the nurſery, a foot and a half aſunder, in rows two feet and a half diſtant, where they may ſtand, if required, till they are ſix or eight feet high, without any other removing.

When they are ſet out finally, they may be planted, with tolerable hopes of ſucceſs; for the Spruce Fir is not ſo nice or difficult in ſhifting its quarters as any of the other ſorts of Pines. But though theſe trees may be tranſplanted at a good height, it is always adviſeable to remove them to the places deſigned for them with all poſſible diſpatch, as they are more certain of growing, and will recover the check occaſioned in all trees by removal in leſs time.

The better the ſoil is, the faſter will the Spruce Fir grow, though it will thrive very well in moſt of our Engliſh lands. In ſtrong loamy earth it makes a ſurpriſing progreſs; and it delights in freſh land of all ſor s, which never has been worn out by ploughing, &c though it be ever ſo poor.

The *Long-coned Corniſh Fir* differs ſcarcely in any reſpect from the Norway Spruce, except that the leaves and the cones are larger.

10. The AMERICAN SPRUCE FIR. This ſpecies includes three *Varieties*: The *White Newfoundland Spruce*; the *Red Newfoundland Spruce*; and the *Black Newfoundland Spruce*. Theſe, however, differ ſo little, that one deſcription is common to them all. They are of a genteel upright growth, though they do not ſhoot ſo freely or grow ſo faſt with us as the Norway Spruce. The leaves are of the ſame green, and garniſh the branches in the ſame beautiful manner as thoſe of that ſpecies, only they are narrower, ſhorter, and ſtand cloſer. The greateſt difference is obſervable in the cones; for theſe are no more than about an inch in length, and the ſcales are cloſely placed. In the cones, indeed, conſiſts the difference of theſe three ſorts: Thoſe of the White ſpecies are of a very light brown colour;

colour; those of the Red species more of a nut-brown or reddish colour; and those of the Black species of a dark or blackish colour. Besides this, there is scarcely any material difference; though it is observable, that this trifling variation seems to be pretty constant in the plants raised from the like seeds. These sorts will often flower, and produce cones when only about five or six feet high; and indeed look then very beautiful: but this is a sign of weakness in the plant, which it does not often fairly overget.

In many parts of England this is a very difficult tree to raise. It spends itself in cones, and becomes stunted and unsightly. Nevertheless, in the vallies of the Highlands, it thrives with full luxuriance and vigour; forming a rich picturable outline, possessing more strength of feature than most of the Pines. As a standard in polished scenery, there are few trees that equal it; as may be seen at ENVILLE and FISHERWICK.

11. The HEMLOCK FIR possesses as little beauty as any of the Fir tribe; though being rather scarce in proportion, it is deemed valuable. It is called by some the Yew-leaved Fir, from the resemblance of the leaves to those of the Yew tree. It is a tree of low growth, with but few branches; and these are long and slender, and spread abroad without order. The leaves do not garnish the branches so plentifully as those of any other sort of Fir. The cones are very small and rounded; they are about half an inch long; and the scales are loosely arranged. We receive these cones from America, by which we raise the plants; though this caution should be given to the planter, that this tree is fond of moist rich ground, and in such a kind of soil will make the greatest progress.

12. The ORIENTAL FIR. This is a low but elegant tree. The leaves are very short, and nearly square. The fruit is exceedingly small, and hangs downward; and the whole tree makes an agreeable variety with the other kinds.

In PROPAGATING the AMERICAN SPRUCE and the HEMLOCK FIR, the seeds being very small, a more than ordinary care should be taken of them, lest they be lost. They should be sown in pots or boxes of fine light mould, and covered over hardly a quarter of an inch.

They ſhould be then plunged up to the rims in a ſhady place, and netted, to ſave them, when they firſt appear, from the birds. If the place in which they ſtand is ſhaded, they will need little or no water all ſummer, unleſs it proves a very dry one; and being all of a very hardy nature, they will not require the trouble of covering in the winter. The beginning of July after that, the Newfoundland Spruce Fir ſhould be pricked out in beds at a ſmall diſtance, though the Hemlock Spruce ſhould remain in the pots a year longer, as they will then be very ſmall. After they are planted, they muſt be well watered, and the beds muſt be hooped, to be covered with mats for ſhade. In hot weather the mats ſhould be put over the beds by nine o'clock in the morning, and conſtantly taken off in the evenings, and remain ſo in cloudy and rainy weather. After they have taken root, they require no farther care, until they are planted out; which, ſays HANBURY, "cuſtom has taught us to do in the autumn or in the ſpring; but I have by much experience found, that July is a good month for planting out all the ſorts of Firs; and if it were done in a wet time, and the weather ſhould continue moiſt or cloudy for two or three weeks, it would be by far the beſt time in the whole year. Whoever, then, plants out Firs in July, unleſs ſuch weather happens, muſt ſhade and water them for a month or ſix weeks; but as ſhade is not to be afforded large trees of this kind, if there be many of them, their removal muſt be at the uſual times, leſt that parching time which often comes in the middle of ſummer burn them up before they can have time to take root. On this account, the planting of trees at Midſummer ſhould be tenderly enforced: though I muſt declare, that I have repeatedly planted Scotch Firs of different ſizes, ſome one yard and more, others ſix feet high, in the ſcorching heat, and left them to Nature, without giving them any aſſiſtance, and they have for the moſt part grown. Let others, if they pleaſe, make the experiment with a few, before they venture to plant out quantities at that ſeaſon." This information, and the manner in which it is conveyed, do Mr. HANBURY great credit as a practical man and a writer.

[*For a farther Account of the* LARCH, *ſee the cloſe of the Article* TIMBER GROVES *in the Firſt Volume.*]

PIS-

PISTACIA.

LINNEAN Clafs and Order, *Dioecia Pentandria:* Male flowers containing five ftamina, and female flowers containing three piftils, upon diftinct plants: There are five SPECIES; four of which will endure our winters, provided they be placed in a warm well fheltered fituation:

1. PISTA'CIA *Terebi'nthus:* The COMMON TURPENTINE TREE; *a fub-evergreen tree or fhrub*; native of Italy, Spain, and fome parts of Africa.

2. PISTACIA *Ve'ra:* The COMMON PISTACIA, or PISTACIA NUT TREE; *a fub-evergreen tree or fhrub*; native of Perfia, Arabia, Syria, and India, from whence we receive the nuts.

3. PISTA'CIA *Trifo'lia:* The THREE-LEAVED PISTACIA, or TURPENTINE TREE; *a low fub-evergreen tree or fhrub*; native of Sicily.

4. PISTA'CIA *Narbone'nfis:* The LARGE-FRUITED TURPENTINE TREE; *a fub-evergreen tree or fhrub*; native of Perfia, Armenia, Mefopotamia, and the fouth of France.

1. The COMMON TURPENTINE TREE will grow to the height of about thirty feet. The bark of the trunk is thick, full of cracks, and of a dark brown colour; whilft that on the young fhoots is thin and fmooth. The leaves are pinnated and large, of a dark green colour, and grow alternately on the branches. The folioles of which each leaf is compofed are oval, fpear-fhaped, and confift of three or four pairs, which are placed on the midrib, befides the odd one with which they are terminated. There will be male and female flowers on different plants. They exhibit their bloom in April: The male flower is nothing but a catkin, and the females make no figure; fo that where philofophy has no view, it is from the defire of having an extenfive collection that we procure thefe trees. In warm countries, the leaves of the Piftacia continue all the

the year; with us, they fall off when attacked by the frosts. From the trunk flows the true turpentine; in the room of which, that taken from some of our Pines is generally substituted.

2. The COMMON PISTACIA is about twenty feet in height. The trunk of this species also is covered with a dark brown bark, full of cracks, whilst the young shoots are smooth, and of a light brown colour. The leaves are likewise pinnated, being composed of about two or three pairs of folioles, which do not always stand exactly opposite on the midrib, terminated with an odd one. These folioles are large, and nearly of an oval figure: Their edges turn backwards, but have nevertheless a noble look. The male flowers are catkins of a greenish colour; and the female flowers are very small, and produced in clusters from the sides of the branches. April is the month of their flowering; and the female flowers are succeeded by the Pistacia nuts we eat.

3. The THREE LEAVED PISTACIA is of about twenty-five feet growth. The bark of the trunk is very rough, and of a dark brown colour; but that of the young shoots is smooth, and lighter. The leaves of this species are trifoliate. The folioles are of an oval figure, of a very dark green colour, and are greatly ornamental to the plant. Different trees will have male and female flowers: The males are greenish catkins; and the females have no petals, are small, and make no show.

4. LARGER-FRUITED TURPENTINE TREE will grow to be about twenty-five feet high. The bark partakes more of a whitish colour, and is smoother than those of the other species. The leaves also are pinnated; but the folioles of which each is composed are not always of the same number: Sometimes there are three, sometimes five pair of folioles to form the compound leaf. These are of a paler green than any of the other sorts, of a roundish figure, and stand on longish footstalks. The male flower of this species also is a catkin; and the females are succeeded by nuts, which by many are liked, being eatable, like the Pistacia nuts. The leaves continue on these trees great part of the year, in warm countries.

The

The PROPAGATION. The ſeeds, which we receive from abroad, ſhould be ſown as ſoon as poſſible after their arrival. A compoſt ſhould be prepared for them, mixed in the following proportions: Six barrows full of earth, from a freſh paſture, taken from thence at leaſt a year before, with the green ſward, and well turned and rotted; three barrows of drift or ſea ſand; and one barrow of old lime rubbiſh, beaten to duſt: theſe ſhould be all well mixed together. The ſeeds ſhould be ſown about half an inch deep in pots, which may then be ſet under a warm wall or hedge, until the hot weather begins to come on, when they ſhould be removed into the ſhade, and plunged up to the rims in ſome mould. At the approach of winter, they may be removed into a warm place, and in ſpring a hotbed muſt be prepared for their reception. As theſe plants rarely come up the firſt year, this will be a better method than to plunge them in a hotbed ſoon after they are ſown; for even with this aſſiſtance, they will be later before they come up, will be very weak and tender plants in the autumn, and will require extraordinary future care to preſerve them; whereas, if they are ſuffered to remain unforced for one turn, they will be preparing to vegetate, and of courſe will come up themſelves the ſecond ſpring; but a hotbed will be neceſſary, as at that time it will make them ſhoot ſtronger. But this forcing muſt by no means be continued; a hitch only is to be given them, and they ſhould immediately be hardened to the air. Watering and ſhade all ſummer muſt be allowed them; and they ought to be made as hardy as poſſible by the autumn. At the approach of winter, when other plants are to be ſet in the greenhouſe, theſe ſhould go with them, or be placed under a hotbed frame. They ſhould be ſet out with them in the ſpring, and in May the pots muſt be plunged up to the rim in the ſhade as before. The next winter they will require the greenhouſe; and in the ſucceeding ſpring they will be two-years-old ſeedlings; at which time they ſhould be ſhaken out of the pots, and each planted in a ſeparate pot, in the ſame ſort of compoſt in which the ſeeds were ſown: This being done, they ſhould be afforded a heat in the bed to ſet them forward. After they have begun ſhooting freely,

freely, the glaſſes ſhould be taken off by degrees; and now they will want no more hotbeds. Watering muſt be given them in dry weather; and in the autumn they muſt be removed into the greenhouſe, with other plants. And thus they ſhould be treated as a greenhouſe plant for four or five, or if even ſix years, it will be ſo much the better; obſerving always, however, in the ſpring, to ſhift them into a freſh and larger pot every other year. The plants being now five or ſix years old, and being become tolerably ſtrong and woody, may be ſet out in the places where they are to remain. Theſe, as was obſerved, muſt be warm well ſheltered places, with a naturally dry ſoil; and if the two or three ſucceeding winters ſhould prove mild and favourable, they will by that time be grown to be very hardy, and may bid defiance to almoſt any weather. The Common Turpentine tree and the Piſtacia Nut tree, when grown old, reſiſt our ſevereſt froſts; and the other ſorts, though rather of a more tender nature, even if not old, will droop to none but the moſt piercing.

P L A T A N U S.

LINNEAN Claſs and Order, *Monoecia Polyandria:* Male flowers containing many ſtamina, and female flowers containing ſeveral piſtils, upon the ſame plant; the males being collected in a globular catkin, and the females digeſted in a roundiſh ball. There are only two SPECIES:

1. PLA'TANUS *Orienta'lis:* The ASIATIC or ORIENTAL PLANE; *a tall deciduous tree*; native of Aſia.

2. PLA'TANUS *Occidenta'lis:* The AMERICAN or OCCIDENTAL PLANE; *a tall deciduous tree*; native of North America.

1. The ORIENTAL PLANE riſes to a very great height, and in its native ſoil grows to a prodigious ſize: The ſtem is covered with a ſmooth bark, which falls off annually. The bark of the young branches is of

of a dark brown, inclining to a purple. The leaves are large and palmated, being deeply cut into five ſegments: their upper ſides are of a deep green, and the under ſides pale. The flowers are very minute: they come out at the ſame time as the leaves, which is in June. This is very late, and is no doubt a blemiſh to the beauty of this nevertheleſs highly *ornamental* tree. The Antients were very partial to this tree; which is not to be wondered at, when we conſider the extenſive canopy it forms, the impenetrable ſhade given by the number and ſize of its leaves, and conſequently the grateful coolneſs it muſt afford in a ſultry climate. EVELYN and HANBURY claſs this and the next ſpecies amongſt Foreſt or Timber trees; and their wood may rank with that of the Sycamore, which bears a conſiderable reſemblance to this genus of plants, and which in the North of England is called the *Plane Tree*.

2. The AMERICAN PLANE. This alſo grows to a great ſize; the ſtem not only ſwells to an immenſe thickneſs, but, riſing erect, ſhoots up perfectly ſtraight and cylindrical to an amazing height. The Hon. Paul Dudley, in a Letter to the Royal Society, ſays, "he obſerved in New England one of theſe Plane Trees nine yards in girth, which continued its bulk very high; containing when felled twenty-two loads of timber." The bark is ſmooth, and, like that of the Aſiatic ſpecies, falls off annually. The leaves are broad, with long footſtalks, and are cut into angles at their edges, but not divided nearly ſo deep as thoſe of the foregoing ſpecies: The upper ſide is of a light green, the under ſide paler: The flowers are ſmall, and come out with the leaves about the ſame time as thoſe of the Oriental Plane. Altogether, this tree is peculiarly refreſhing to the eye, and truly *ornamental*.

Beſides theſe two diſtinct ſpecies, there are two *Varieties*:

The Maple-leaved Plane.
The Spaniſh Plane.

The *Maple-leaved Plane*, ſays MILLER, is certainly a ſeminal variety of the Eaſtern Plane. It differs from the two ſorts beforementioned, in having its leaves not ſo deeply cut as thoſe of the Eaſtern Plane, but much more deeply than thoſe of the Occidental Plane. The foot-

footſtalks of the leaves are much larger than thoſe of either of the former, and the upper ſurface of the leaves is rougher.

The *Spaniſh Plane* has larger leaves than either of the other ſorts. They are divided in a ſimilar manner to thoſe of the Maple-leaved Plane. Some of them are cut into five, and others into three lobes: Theſe are ſharply indented on their edges, and are of a light green. This is by ſome called the *Middle Plane*, from its leaves being ſhaped between thoſe of the firſt two ſorts. This too is probably a Variety of the Oriental Plane.

The method of PROPAGATING the ORIENTAL PLANE is from ſeeds, when they can be eaſily procured; but whoever enjoys not this convenience muſt have recourſe to layers. 1. The ground proper for the ſeminary ſhould be moiſt and ſhady, well dug, and raked until the mould is fine; then in the autumn, ſoon after the ſeeds are ripe, let them be ſcattered over this ground, and the ſeeds raked in, in the ſame manner as turnep ſeeds. In the ſpring, many of the young plants will come up, though you muſt not expect the general crop until the ſecond year; the ſpring after which they may be taken out of the ſeminary, and planted in the nurſery in rows one yard aſunder, and at one foot and a half diſtance in the rows. Here they may remain, with the uſual care of digging between the rows, and keeping them clean, till they are of ſufficient ſize to be planted out. 2. Where the ſeeds of theſe trees cannot be procured, layering muſt be the method of propagation. For this purpoſe, a ſufficient number muſt be planted out for ſtools, on a ſpot of earth double dug. After they have ſtood one year, they ſhould be cut down, in order to make them throw out young wood for layering. The autumn following, theſe ſhould be laid in the ground, with a little nick at the joint; and by the ſame time twelve months after, they will be trees of a yard high, with a good root, ready to be planted out in the nurſery, where they may be managed as the ſeedlings; and as the ſtools will have ſhot up freſh young ſhoots for a ſecond operation, this treatment may be continued at pleaſure.

The AMERICAN PLANE is PROPAGATED by cuttings;

tings; which, if they be taken from ſtrong young wood, and planted early in the autumn, in a moiſt good mould, will hardly fail of ſucceeding. They are generally planted thick; and then removed into the nurſery ground, as the layers of the other ſort: But if a large piece of ground was ready, the cuttings might be placed at ſuch a diſtance as not to approach too cloſe before they were of a ſufficient ſize to be planted out to ſtand; and this would ſave the expence and trouble of a removal. The Oriental Plane tree will grow from cuttings, but not ſo certainly as this; and whoever has not the convenience of proper ground for the cuttings, muſt have recourſe to layers, which, indeed, is for either ſort the moſt effectual and ſure method.

Plane trees delight in a moiſt ſituation, eſpecially the Occidental ſort. Where the land is inclined to be dry, and Plane trees are deſired, the others are to be preferred. But in moiſt places, by the ſides of rivulets, ponds, &c. the Occidental makes ſuch ſurpriſing progreſs, that it might be ranked among the Aquatics. The bright colour of the Planes gives variety to groves and maſſes of wood; in groups and ſingle trees they are ſingularly elegant; as may be ſeen at Fiſherwick.

POPULUS.

LINNEAN Claſs and Order, *Dioecia Octandria:* Male flowers containing eight ſtamina, and female flowers containing one piſtil, upon diſtinct plants; the males and females being ſimilarly ſituated in long, looſe catkins. There are five SPECIES:

1. PO'PULUS *A'lba:* The ARBEEL, or WHITE POPLAR; *a deciduous aquatic tree*; growing common in England and moſt parts of Europe.

2. PO'PULUS *Ni'gra:* The COMMON POPLAR; *a deciduous aquatic tree*; this alſo grows common in England and moſt parts of Europe.

3. Po'pulus *Tre'mula:* The Aspen; *a deciduous tree*; native of England and the colder parts of Europe.

4. Po'pulus *Balsami'fera:* The Balsam Poplar, or Tacamahaca; *a deciduous tree*; native of Carolina and many parts of North America.

5. Po'pulus *Heterophy'lla:* The Virginia Poplar; *a deciduous tree*; native of Virginia.

1. The Arbeel. This is a tall ſpreading tree, one of the largeſt of the aquatic tribe. The trunk is covered with a ſmooth whitiſh bark. The leaves are about three inches long, and ſtand upon footſtalks about an inch in length: they are indented at the edges; and are of a dark green on the upper ſurface, but white and woolly underneath.

2. The Common Poplar. This tree will alſo grow to a large ſize. Its leaves are not ſo large as thoſe of the former. Their colour is a pleaſant green; they are heart-ſhaped, and appear about the middle of April.

The *Lombardy Poplar*, or the *Po Poplar*, ſeems to be a *Variety* of the Common Poplar: the chief difference is, the Common Poplar throws out a large ſpreading head, whilſt the Lombardy aſpires with a remarkably cloſe one, riſing like an obeliſk. There is a peculiar elegance in this plant when young, and in a moiſt rich ſituation it flouriſhes extraordinarily; but it will not thrive in a dry barren ſoil, ſoon growing ſtunted and moſſy.

Poplars in general, whilſt young, are elegant and *ornamental*; but the litter which is made by their catkins renders them in the ſpring diſagreeablle neighbours to kept walks and ſhruberies. The *uſes* of the wood are not many: it makes good boards, which have one peculiar good quality for flooring: they can with difficulty be made to burn, and will never blaze out like thoſe of other wood: it would be needleſss to add that the wood of the Poplar is the worſt of fuel. It is a quick-growing tree, and may frequently be made uſe of as a ſkreen to hide ſwamps or other deformities.

3. The Aspen. This tree will grow to a great height, and takes a good outline. Its leaves are ſmaller than thoſe of the Common Poplar; they ſtand on long ſlender *flat* footſtalks, which render it of all the other ſorts the moſt tremulous; they are roundiſh, and ſmooth,

ſmooth on both ſides; but do not make their appearance before the beginning of May. The profuſion of ſuckers thrown up by this tree renders it altogether unfit for kept grounds.

The PROPAGATION of theſe three ſpecies of Poplar is very eaſy: they will grow from cuttings, ſets, truncheons, &c.; but, ſays HANBURY, "I by no means approve of the planting of truncheons, as has been often practiſed on boggy places; becauſe I have always obſerved, that plantations of theſe luxuriant trees, attempted to be raiſed in this manner, have been frequently ſtunted, and very unpromiſing; and that the moſt promiſing trees have never equalled, in goodneſs or beauty, thoſe planted with regular trees raiſed in the nurſery. In order, therefore, to obtain a quantity of Poplars, proper to be planted in avenues or clumps, by the ſides of rivulets, bogs, or any other places where they are deſired, you muſt get a piece of ground double dug for the nurſery. If the trees wanted are to be planted for good in a watery ſituation, this nurſery ground ſhould be pretty near it; but if they are deſigned for paſture grounds, fields, or ſuch as have no more than a common degree of moiſture, the ſoil of the nurſery ſhould be proportionably drier. The latter end of October is the beſt ſeaſon for planting the cuttings; though they will grow if planted in any of the winter months. They ſhould be all of thoſe laſt year's ſhoots which have been vigorous, or at leaſt not older than two years wood. Theſe cuttings ſhould be one foot and a half in length; and muſt be planted in the nurſery ground in rows a yard aſunder, and at a foot and a half diſtance from one another. They ſhould be planted a foot in the ground, while the other half muſt remain to ſend forth the leading ſhoot. Now in order to have one leading ſhoot only, in ſummer theſe plants ſhould be carefully looked over, and all young ſide branches nipped off, in order to encourage the leading branch. After this, no farther care need be taken of them than keeping them clean from weeds, and digging between the rows in the winter, till they have attained a proper ſize to be planted out.

4. The BALSAM POPLAR will grow to be a large timber tree, "and has a majeſty both enchanting and

 peculiar."

peculiar." It is an exceedingly ſwift grower, inſomuch that it has been known to ſhoot ten feet in the ſpace of one ſummer, and to be in thickneſs, neareſt the baſe, an inch in diameter. The bark is ſmooth, and of a whitiſh colour; though that on the young ſhoots is of a fine green. The young ſhoots are cornered, having five angles; and the bark of which theſe are compoſed, being extended by the future growth, leaves only the traces on the older branches of theſe angles. "This (continues HANBURY) gives the tree in winter a particular look; for at the baſe of each bud they curve over and meet. Thus there will be between every bad formed by the bark, figures like niches, as it were, of public buildings, though with an upright in the middle, at the top of each of which, like an ornament, is ſeated the bud, for the future ſhoot or leaf. Theſe buds are only to be found on the younger branches; but the figure is retained on the bark of the older without thoſe ornaments. But of all the trees in a collection, none more agreeably by its leaves entertains us than this, whether we conſider their colour, figure, or ſize. The colour is a light ſhining green, which is heightened in the autumn by the ſtrong midrib, and the large veins that iſſue from it, turning to a red colour; the leſſer veins alſo being in ſome degree affected, occaſions upon the ſame leaf a ſweet contraſt. Their figure nearly reſembles that of a heart, and they are notched at their edges. But the chief majeſty this tree receives is from the ſize of the leaves: I have meaſured ſome of the younger trees, and found the leaves ten inches long and eight broad, with a ſtrong footſtalk of four inches in length. Theſe majeſtic leaves are placed alternately on the branches; though, as the tree advances in height, they diminiſh in ſize. This ſpecies ſhoots late in the autumn; and theſe young ſhoots have their ends often killed in hard winters; which is an imperfection, as it cauſes the tree to have a very bad look in the ſpring, before and when the leaves are putting out: However, theſe laſt will not fail afterwards to make ample amends for the former defect. The flowers afford no pleaſure to the Gardener: They are only catkins, like other Poplars, and fit only for the curious Botaniſt's inſpection."

5 The

5. The VIRGINIA POPLAR grows to be a large timber tree. The branches are numerous, veined, and angular. The leaves are heart-ſhaped, broad, ſlightly ſerrated, and downy on their firſt appearance. The flowers come out in looſe catkins, and make little ſhow: They appear early in the ſpring; and are ſucceeded by numerous downy ſeeds, which are diſperſed all about to a conſiderable diſtance.

Theſe two ſpecies are PROPAGATED, 1. By cuttings. In order to obtain proper cuttings for the purpoſe, the plants ſhould be headed the year before, and a foot and a half of the thickeſt part of the former ſummer's ſhoots ſhould be taken. The month of October is the ſeaſon; and theſe cuttings ſhould be planted in a moiſt ſhady ſoil, one foot deep, with the other half foot above ground. Many of them will grow; though it is generally allowed to be a good crop if half ſucceed. 2. By layers. Theſe muſt be of the laſt ſummer's ſhoots; and the operation ought to be performed in the autumn, before they have done growing; for the ſap being then in motion, they may readily be brought down; whereas, if it is deferred until winter, the young ſhoots are then ſo exceedingly brittle, that though all poſſible care be taken, many of them, in attempting to bring them down, will be broken. A ſmall ſlit with the knife muſt be given to each; and after the operation is performed, ſome furze buſhes ſhould be ſtuck round each ſtool, to break the keen edge of the black froſts, and preſerve the ends of the layers from being killed. In the ſpring they ſhould be cut down to within one eye of the ground; and by the autumn they will have ſtruck root, and be good plants, either for the nurſery ground, or where they are intended to be ſet out to ſtand.

POTENTILLA.

LINNEAN Claſs and Order, *Icoſandria Polygynia*: Each flower contains twenty males and many females.

 There

There are numerous SPECIES of this genus of plants; but only one of a ligneous nature:

POTENTI'LLA *Frutico'sa:* The CINQUEFOIL SHRUB; *a low deciduous ſhrub*; native of Yorkſhire.

The CINQUEFOIL SHRUB grows to the height of three feet; exceedingly buſhy and full of wood; which renders it the leſs ornamental; but the ſize and luſtre of its flowers do away that objection. The leaves are five-fold, and of a pale green colour. The flower is of a bright yellow.

The PROPAGATION of this ſhrub is by layers or by cuttings; the ſeeds, though a native of the mountainous parts of this iſland, ſeldom ripen in gardens. If the Propagation is by cuttings, the ground ſhould be in the ſhade, naturally moiſt and well dug; and if they are planted any time in the autumn, winter, or early in the ſpring, they will readily grow. In the autumn or winter following, the ſtrongeſt plants may be removed to the places where they are deſigned to remain. If the operation is performed by layering, the trouble then is only pegging down the young branches, and drawing a little mould over them: they will then readily ſtrike root, and in the autumn or winter following, they muſt be taken up trimmed and planted, in the nurſery way, to remain there for a year, before they are removed to the places of their final deſtination.

PRINOS.

LINNEAN Claſs and Order, *Hexandria Monogynia:* Each flower contains ſix males and one female. There are only two SPECIES:

1. PRINOS *Verticilla'tus:* The VIRGINIA WINTERBERRY, or DESIDUOUS PRINOS; *a deciduous ſhrub*; native of Virginia, Pennſylvania, and other parts of North America.

2. PRI'NOS *Gla'ber:* The CANADA WINTERBERRY, or EVERGREEN PRINOS, *an evergreen ſhrub*; native of Canada.

1. The

1. The DECIDUOUS WINTERBERRY is a ſhrub of about ſix or eight feet in growth, ſending forth many branches from the bottom to the top, which are covered with a browniſh bark. The leaves are ſpear-ſhaped, pretty large, of a ſtrong green colour, lengthways ſerrated, and placed alternately on ſlender footſtalks on the branches. The flowers are produced at the ſides of the branches, growing one or two together at the joints; but make no ſhow. They appear in July; and are ſucceeded by purple-coloured berries, which remain on the trees all winter, and look well.

The beſt way of PROPAGATING this plant is from its ſeeds. Theſe ſhould be ſown, ſoon after they are ripe, in beds of fine ſandy earth; and if the garden does not naturally afford ſuch, a few barrows full of drift ſand muſt be brought to mix with the common mould. The beds being thus prepared, and made ready for ſowing, the ſeeds ſhould be ſown about three quarters of an inch deep. It is very ſeldom that any of the ſeeds come up the firſt ſpring after; if any do, there will be but few; ſo that all the ſummer they muſt be kept clean from weeds. The ſpring following the plants will come up; though many will lie until the third ſpring before they make their appearance. After they are come up, weeding and watering muſt be afforded them in the ſummer; and with this care they may remain in the ſeed bed two years. In March, being then two-years-old ſeedlings, they ſhould be taken up, and planted in the nurſery, at very ſmall diſtances; and here they may remain, with the uſual nurſery care, until they are ſet out.

2. The EVERGREEN WINTERBERRY grows to about eight or ten feet high, ſends forth many branches from the bottom to the top, and the whole plant aſſumes the appearance of an *Alaternus*. The leaves are oblong, ſpear-ſhaped, acute, ſerrated, of a ſtrong green colour, and placed alternately on the branches. The flowers come out from the wings of the leaves, two or three together on a footſtalk. They are ſmall, white, appear in July, and are ſucceeded by red or purple berries, which remain on the trees all winter.

The PROPAGATION of this ſhrub is exactly the ſame as that of the deciduous ſpecies, except that this ſpecies is of a more tender natute; and instead of setting out the seedlings in the nursery ground, each should be set in a separate pot, to be placed under shelter in winter for a few years, until they are grown strong plants, and after that to be turned out, with the mould at the roots, into the places where they are designed to remain, which ought always to be in a dry sandy soil, and a well sheltered situation.

PRUNUS.

LINNEAN Class and Order, *Icosandria Monogynia*: Each flower contains about twenty males and one female. There are fifteen SPECIES; twelve of which are here treated of; most of them well known species; including a numerous tribe of orchard, garden, and shrubery plants.

1. PRU'NUS *Pa'dus*: The PADUS, or COMMON BIRD CHERRY; *a low deciduous tree*; native of England, Scotland, and most parts of Europe.

2. PRUNUS *Virginia'na*: The VIRGINIA PADUS; *a low deciduous tree*; native of Virginia, Pennsylvania, and Carolina.

3. PRU'NUS *Canadé'nsis*: The CANADIAN PADUS; *a deciduous shrub or tree*; native of Canada and many other parts of America.

4. PRU'NUS *Maha'leb*: The MAHALEB, or PERFUMED CHERRY; *a tall deciduous shrub*; native of Switzerland and the North of Europe.

5. PRU'NUS *Armeni'aca*: The APRICOT; *a low deciduous tree*; whose native country is unascertained.

6. PRU'NUS *Ce'rasus*: The CHERRY, or the CULTIVATED CHERRY; *a deciduous tree*; native of England and most parts of Europe.

7. PRU'NUS *A'vium*: The WILD CHERRY; *a deciduous tree*; native of England and the North of Europe.

8. PRU-

8. PRU'NUS *Dome'stica:* The PLUM; *a deciduous tree*; native of many parts of Europe.

9. PRU'NUS *Insiti'tia:* The BULLACE; *a deciduous tree or shrub*; native of England and Germany.

10. PRU'NUS *Spino'sa:* The SLOE THORN, or BLACK THORN; *a deciduous shrub*; native of England and most countries of Europe.

11. PRU'NUS *Lauro-Ce'rasus:* The LAUREL, or the COMMON LAUREL; *an evergreen shrub or tree*; native of Trebisond, near the Black Sea.

12. PRU'NUS *Lusita'nica:* The PORTUGAL LAUREL; *an evergreen shrub*; native of Portugal; also of Pennsylvania and other parts of America.

1. The COMMON PADUS, or BIRD CHERRY, is a tree of about twenty feet growth; oftentimes it rises higher. It grows with an upright stem, and makes a handsome appearance. The bark of the older shoots is of a dark brown, inclined to a purple colour, and is besprinkled with a few grayish spots; while the preceding summer's shoots are smoother and of a reddish cast. The buds early in the winter will begin to swell, for the future shoots. The leaves are large, and grow alternately on the branches. Their figure is nearly oblong. They are rough, and have their edges serrated. Their under surface is of a lighter colour than their upper, and they have two glandules at their base. The flowers are white, and produced in May, in long bunches. A kind of spike of white flowers grows from the sides of the branches; and these waving about on every side, in a loose and easy manner, have a genteel and pleasing effect. The flowers of which these spikes are composed stand each on their own proper pedicles, and are all arranged alternately along the main stalk, which is tolerably long. These flowers are succeeded by fruit, which is a small berry, that ripens in August, at which period it will be black; but besides this, it will undergo the changes of being first green and afterwards red. When these berries are ripe, they are of a sweet disagreeable taste, but so liked by the birds (which will flock from all parts to feed on them) as to occasion its being called the Bird Cherry; and for their sake purely many persons plant a more than

than common quantity of these trees, that they may have these feathered songsters in greater plenty.

There is a *Variety* of this tree, called the *Cornish Bird-Cherry*, which differs from it in some respects; but these differences are inconsiderable.

2. The VIRGINIA PADUS will grow to thirty or forty feet high, and is said to afford wood of great value. The bark is of a dark brown, inclined to a purple colour, and spotted irregularly with some grayish blotches. The young shoots are of a lighter colour, and very smooth; and the whole tree is more ramose than the former sort. The leaves are oval, and of a shining green colour. Their edges are serrated, and placed alternately on the branches. They stand on short footstalks, and continue on the trees late in the autumn. Their flowers are white, and produced in May, in the same sort of long bunches as the other; and are succeeded by black berries, which are equally coveted by the birds, for whose sake only this species also is frequently planted. HANBURY classes this tree amongst his Forest Trees, and says, "The wood is very valuable; is much used by the cabinet makers; will polish very smooth, and display beautiful veins, both black and white." MILLER tells us, "that the *Padus Virginiana* will grow to be a large tree when it is planted in a moist soil, but in dry ground it rarely rises more than twenty feet high." He has also a *Padus Caroliniana* (probably a *Variety* of this species), the seeds of which he says "were sent from Carolina by the title of BASTARD MAHOGANY, from the colour of the wood, which is somewhat like Mahogany." He adds, however, that "this seems to be little more than a shrub, if we may judge from the growth here."

3. The CANADA-PADUS is of much lower growth than the former sorts. The branches are smooth, The leaves are broad, spear-shaped, rough, downy, and destitute of glands, like those of the former species. The flowers grow in long, branching bunches: Their colour is white; they come out in May, and are succeeded by small, round, black berries, which will be ripe in the autumn.

4. The

4. The PERFUMED CHERRY ſeldom grows to be more than ten or twelve feet high. The branches are covered with a ſmooth, whitiſh gray bark. The leaves are ſmall, of a lucid green colour, of an oval figure, and ſtand alternately on the branches. The flowers are white, produced in May in roundiſh cluſters, and are ſucceeded by berries, of which the birds alſo are very fond. The wood of all theſe ſorts is much eſteemed by the cabinet makers, particularly amongſt the French, as it always emits a very agreeable odour.

5. The APRICOT TREE is often planted as a flowering-ſhrub; for though it will grow to be thirty feet high, it may nevertheleſs be kept down to what height the owner deſires. "This tree, ſays HANBURY, as well as moſt ſorts of fruit-trees, is exceeded by few in ornament; for being permitted to grow in its natural ſtate to twenty or thirty feet high, with all its luxuriancy of branches, covered with their delightful heart-ſhaped leaves, what a glorious figure will it preſent! But when we reflect on the fine appearance ſuch a tree muſt make, early in the ſpring, when covered all over with the bloom of ſuch fine flowers as thoſe of the Apricot are known to be, this enhances the value; and either of theſe motives is ſufficient for introducing theſe trees into plantations of this kind. Add to this, ſome of the ſorts, in warm well ſheltered ſituations, will produce fruit when growing in this manner, as well as if planted and trained againſt walls; ſo that additional returns will be made by the fruit to the curious planter of theſe trees."

6. The CHERRY TREE of our orchards is too well known, with all its Varieties, to need any deſcription. HANBURY obſerves, "were the tree ſcarce, and with much difficulty propagated, every man, though poſſeſſed of a ſingle tree only, would look upon it as a treaſure. For beſides the charming appearance theſe trees have, when beſnowed, as it were, all over with bloom in the ſpring, can any tree in the vegetable tribe be conceived more beautiful, ſtriking, and grand, than a well grown and healthy Cherry Tree, at that period when the fruit is ripe?"

The many kinds of Cherry Trees afford an almoſt endleſs *Variety*; all differing, in ſome reſpect, in their

their manner of ſhooting, leaves, flowers, or fruit: Two in particular demand admiſſion into the pleaſure-garden; the Double-bloſſomed and the Red-flowering.

The *Double-bloſſomed Cherry.* The pleaſing ſhow the common Cherry Tree makes when in blow is known to all; but that of the Double-bloſſomed is much more enchanting. It bloſſoms, like the other, in May; the flowers are produced in large and noble cluſters; for each ſeparate flower is as double as a roſe, is very large, and placed on long and ſlender footſtalks, ſo as to occaſion the branches to have an air of eaſe and freedom. They are of a pure white; and the trees will be ſo profuſely covered with them, as to charm the imagination. Standards of theſe trees, when viewed at a diſtance, have been compared to balls of ſnow; and the nearer we approach, the greater pleaſure we receive. Theſe trees may be kept as dwarfs, or trained up to ſtandards; ſo that there is no garden or plantation to which they will not be ſuitable. By the multiplicity of the petals the organs of generation are deſtroyed; ſo that thoſe flowers which are really full are never ſucceeded by any fruit.

The *Red-flowering Cherry Tree* differs in no reſpect from the Common Cherry Tree, only that the flowers are of a pale red colour, and by many are eſteemed on that account.

Beſides the *ornament* and utility afforded us by the flowers and fruit of the Cherry, its *timber* is a farther inducement for propagating it; more eſpecially that of the ſmall *Black Wilding* ſort; which may perhaps with propriety be conſidered as the genuine ſpecies, and a native of this iſland. Be this as it may, it will grow, in a ſoil and ſituation it affects, to be a large timber tree, which, if taken in its prime before it become tainted at the heart, will turn out perhaps not leſs than a ton of valuable materials, peculiarly adapted to the purpoſes of furniture. The grain is fine, and the colour nearly approaching to that of mahogany, to which valuable wood it comes nearer than any other which this country produces.

7. The WILD RED CHERRY is a very large growing tree, and may like the Black Wilding be an object for

for timber. The leaves are oval, ſpear-ſhaped, and downy underneath. The flowers come out from the ſides of the branches in ſeſſile umbels. They appear rather later than the cultivated ſorts; and are ſucceeded by ſmall red fruit, which ripens late in the autumn. This is often called the Wild Northern Engliſh Cherry.

8. The PLUM TREE, with all its Varieties, is ſo well known as to require no deſcription. No one need be told, that the Plum Tree is a large growing tree, and that it has a beautiful appearance in ſpring when in blow. The fruit that ſucceeds the bloſſom is of many colours, ſhapes, and ſizes; and the trees of the variety of ſorts will be ſo adorned with them in the autumn, as to have a noble and delightful effect, being hardly exceeded by the Cherry itſelf. Theſe are ſeldom planted anywhere except in orchards; but let them be ſet where they will, they never fail to repay the owner with pleaſure and profit.

The *Varieties* which are principally eligible for Ornamental Plantations are, The Cherry Plum Tree, the Double-bloſſomed, the Stoneleſs, the Gold-ſtriped, and the Silver-ſtriped Plum.

The *Cherry Plum Tree* is always planted among flowering ſhrubs, on account of its early flowering. It may be kept down to any height; and the flowers will be produced in March, in ſuch plenty, and ſo cloſe, as almoſt to cover the branches. It is admired by all for the early appearance of its flowers, which are ſucceeded, after a mild ſpring, by a round reddiſh plum, on a long ſlender footſtalk, that has the reſemblance of a Cherry. Unleſs there is little or no froſt after theſe trees have been in blow, it rarely happens that any fruit ſucceeds the flowers.

The *Double-bloſſomed Plum Tree* is another Variety. The flowers of this ſort are exceedingly double, and the twigs will be richly furniſhed with them in the month of May. Their petals, like thoſe of the Cherry, are of a pure white, though amongſt theſe ſome filaments with darkiſh antheræ appear. As ſoon as the ſhow of flowers is over, we are not to give up all expectations from this tree; for many of them will be ſucceeded by fruit, which is of the ſame colour, ſhape,

ſhape, and taſte, with the common Damaſcene, though ſmaller, and is liked by many.

The *Stoneleſs Plum*. This is a Variety that ſhould be admitted on no other account than becauſe the pulp ſurrounds a kernel, without having any ſtone. It is a ſmall blue plum; and thoſe people who have it in poſſeſſion, take a pleaſure in ſhewing it as a curioſity.

The *two Striped ſorts* make a Variety by their variegated leaves; on which account they are frequently ſought after by the curious.

9. The BULLACE TREE is ſometimes planted in ſhrubery quarters, for the ſake of the fruit; which by many perſons is deemed very agreeable, being poſſeſſed of a fine acid. It ought to be pulled and eaten immediately from the tree.

The *Varieties* of this ſpecies are, The *Black*, the *White*, and the *Red Bullace*.

10. The SLOE TREE. The Sloe Buſh is, without all doubt, a ſpecies diſtinct from either Plum or Bullace. And indeed it is ſuch a ſpecies, that, were it not for its commonneſs, it would be thought inferior in beauty to none of our ſhrubs. The commonneſs of this tree, however, cauſes its beauties to be unnoticed, and forbids us to admit too many into our collection.

11. The COMMON LAUREL will grow to about thirty feet, and the leaves, which are ſometimes five or ſix inches long and three broad, being likewiſe of a firm ſtructure even at the edges, garniſh the branches of the tree in ſuch a manner as would excite our admiration, did not the frequency of this noble plant diminiſh our reſpect. The Laurel however will, we apprehend, ever preſerve its rank as a *Stock Plant* in ſhruberies and other *ornamental* grounds. EVELYN places the Laurel among Foreſt Trees; and HANBURY ſpeaks of the *uſes* of its timber to the turners and cabinet makers. It ſeems peculiarly adapted to the purpoſe of ORNAMENTAL UNDERWOOD; as it is of quick growth, and will flouriſh under the drip and ſhade of other trees.

The Common Laurel affords two *Varieties*: The *Gold ſriped* and *Silver-ſtriped* Laurel.

12. The

12. The PORTUGAL LAUREL is a lower growing tree than the former; and though its leaves, flowers, &c. are proportionally ſmaller, it is thought by many to be much the moſt beautiful; the commonneſs of the one, and ſcarcity of the other, may perhaps not a little contribute to this opinion. The Portugal Laurel will grow to be fifteen or twenty feet high, accordingly as the ſoil in which it is placed contributes to its increaſe. The branches are produced in an agreeable manner, being chiefly inclined to an upright poſture; and the young ſhoots are clothed with a ſmooth reddiſh bark. The leaves are ſmooth, and of a fine ſtrong green colour, though their under ſurface is rather paler than the upper. They are much ſmaller than thoſe of the Common Laurel, are of an oval figure, and have their edges ſerrated; they are of a thick conſiſtence, and juſtly entitle the tree to the appellation of a fine Evergreen. The flowers are produced in the ſame manner as thoſe of the Common Laurel; but are ſmaller. They are white, appear in June, and are ſucceeded by berries, which when ripe are black; though before they will undergo the different changes of being firſt green, and then red.

The PROPAGATION of the ſeveral tribes of *Prunus* varies with the reſpective ſpecies.

The ſpecies and varieties of PADUS, or BIRD CHERRY, may be raiſed, 1. From ſeeds, ſown in autumn, in beds of light earth, about half an inch deep. The plants will appear the firſt ſpring, and the ſpring following they may be planted out into the nurſery, at the diſtance of two feet by one; in which ſituation they may remain until wanted for planting out. 2. This claſs may alſo be propagated by layers; the young twigs, being ſimply laid in the ground, will without any other trouble ſtrike root in one year, and may be taken up and tranſplanted into the nurſery, or be planted where they are to remain, as circumſtances may ſuit. 3. Theſe trees will alſo grow from cuttings, planted in October, in a moiſt ſituation: if the ſpring and ſummer prove dry, they will require to be watered.

This claſs of *Prunus* affects a moiſt ſituation.

The

The PERFUMED CHERRY is PROPAGATED by grafting, or by budding upon any of our Cherry ſtocks.

The APRICOT TREE is PROPAGATED by budding it upon the Plum ſtock.

The FRUIT-BEARING, DOUBLE-BLOSSOMED, and RED-FLOWERING CHERRY TREES are PROPAGATED by grafting upon ſtocks raiſed from the ſtones of the Black Cherry Tree; though it may be proper to obſerve here, that when the Double-bloſſomed Cherry is wanted to be kept very low, in its dwarf ſtate, the Common Bird Cherry will be a much more proper ſtock to work it upon, as that ſort is naturally of much lower growth than the Black Cherry Tree.

The PLUM TREE, in all its varieties, and the Bullace Tree, the Cherry Plum, the Double-bloſſomed Plum, and the Stoneleſs Plum, are PROPAGATED by grafting upon Plum ſtocks raiſed from ſeeds; though it is obſervable, that ſuckers of the Bullaces will grow to be trees, and produce plenty of good fruit; but theſe will not be ſo good as thoſe grafted on the Plum ſtocks.

The SLOE BUSH may be obtained from the places where they grow; for from thence a ſucker or two may be taken, and planted for the conveniency of obſervation; but theſe will not be ſo good as thoſe raiſed from the ſtones.

The COMMON LAUREL is PROPAGATED either from ſeeds or from cuttings. 1. If the former method is practiſed, the ſeeds muſt be gathered from the trees when they are full ripe: This will be known by their being quite black, which is generally about the beginning of October. Theſe ſeeds ſhould be ſown directly in beds of light earth, half an inch deep, which muſt be afterwards hooped over, to be covered in very ſevere froſts. A hedge of furze buſhes alſo ſhould be made around them, to break the force of the freezing black winds, and ſecure the ſeeds, together with the mats, from being deſtroyed. This is a much ſafer method than covering the beds with litter, which, if neglected to be taken off when the froſt is over, will retain the rains which generally ſucceed ſuch weather, ſodden the beds, and make them ſo wet as frequently to to deſtroy the whole of the expected crop. The ſeeds being ſown, and preſerved with the above care, will appear in the ſpring. During the ſummer they ſhould be kept clear of weeds, as well as watered in dry weather; and all the enſuing winter they muſt remain untouched in their beds, the furze hedge ſtill ſtanding till the froſty weather is paſt; for if theſe young ſeedlings are planted out in the autumn, the major part of them will be in danger, before the winter be expired, of being thrown out of the ground by the froſt; and not only ſo, but of being really killed by it, as they are not very hardy at one year old. In the ſpring, therefore, when the bad weather is ceaſed, let them be planted out in the nurſery ground, in rows two feet aſunder, and the plants a foot and a half diſtant in the rows; where they may ſtand till they be finally planted out. 2. Trees raiſed from ſeeds generally grow more upright, and ſeldom throw out ſo many lateral branches as thoſe reared from cuttings; nevertheleſs, as the expectation of a crop from ſeeds has ſo often failed, notwithſtanding great care has been uſed; and as the difficulty of procuring the ſeeds, and preſerving them from the birds, has been very great; the moſt certain and expeditious method of raiſing quantities of theſe trees is by cuttings, and is as follows: In the month of Auguſt the cuttings ſhould be gathered, about a foot and a half in length. They will thrive the better for having a bit of the laſt year's wood at the end, though without this they will grow exceedingly well. The under leaves ſhould be cut off a foot from the thick end of the cuttings, which muſt all be planted about a foot deep in the ground; the other half foot, with its leaves, being above it. No diſtance need be obſerved in planting theſe cuttings, which may be ſet as thick as you pleaſe, though the ground for raiſing them ſhould be ſheltered, leſt the winds, which are frequently high at this time of the year, or ſoon after, looſen the plants juſt when they are going to ſtrike root The weather when the cuttings are to be planted ſhould be either rainy or cloudy; and if no ſhowers ſhould fall in Auguſt, the work muſt be deferred till they do; for if cuttings are planted in Auguſt, when the weather is parching and dry, they will be burnt up, without great care and trouble in ſhading and watering. Neither is cloudy or rainy weather only to be recommended in planting theſe cuttings, but a ſhady ſituation alſo, either under a north wall, or in beds which are covered the greateſt part of the day with the umbrage of large trees. This ſhady ſituation is very neceſſary for them; ſince, though the weather be rainy and cloudy when they are planted, yet ſhould it prove fair afterwards, the ſun will ſoon dry up the moiſture at that ſeaſon, and endanger the plants, if they are not conſtantly watered and protected with a ſhade; which at once ſhews the expediency of pitching on a ſpot where ſuch a conveniency is natural. If theſe cuttings are planted in Auguſt, they will have taken root before winter, eſpecially if they have ſhade, and water in dry weather: but they ſhould remain undiſturbed till the ſpring twelvemonth following, in order to acquire ſtrength to be planted in the nurſery. During the ſummer, they will require no other trouble than watering in dry weather, and being kept clean from weeds; and by the autumn they will have made a ſhoot of perhaps a foot or more in length. In the beds nevertheleſs, they may remain until the ſpring, when they ſhould be all carefully taken out, and planted in the nurſery, as was directed for the ſeedlings.

The Common Laurel may be tranſplanted, at Midſummer, with ſafety, and great ſucceſs.

The PORTUGAL LAUREL is to be raiſed the ſame way as the Common Laurel, by *ſeeds* and *cuttings*; but the cuttings of the Portugal Laurel do not take ſo freely as thoſe of the Common ſort; and the young practitioner, out of a good bed of cuttings, muſt expect to ſee but a few real plants ſucceed. If they are planted in July or Auguſt, they muſt be ſhaded, and kept moiſt during the hot weather; and that will be the moſt probable way to enſure ſucceſs. If a perſon has the conveniency of a good ſtove, the beſt method is not to plant them until the ſpring; and then many cuttings may be planted in one pot, and afterwards plunged into the bark bed; and by this means numerous plants may eaſily be obtained.

Neither of the Laurels can bear a very wet ſituation; and both of them are liable to be injured by ſevere froſt.

PTELEA.

P T E L E A.

LINNEAN Clafs and Order, *Tetrandria Monogynia*. Each flower contains four males and one female. There are only two SPECIES; one of them introduced into our fhruberies; the other a late difcovery.

PTELEA *Trifolia'ta:* The THREE-LEAVED PTELEA, or the TREFOIL SHRUB; *a deciduous fhrub*; native of Virginia and Carolina.

The PTELEA will grow to the height of ten feet. The branches are not very numerous; when broken, they emit a ftrong fcent: They are brittle, full of pith, and covered with a fmooth purplifh bark. The leaves are trifoliate, and grow irregularly on the branches, on a long footftalk. The folioles are oval, fpear-fhaped, of a delightful ftrong green colour on their upper fide, lighter underneath, fmooth, and pretty large when they are fully out, which will not be before part of the fummer is elapfed; for they put out late in the fpring. The flowers are produced in bunches, at the ends of the branches: Their colour is a greenifh white. They come out in June; and are fucceeded by roundifh bordered capfules; but the feeds feldom ripen in England.

This fhrub may be PROPAGATED either by feeds, layers, or cuttings. 1. By feeds. Thefe fhould be fown in a warm border, in the fpring, in common garden mould made fine; and if the feeds are good, they will grow, and come up the firft fummer. We generally receive the feeds from abroad; though they will in fome warm feafons ripen here with us. When the young plants begin to come up, which will be, if the feeds are good, by the end of May, they fhould be fhaded, and every fecond evening duly watered; and this, together with conftant weeding, will be all the care they will require until the autumn. At the approach of winter it will be proper to prick fome furze bufhes round the bed, to break the keen edge of the black frosts. They will then require no other trouble until the second spring after they are come up; when they should be all taken out of the seed bed, and planted in the nursery, a foot asunder; and in two or three years they will be fit to be finally planted out. 2. By layers. For this purpose a number of plants must be planted for stools; and, after they have stood a year or two, these should be cut down pretty near the ground. By the autumn they will have made shoots, some of which will be five or six feet, or more, in length; and these are the shoots for layering. October is the best month for the work; and the operation is to be performed by cutting the twig half through, and making a slit half an inch long. Any thing may be put into this slit, to keep it open; and after the mould is levelled all round, the longest ends should be taken off. By this method they will generally have good roots by the autumn following: and the stools will have shot out fresh wood for a second layering. At this time they should be taken up, and the weakest planted in the nursery, to get strength; whilst the stronger layers will be good plants to set out to stand. After this, the operation may be again repeated, and so continued annually, at pleasure. 3. By cuttings. In order to obtain plenty of good cuttings, the plants should be headed as for layering. In October the young shoots should be taken off, and cut into lengths of a little more than a foot, two thirds of which should be set in the ground. Some of these cuttings will grow; though, says HANBURY, I ever found this way very uncertain, and not worth the practising: But if the cuttings are planted in pots, and assisted by artificial heat, they will grow readily. This, however, is not a good method; for they will be tender the first winter, as well as require to be protected in the greenhouse, or under some cover, which will occasion more trouble than if they had been layered. By layers and seeds, therefore, are the best and most eligible methods of encreasing these trees.

PUNICA.

LINNEAN Claſs and Order, *Icoſandria Monogynia*: Each flower contains twenty males and one female. There are two SPECIES; the one a greenhouſe plant; the other

PU'NICA *Grana'tum*: The COMMON POMEGRANATE; *a tall deciduous ſhrub*; native of Spain and the South of Europe.

"The POMEGRANATE," ſays MILLER, "riſes with a woody ſtem eighteen or twenty feet high; ſending out branches the whole length, which likewiſe put out many ſlender twigs, ſo as to render them thick and buſhy."

There is a *Variety* with *double flowers*.

Theſe plants may be PROPAGATED by laying down their branches in the ſpring, which, ſays MILLER, in one year's time will take good root, and may be tranſplanted where they are deſigned to remain.

PYRUS.

LINNEAN Claſs and Order, *Icoſandria Pentagynia*: Each flower contains about twenty males and five females. There are eleven SPECIES; four of which are as follow:

1. PY'RUS *Commu'nis*: The PEAR; a well known *deciduous tree*; native of moſt parts of Europe.

2. PY'RUS *Ma'lus*: The APPLE; an equally well known *deciduous tree*; native alſo of moſt parts of Europe.

3. PY'RUS *Corona'ria*: The SWEET-SCENTED CRAB; *a deciduous tree*; native of Virginia.

4. PY'RUS *Cydo'nia*: The QUINCE; *a deciduous shrub or tree*; native of the Banks of the Danube.

1. The PEAR. Of the numerous *Varieties* of this species of *Pyrus* there are two admissible into ornamental grounds:

The Double-blossomed Pear,
The Twice-flowering Pear.

The *Double-blossomed Pear* differs from the other sorts only in that the flowers are double. The leaves, indeed, are not so much serrated as some of the other Pears; nay, scarcely any serratures appear, excepting on the oldest leaves; for the younger are perfectly entire and downy. The multiplicity of the petals of this flower is not sufficient to entitle it to the appellation of a full flower; for it consists only of a double row of petals; but as these are all large, produced in clusters, and of a pure white, they entitle the tree to be called a flowering tree, with greater propriety than the ordinary Pears can be so styled. The planter of this species is rewarded in a double respect; for as the petals are not multiplied in so great a degree as to destroy the stamina, the flowers are succeeded by a good fruit, whose properties are such as entitle it to the rank of a good baking Pear.

The *Twice-flowering Pear*. This species is sufficiently described by the title; it being a Pear that often produces flowers in the autumn, when the fruit that succeeded those of the spring are nearly ripe This tree deserves to be planted both for its beauty and singularity; for it sometimes happens, though by no means constantly, that it is covered over in September with bloom and fruit. This autumnal bloom falls away, and the chilling cold often prevents its coming to any embryo fruit.

2. The APPLE. This species likewise affords us two ornamental *Varieties*.

The Paradise Apple,
The Fig Apple.

The *Paradise Apple* is rather a shrub than a tree. There are two sorts of it, which Gardeners distinguish by the names of the French and the Dutch Paradise Apple. They are both low growing trees; and the only difference between them is, that the Dutch sort is rather

rather the ſtrongeſt ſhooter. They are chiefly uſed for ſtocks to graft apples upon, in order to make them more dwarfiſh; ſo that a plant or two in a collection, for the ſake of variety, will be ſufficient.

Fig Apple has a place here for no other reaſon than its being deſtitute of the moſt beautiful parts of which the flowers are compoſed; viz. the petals: They have all the ſtamina, &c. but no petals, which is a ſingular imperfection; though by many they are coveted on that account. As the ſtamina and other parts are all perfect, the flowers are ſucceeded by a tolerably good eating Apple; for the ſake of which this tree deſerves to be propagated.

3. The Sweet-scented Crab of Virginia differs from our Crab in the leaves, flowers, and fruit. The leaves are angular, ſmooth, of a fine green colour, and have a look entirely different from any of our Crabs or Apples. The flowers ſtand on larger footſtalks than thoſe of the generality of our Crabs, and are remarkable for their great fragrance. This tree is ſeldom in full blow before the beginning of June. The flowers, when they firſt open, are of a pale red, though the petals ſoon after alter to a white colour. They are ſucceeded by a little round Crab, which, of all others, is the ſoureſt, rougheſt, and moſt diſagreeable, that can be put into the mouth.

There is a *ſub-evergreen Crab* of America, ſuppoſed to be a *Variety* of this Species. Its natural growth ſeems to be not more than twelve feet; and the branches are covered with the ſame kind of ſmooth brown bark as our common Crab Tree. The leaves are long and narrow, and will often be found of different figures; for though ſome will be angular, others again are oblong, or of a lanceolate figure. They are fine, ſmooth, of a ſtrong dark green colour, and have their edges regularly ſerrated. They will remain until late in the ſpring, which rather entitles this ſhrub to a place here; though in an expoſed ſituation, the ends of the branches will be often ſtripped of thoſe ornaments, after a few ruffian attacks of the piercing northern blaſts: So that this tree, when conſidered as an evergreen, ſhould always be planted in a well ſheltered place, where it will retain its leaves, and look very well all winter.

4. The QUINCE. There are many *Varieties* of the Quince Tree, which are chiefly raiſed for the fruit. The Quince Tree ſeldom grows to be higher than eight or ten feet; and the bark on the branches is often of a kind of iron colour. The leaves are large and oval: Their upper ſurface is of a pleaſant green colour, though often poſſeſſed of a looſe downy matter, and their under ſide is hoary to a great degree. The flowers are produced in May, all along the branches: They grow upon young ſhoots of the ſame ſpring, and are very large and beautiful; for although each is compoſed of about five petals only, yet theſe are often an inch long, are broad and concave, and of a fine pale red as they firſt open, though they afterwards alter to a white; and thoſe flowers being produced the whole length of the branches, and beſpangling the whole tree in a natural and eaſy manner, juſtly entitle this ſpecies to no mean place among the flowering kinds. They are ſucceeded by that fine large yellow fruit which is ſo well known, and which at a diſtance, on the tree, appears like a ball of gold. Indeed, theſe trees ſhould always be planted at a diſtance from much frequented places; for the fruit, valuable as it is when properly prepared for uſe, has a ſtrong diſagreeable ſcent, that will fill the air all around with its odour, which to moſt people is offenſive.

PROPAGATION. HANBURY ſays, all theſe ſorts will take by grafting or budding upon one another, notwithſtanding what MILLER has alledged to the contrary. He continues, "I have a tree that bears excellent Apples grafted upon a Pear ſtock; and Pears grafted upon Crab ſtocks that have not yet borne." The uſual way is to graft the PEARS on ſtocks raiſed from the kernels of Pears, and the APPLES on Crab ſtocks. Theſe ſhould be ſown, ſoon after the fruit is ripe, in beds half an inch deep, and carefully guarded from mice, which will ſoon deſtroy the whole ſeminary, if once found out. In the ſpring the plants will come up; and in the winter following they ſhould be planted out in the nurſery, in rows two feet aſunder. In a year or two after this they will be fit for working;

and

and by this method all the ſorts of Pears and Apples are propagated.

The PARADISE APPLE is generally raiſed by layers or cuttings; and all the ſorts of QUINCES grow readily by cuttings, planted any time in the winter; though the early part of that ſeaſon is to be preferred.

The *Evergreen Crab* will take by grafting or budding on the common Crab or Apple ſtock: but great care and nicety of execution is requiſite in performing the operation: Budding towards the latter end of July, HANBURY ſays, he has always found to be the moſt certain method.

QUERCUS.

LINNEAN Claſs and Order, *Monoecia Polyandria*: Male flowers containing many ſtamina, and female flowers containing one piſtil, upon the ſame plant. There are thirteen SPECIES.

1. QUE'RCUS *Ro'bur*: The ENGLISH OAK; a well known *tall deciduous tree*; native of England; and is found in moſt parts of Europe.

2. QUE'RCUS *Phe'llos*: The WILLOW-LEAVED OAK; *a deciduous tree*; native of moſt parts of North America.

3. QUE'RCUS *Pri'nus*: The CHESNUT-LEAVED OAK; *a deciduous tree*; native of moſt parts of North America.

4. QUE'RCUS *Ni'gra*: The BLACK OAK; *a low deciduous tree*; native of North America.

5. QUE'RCUS *Ru'bra*: The RED OAK; *a tall deciduous tree*; native of Virginia and Carolina.

6. QUE'RCUS *A'lba*: The WHITE OAK; *a deciduous tree*; native of Virginia.

7. QUE'RCUS *E'ſculus*: The ITALIAN OAK, or the CUT-LEAVED ITALIAN OAK; *a low deciduous tree*; native of Italy, Spain, and the South of France

8. QUE'RCUS *Æ'gilops*: The SPANISH OAK, or OAK WITH LARGE ACORNS AND PRICKLY CUPS; *a tall deciduous tree*; native of Spain.

9. QUERCUS

9. QUE'RCUS *Ce'rris:* The AUSTRIAN OAK, or the OAK WITH PRICKLY CUPS AND SMALLER ACORNS; *a deciduous tree*; native of Auſtria and Spain.

10. QUE'RCUS *Su'ber:* The CORK TREE; *a low evergreen tree*; native of the ſouthern parts of Europe.

11. QUE'RCUS *I'lex:* The ILEX, or COMMON EVERGREEN OAK; *a low evergreen tree*; native of Spain and Portugal.

12. QUE'RCUS *Cocci'fera:* The KERMES OAK; *a tall evergreen ſhrub*; native of France and Spain.

13. QUERCUS *Molu'cca:* The LIVE OAK; *a low evergreen tree*; native of America.

1. The ENGLISH OAK will grow to great ſtature, and live to a great age. EVELYN, whoſe learning and induſtry are evident in every page of his elaborate work, fatigues us with a tedious account of large trees which either were growing in his time, or which he found in the mouth of tradition, or in the pages of learning and hiſtory. We would rather however refer our readers to his detail than either copy or abridge it; confining ourſelves to a few individuals of our own time, which now are (or were very lately) actually ſtanding in this kingdom. The COWTHORP OAK, now growing at Cowthorp, near Wetherby in Yorkſhire, has been held out as the *father* of the foreſt. Dr. HUNTER of York, in his brilliant edition of Mr. EVELYN's book, has favoured us with an engraving of this tree; the dimenſions of which, as he juſtly obſerves, "are almoſt incredible." Within three feet of the ſurface, the Doctor tells us, "it meaſures ſixteen yards, and cloſe to the ground, twenty-ſix yards. Its height in its preſent ruinous ſtate (1776) is about eighty-five feet, and its principal limb extends ſixteen yards from the bole. Throughout the whole tree the foliage is extremely thin, ſo that the anatomy of the antient branches may be diſtinctly ſeen in the height of ſummer. When compared to this, all other trees (the Doctor is pleaſed to ſay) are but *children* of the foreſt." If indeed the above admeaſurement might be taken as the dimenſion of the *real ſtem*, its ſize would be truly enormous, and far exceed that of any other Oak in the kingdom; but the Cowthorp Oak has a ſhort ſtem, as moſt *very* large trees it is obſervable have, ſpreading wide

wide at the baſe, the roots riſing above the ground like ſo many buttreſſes to the trunk, which is not like that of a tall ſtemmed tree, a cylinder, or nearly a cylinder, but the fruſtum of a cone. Mr. MARSHAM gives us a plain and accurate account of this tree: He ſays, "I found it in 1768, at four feet, forty feet ſix inches; at five feet, thirty ſix feet ſix inches; and at ſix feet, thirty-two feet one inch." Therefore in the principal dimenſion, *the ſize of the ſtem*, it is exceeded by the BENTLEY OAK; of which the ſame candid obſerver gives the following account: "In 1759 the Oak in Holt Foreſt, near Bentley, was, at ſeven feet, thirty-four feet. There is a large excreſcence at five and ſix feet that would render the meaſure unfair In 1778 this tree was increaſed half an inch, in nineteen years. It does not appear to be hollow but by the trifling increaſe I conclude it not ſound." Extraordinary, however, as theſe dimenſions may appear, they are exceeded by thoſe of the BODDINGTON OAK; a tree which we believe does not appear anywhere upon record, except it be alluded to in Mr. EVELYN's Liſt. This Oak grows in a piece of rich graſs land, called the Old Orchard Ground, belonging to Boddington Manor Farm, lying near the turnpike road between Cheltenham and Tewkſbury, in the Vale of Glouceſter. The ſtem is remarkably collected and ſnug at the root, the ſides of its trunk being more upright than thoſe of large trees in general; nevertheleſs its circumference at the ground, as near to it as one can walk, is twenty paces: meaſuring with a two foot rule, it is ſomewhat more than eighteen yards. At three feet high it meaſures forty-two feet, and at its ſmalleſt dimenſions, namely, from five to ſix feet high, it is thirty ſix feet. At about ſix feet it begins to ſwell out larger; forming an enormous head, which heretofore has been furniſhed with huge, and in all probability extenſive arms. But age and ruffian winds have robbed it of a principal part of its grandeur; and the greateſt extent of arm at preſent (1783) is eight yards, from the ſtem. From the ground to the top of the crown of the trunk is about twelve feet; and the greateſt height of the branches, by eſtimation, forty-five feet. The ſtem is quite hollow; being, near the ground, a perfect ſhell; forming a capacious

capacious well fized room; which at the floor meafures, one way, more than fixteen feet in diameter. The hollownefs, however, contracts upwards, and forms itfelf into a natural dome, fo that no light is admitted except at the door, and at an aperture or window in the fide. It is ftill perfectly alive and fruitful, having this year a fine crop of acorns upon it. It is obfervable in this (as we believe it is in moft old trees), that its leaves are remarkably fmall; not larger, in general, than the leaves of the Hawthorn.

In contemplating thefe wonderful productions of nature we are led to conjecture the period of their exiftence. Mr. MARSHAM in his Paper publifhed in the Firft Volume of the Tranfactions of the Bath Agriculture Society, has given us fome very ingenious calculations on the age of trees; and concludes that the Tortworth Chefnut is not lefs than eleven hundred years old. We have however fhewn under the Article CHESNUT, that Mr. MARSHAM is miftaken in the dimenfions of that tree. Neverthelefs, if it ftood in the days of King John, fix centuries ago, and was then called the Great Chefnut *, we may venture to fuppofe it not much lefs than one thoufand years of age; and farther, if we confider the quick growth of the Chefnut compared with that of the Oak, and at the fame time the inferior bulk of the Tortworth Chefnut to the Cowthorp, the Bentley, and the Boddington Oaks; may we not venture to infer, that the exiftence of thefe truly venerable trees commenced fome centuries prior to the era of Chriftianity?

The root of the Oak ftrikes deep, efpecially the middle or tap root, which has been traced to a depth nearly equal to the height of the tree itfelf: nor do the lateral roots run fo fhallow and horizontal as thofe of the Afh and other trees; but perhaps the roots of very few trees range wider than thofe of the Oak. The ftem of the Oak is naturally fhort, and if left to itfelf, in an open fituation, it will generally feather to the ground. It has not that upright tendency as the Afh, the Efculus, and the Pine tribe: neverthelefs, by judicious pruning, or by planting in clofe order, the Oak

* As Tradition fays it was.

will

will acquire a great length of ſtem; in this caſe, however, it rarely ſwells to any conſiderable girt. Mr. MARSHAM indeed mentions one in the Earl of Powys's Park near Ludlow, which in 1757 meaſured, at five feet, ſixteen feet three inches, and which ran quite ſtraight and clear of arms near or full ſixty feet. But, as has before been obſerved, Oaks which endure for ages have generally ſhort ſtems; throwing out, at ſix, eight, ten, or twelve feet high, large horizontal arms; thickly ſet with crooked branches; terminating in clubbed abrupt twigs; and cloſely covered with ſmooth gloſſy leaves; forming the richeſt foliage, irregularly ſwelling into the boldeſt outline we know of in nature. The Pine tribe and the Eſculus may be called elegant or beautiful; but the general aſſemblage of a lofty full furniſhed Oak is truly ſublime.

It is ſomewhat extraordinary, that the moſt *ornamental* tree in nature ſhould, at the ſame time, be the moſt *uſeful* to mankind. Its very leaves have been lately found to be of eſſential uſe to the Gardener; the Huſbandman is well acquainted with the value of its acorns; and every Engliſhman experiences daily the uſeful effects of its bark. It is wholly unneceſſary to mention the value of its timber: it is known to the whole world. The Oak raiſed us *once* to the ſummit of national glory: and *now* we ought to hold in remembrance that our exiſtence as a nation depends upon the Oak. If therefore our forefathers, merely from the magnitude and majeſty of its appearance, the veneration due to its age, and gratitude perhaps for ſome few economical uſes they might apply it to, paid divine honours to this tree; how much more behoves it us, circumſtanced as we are, to pay due homage to this our national ſaviour? How could our Kings be inveſted with the enſigns of royalty, or our Creator receive at ſtated times the gratitude and praiſe which we owe to him, with greater propriety than under the ſhadow of this ſacred tree? Acts like theſe would ſtamp it with that reſpectability and veneration which is due to it. To corroborate theſe ideas as well as to inſtitute ſuch laws as might be found neceſſary, the ſtate of the growth of Oak in Great Britain ought to be a ſtanding enquiry of the Britiſh Legiſlature. It is far from being impracticable

ticable to have annual returns of Oak fit for ship-building in every parish in the kingdom; with the distance it stands from water carriage. It avails but little our making laws of police, or forming foreign alliances, unless we take care to secure in perpetuity the defence of our own coast. It is idle to think of handing down to posterity a national independency, if we do not at the same time furnish them with the means of preserving it.

The PROPAGATION of the ENGLISH OAK. Having, under the Title WOODLANDS, given directions for raising *Woods* and *Groves* of OAK, it remains to treat of it, here, merely as a *Nursery plant*. There are various opinions about the *choice of acorns*: *Authors* in general recommend those of "fair, straight, large and shining trees;" but *Nurserymen*, we believe, pay little attention as to the tree from which the acorns are gathered. And indeed, when we consider that the seeds of the distinct *Varieties* of any individual species of plants produce one and the same seedling stock, or a similar Variety of seedling plants, we must conclude that little attention is due. If however it be true, that the seeds of some *Varieties* produce *more* of its own kind than those of other *Varieties* of the same species, it may be worth the trouble, when only a small quantity of seed is wanted, to gather it from the most valuable tree. The *preservation of Acorns* is extremely difficult: if we sow them in autumn, they become obnoxious to vermin and birds: if we keep them above ground, it is very difficult to prevent their sprouting, and at the same time preserve their vegetating power. Upon the whole, the fall of the Acorn seems the properest *time of sowing*. For spring sowing, February and March are the proper months. The usual *method of sowing* is either in drills, or promiscuously in beds, covering them about two inches deep. But we would rather recommend placing them in beds in the quincunx manner, from four to six inches apart, covering them one half to two and a half inches deep, according to the stiffness or lightness of the soil. Sowing seeds in drills renders them peculiarly obnoxious to mice and rooks; and by scattering them promiscuously the plants are liable to come up double and irregularly, and the use

uſe of the hoe is precluded. The oakling riſes the firſt ſpring after ſowing. The ſeedling plants, having ſtood two years in the ſeed bed, ſhould be removed into the nurſery, placing them in rows from two and a half to three feet aſunder, and the plants from nine to twelve inches in the rows; the tap root and all long ſprawling fibres having been firſt taken off, and the top trimmed to a ſwitch, if tolerably ſtraight, or, if deformed or maimed, cut down within two or three inches of the ground; remembering to ſort the plants as directed in the Introductory part of our work. Having remained two or three years in the nurſery, they will be ready to be planted out into fenced plantations. Such as are wanted to be trained for ſtandards, may be removed into ſome vacant ground; firſt pruning them in the conoidic manner, and afterwards remembering from time to time to pay proper attention to their leaders.

The Engliſh Oak admits of ſome *Varieties:* indeed, if we attend minutely to particulars, we ſhall find them almoſt infinite. There is one Variegation under the name of the *Stripe-leaved Oak:* But the moſt intereſting Variety of the Engliſh Oak is the *Lucombe* or *Devonſhire Oak.* In the Sixty-ſecond Volume of the Philoſophical Tranſactions, a particular account is given of this Oak; ſetting forth that Mr. Lucombe, a Nurſeryman near Exeter, having, about the year 1765, ſowed a parcel of acorns ſaved from a tree of his own growth, and obſerving that one of the ſeedling plants preſerved its leaves through the winter, he paid particular attention to it, and propagated, by grafting, ſome thouſands from it. Its being a ſub-evergreen is not the only peculiarity of this Variety; it has a ſomewhat more upright tendency, and ſeems to be of a quicker growth, than Oaks in general. The plants however, which we have ſeen, do not anſwer altogether the deſcription given in the account abovementioned; but as they are now in the hands of almoſt every Nurſeryman, we forbear ſaying any thing further reſpecting them.

2. The Willow-leaved Oak will grow to be a large timber tree. It receives its name from its leaves reſembling very much thoſe of the Common Willow. Theſe long narrow leaves have their ſurface ſmooth,

and their edges entire; and their acorns will be almoſt covered with their large cups.

There are ſeveral *Varieties* of this ſort; ſome having ſhorter leaves, others broader, and hollowed on the ſides; ſome large acorns, others ſmaller, &c. all of which are included under the appellation of Willow-leaved Oaks.

3. The CHESNUT-LEAVED OAK. This alſo will grow to be a large timber tree; and in North America, where it grows naturally, the wood is of great ſervice to the inhabitants. It is ſo called, becauſe the leaves greatly reſemble thoſe of the Spaniſh Cheſnut Tree. They are about the ſame ſize, ſmooth, and of a fine green colour.

There are two or three *Varieties* of this ſort; but the leaves of all prove that they are of the ſpecies called the Cheſnut-leaved Oak; ſo that nothing more need be obſerved, than that the leaves of ſome ſorts are larger than thoſe of others; that the acorns alſo differ in ſize, and grow like thoſe of our Engliſh Oak, on long or ſhort footſtalks as it ſhall happen.

4. The BLACK OAK is a tree of lower growth, it ſeldom riſing to more than thirty feet high. The bark of this tree is of a very dark colour, which occaſioned its being named the Black Oak. The leaves are ſmooth, very large, narrow at their baſe, but broad at their top, being in ſhape like a wedge: They have indentures at the top, ſo as to occaſion its having an angular look; they are of a ſhining green colour, and grow on ſhort footſtalks on the branches.

There is a *Variety* or two of this ſort, particularly one with *trifid leaves*, and another ſlightly trilobate, called *The Black Oak of the Plains*, the leaves and cups of all which are ſmall.

5. RED VIRGINIA OAK. The Red Oak will grow to be a timber tree of ſixty or ſeventy feet high, and the branches are covered with a very dark coloured bark. It is called the Red Oak from the colour of its leaves, which in the autumn die to a deep red colour.

There are ſeveral *Varieties* of this ſpecies, the leaves of which differ in ſize and figure; but thoſe of the larger ſort are finely veined and exceedingly large, being often

often found ten inches long, and five or fix broad: They are obtufely finuated, have angles, and are of a fine green colour in the firft part of the fummer, but afterwards change by degrees to red, which is mark enough to know thefe trees to be of this fpecies. There are feveral Varieties of this tree, which exhibit a manifeft difference in the fize of the leaves, acorns, and cups. That is the beft which is commonly called the *Virginian Scarlet Oak*; and the bark is preferred for the tanners ufe before that of all the other forts.

6. The WHITE OAK. The White Oak will not grow to the fize of the former, it feldom being found higher than forty feet even in Virginia, where it grows naturally. But though the timber is not fo large, yet it is more durable, and confequently of greater value for building to the inhabitants of America, than any of the other forts. The branches of this tree are covered with a whitifh bark; the leaves alfo are of a light colour. They are pretty large, being about fix inches long and four broad. They have feveral obtufe finufes and angles, and are placed on fhort footftalks.

There is a *Variety* or two of this fpecies; and the acorns are like thofe of our Common Oak.

7. The ITALIAN OAK will grow to about the height of thirty feet. The branches are covered with a dark purplifh bark. The leaves are fmooth, and fo deeply finuated as to have fome refemblance of pinnated leaves; and each has a very fhort footftalk. The fruit of this fpecies fits clofe to the branches. The cups are in fome degree prickly and rough, and each contains a long flender acorn, that is eatable. This (fays HANBURY) is the true *Phagus* of the Greeks, and the *Efculus* of Pliny. In the places where thefe trees grow naturally the acorns are, in times of fcarcity, ground into flour, and made into bread.

8. The SPANISH OAK will grow to be as large a tree as our Common Oak, and is no way inferior to it in ftatelinefs and grandeur; for the branches will be far extended all around, caufing, with the leaves, a delightful fhade. Though the bark of thefe branches is of a whitifh colour, yet they are neverthelefs fpotted with brownifh fpots. The leaves are of an oblong oval figure, but not very long, feldom being longer than

 three

three inches, and two broad. They are ſmooth, and have their edges deeply ſerrated: Theſe ſerratures are acute, and chiefly turn backwards. Their upper ſurface is of a fine light green colour, and their under of an hoary caſt; and with theſe beautiful leaves each branch is plentifully ornamented all over the tree. The cups are moſt peculiar and ſingular; for they are very large, and compoſed of ſeveral rough, black, large ſcales, that lap over one another like the ſcales of a fiſh. They almoſt cover the acorn, though they are pretty large, narrow at the bottom, but broader higher, and have their tops flat. The Greeks call the acorns *Velani*, and the tree itſelf *Velanida*. The acorns are uſed in dyeing.

9. The AUSTRIAN OAK is of lower growth than the preceding ſpecies, it ſeldom riſing to more than forty feet high. The leaves are of two colours; their upper ſurface being of a fine green colour, and their under downy. Their figure is oblong; but they are ſo indented about the middle, as to make them have the reſemblance of a lyre. They are wing-pointed, tranſverſely jagged, and ſtand on ſlender footſtalks on the branches. The cups of this ſort alſo are ſmaller and prickly, and the acorns alſo proportionally ſmaller than thoſe of the preceding ſpecies.

All theſe foreign deciduous ſorts may be PROPAGATED from the acorns, which muſt be procured from the places where the trees naturally grow. They ſhould be ſown as ſoon as poſſible after they arrive; and if any of them have ſprouted, great care muſt be uſed in taking them out of the boxes in which they were conveyed. Any ſort of our common garden mould, made fine, will ſuit them; and they ſhould be ſown in drills, in beds an inch deep. The firſt ſpring after ſowing, the plants will come up; they ſhould be always kept clean from weeds, and if they are watered in dry weather, it will be the better. They will want no preſervation in winter, for they are all very hardy, even when young. In March they ſhould be all taken out of the ſeed bed, have their tap roots ſhortened, and be planted in the nurſery ground a foot aſunder, and two feet diſtant in the rows, where they may ſtand, with the uſual nurſery care, until they are to be planted out.

The STRIPED-LEAVED OAK is uſually PROPAGATED

GATED by inarching into the Common Oak; but it is beſt increaſed by grafting. In the ſame manner, alſo, any particular Variety belonging to the other ſpecies may be continued and multiplied.

EVELYN ſays, "the Oak will endure the LAYING, but never to advantage of bulk or ſtature."

10. The CORK TREE admits of two *Varieties:*

The Broad-leaved Cork Tree.

The Narrow-leaved Cork Tree.

The *Broad-leaved Cork Tree* is a timber tree in Portugal and Spain, and other ſouthern parts of Europe, where it grows naturally. In our preſent plantations, it ſhould be placed near the middle of our largeſt quarters, among others of about forty feet growth; and a few alſo ſhould be planted ſingly in opens, that its fungous bark may be in view: not that there is any great beauty merely in the ſight, but with us it is a curioſity; being the true Cork, and is of the ſame nature with what comes from abroad, and we uſe for bottles, &c. Thus rough and ſpongy is the bark on the trunk and main branches; but the bark on the young ſhoots is ſmooth and gray, and that on the youngeſt white and downy. The leaves are of an oblong, oval figure, with ſawed edges. Their upper ſurface is ſmooth, and of a ſtrong green colour, but their under is downy. They grow alternately on the branches, on very ſhort though ſtrong footſtalks, and indeed differ in appearance very little from many ſorts of the *Ilex*. As the flowers of the *Quercus* make no ſhow, we ſhall proceed to the next ſort, after obſerving, that the acorns of the Cork Tree are longiſh, ſmooth, and brown when ripe, and of the ſize and ſhape of ſome of our common acorns, to which they are ſo much alike, as not to be diſtinguiſhed, if mixed together.

The *Narrow-leaved Cork Tree* is a Variety only of the common and moſt general ſort; ſo that, as this article requires nothing more than obſerving that the leaves are ſmaller, and as ſuch make a variety in plantations, it may not be amiſs to ſay ſomething of the Cork, which we receive from abroad, and which is collected from theſe trees. The beſt cork, then, is taken from the oldeſt trees, the bark on the young trees being too porous for uſe. They are, nevertheleſs, barked before they are twenty years old; and this barking is necessary, to make way for a better to succeed; and it is observable, that after every stripping the succeeding bark will increase in value. They are generally peeled once in ten years, with an instrument for the purpose; and this is so far from injuring the trees, that it is necessary, and contributes to their being healthy; for without it they thrive but slowly: may, in a few years they will begin to decay, and in less than a century a whole plantation will die of age; whereas those trees that have been regularly peeled will last upwards of two hundred years. "Wonderful, then, is the wisdom and goodness of Almighty God, and calls for our profoundest admiration, that he should not only provide for us his creatures such variety of things for use, but cause, as in this instance, what would be death to one tree, to be refreshment to another, for the supply of our necessaries; and in the formation of this tree, not only causing the cork to grow, but providing also an interior bark sufficient to nourish the tree, and even in a manner exhilarate it, as the loaded wool is shorn from the fleecy kind. To make our gardening to the utmost degree useful, we should be always exercised in these considerations, and this will inspire us with acts of gratitude and obedience." HANBURY.

11. The ILEX is a well known Evergreen, of which there are many *Varieties*; all of which add great beauty to the large quarters of Evergreen trees. The bark of all these sorts is entire, and that of the younger sorts smooth; but the leaves are of different shapes and composition, according to the nature of their Variety. Some of them are nearly like those of both sorts of the Cork Tree; others again are nearly round and prickly; some are long, smooth, and narrow, with few indentures; whilst others are broad, and much serrated. All these *Varieties* will often proceed from acorns gathered of the same tree; nay, the leaves of the same tree will not be always alike, being often found very different on the same plant; so that a quantity of plants of this species raised from seeds, will of themselves afford considerable variety. The acorns of all these sorts are of different sizes, though their shape is nearly the same, which is like that of some sorts of our Common

mon Oak, but ſmaller. The moſt ſtriking *Variety* of the *Ilex* is the *Holly-leaved Oak*, which differs from the other ſorts only that the leaves are ſhaped like thoſe of the Holly Tree. They are of an oblong, oval figure, ſinuated, prickly, and downy underneath; but many ſorts raiſed from ſeeds of the *Ilex* will have ſuch kind of leaves; and it conſtitutes no farther a *Variety*, than what may reaſonably be expected from a quantity of the acorns of the *Ilex* ſown.

12. KERMES OAK. This is a low growing tree, and a fine Evergreen: It ſeldom grows to be twenty feet high, and it may be kept down to what height is required. It has the appearance of ſome of the ſorts of the *Ilex*, from which it looks to be a Variety only, though doubtleſs this is of itſelf a diſtinct ſpecies. The leaves are ſmooth, and of an oval figure. They are of a thickiſh conſiſtence, and larger than moſt ſorts of the *Ilex*. Their verge is indented, and many of them are poſſeſſed of ſmall ſpines; and they are placed on ſhort ſtrong footſtalks on the branches. The acorns of this ſort are ſmall, though there are to be found in our woods acorns of about the ſame ſize and ſhape.

MILLER ſays, "this is the Oak from which the Kermes or Scarlet Grain is collected, which is an inſect that harbours on this tree."

13. The LIVE OAK is common in America, where it grows to timber. The leaves are large, ſpear-ſhaped, oval, of a fine dark green colour, entire, and placed on ſhort footſtalks on the branches. The acorns of this ſort are ſmall, though they grow in cups with footſtalks like the other ſorts. The wood of this tree is very uſeful to the inhabitants of Carolina, Pennſylvania, and Virginia, where it grows naturally, being very tough and hard, and ſerves for many purpoſes that require ſuch a ſort. The acorns ſerve for food for the meaneſt people, who not only eat them as ſuch, but, being of a very ſweet nature, they are liked by perſons of all ranks. From theſe acorns a ſweet oil alſo is extracted, which is very good.

There are many other *Varieties* of the different ſpecies of Evergreen Oaks, which it will not be ſo neceſſary to ſearch for here, as the ſorts mentioned are the bulk of the tribe, and of themſelves afford much variety; and indeed, if much coſt and trouble were beſtowed in procuring others, the variety would be little heightened, particularly as the pleaſure received from the variation ariſes principally from the different forms of the leaves; for none of theſe trees produce flowers for ornament, and the acorns afford too minute a variety to require dwelling long on here.

All the ſpecies of EVERGREEN OAKS are to be raiſed from *acorns*, in the manner which has been directed for the foreign deciduous ſorts. The beſt acorns we receive from abroad; for they ſeldom ripen well with us. Theſe acorns often ſprout in the paſſage; ſo that care muſt be uſed in taking them out of what they are incloſed in, and they ſhould be put into the mould as ſoon as convenience will permit. Traps for mice, &c, muſt be ſet: and after they come up, they will want nothing but weeding for at leaſt three years; for I would not have them taken out of the ſeed beds ſooner; eſpecially the ſorts of the *Ilex*; for when theſe have been pricked out of the ſeed beds at one year old, they have ſeldom grown; and though ſometimes ſome of them will be green, and have the appearance of growing during one ſummer, they will ofteneſt turn brown, and gradually go off afterwards. "After theſe plants have ſtood to be two or three feet high, I always found them more ſure of growing when moved. I have tranſplanted ſuch plants at moſt times of the year with ſucceſs; in the ſpring, in the depth of winter, and in the autumn, and have had them grow well when moved in July; and indeed I am pretty well perſuaded there is no month in the year more proper than that for the removing of moſt ſorts of Evergreens, provided the weather be rainy or hazy at their planting, and ſhade can be afforded them for ſome time after." HANBURY.

Theſe trees may be alſo increaſed by *inarching*, for they will grow very readily this way on ſtocks of our Common Oak; ſo that having a tree or two of any of the ſorts, if young Oaks are planted round each of them, after they have grown a ſummer or two, they will be ready to embrace the young ſhoot. After they are well joined, they may be cut off from the mother tree, and tranſplanted into the nurſery ground, or where they are to remain, and freſh Oaklings planted

round

round the trees to be multiplied; and the continuance of the repetition of this may be at pleaſure. In removing of the inarched plants, the time ſhould be obſerved as in removing young plants of our Common Oak, the roots ſtill remaining of that kind and nature.

Theſe trees will take by *grafting* on the young ſtocks of our Common Oak. The ſtocks ſhould be young and healthy, the cuttings ſtrong and good, and great care muſt be taken in properly joining and claying them, or they will not grow; which makes the inarching more neceſſary, as by that practice no cutting is in danger of being loſt.

RHAMNUS.

LINNEAN Claſs and Order, *Pentandria Monogynia*: Each flower contains five males and one female. There are twenty ſeven SPECIES; ſeven of which may be admitted into our collection.

1. RHAMNUS *Catharticus*: The COMMON BUCKTHORN; *a tall deciduous ſhrub*; native of England, and (one of its Varieties) of Spain, Italy, and France.

2. RHA'MNUS *Frangula*: The FRANGULA, or the BERRY-BEARING ALDER; *a tall deciduous ſhrub*; native of England and moſt of the northern parts of Europe.

3. RHA'MNUS *Alpi'nus*: The ALPINE RHAMNUS, or the ROUGH-LEAVED FRANGULA; *a deciduous ſhrub*; native of the Alps.

4. RHA'MNUS *Paliu'rus*: The PALIURUS, or THORN OF CHRIST, or CHRISTI THORN; *a deciduous ſhrub*; native of Paleſtine, alſo of Spain, Portugal, and Italy.

5. RHA'MNUS *Alaternus*: The COMMON ALATERNUS; *an evergreen tree or ſhrub*; native of the South of Europe.

6. RHA'MNUS *inſecto'rius*: The NARROW-LEAVED EVERGREEN BUCKTHORN; *an evergreen ſhrub or tree*; native of Spain.

7. RHAMNUS *Oleo'ides:* The OLIVE-LEAVED EVERGREEN BUCKTHORN; *an evergreen ſhrub*; native of Spain.

1. The COMMON BUCKTHORN. Of this ſpecies there are the following *Varieties:* Dwarf Buckthorn, Long-leaved Dwarf Buckthorn, and the Common Buckthorn of our hedges. Variety is the ſole motive for admitting theſe ſorts into a collection. The flowers have no beauty to catch the attention; though their berries, their manner of growing, the colour of their bark in winter, and verdure of their leaves in ſummer, court us to admit a few of them.

Dwarf Buckthorn is a ſhrub of about a yard high. The branches grow irregular, and are covered with a blackiſh coloured bark. The leaves are nearly oval, though they end in a point. They are ſcarcely an inch long, about half that breadth, and ſtand oppoſite by pairs for the moſt part. The flowers grow on ſhort footſtalks, on ſpurs, by the ſides of the branches. They are of a greeniſh colour, and make little ſhow.

Long-leaved Dwarf Buckthorn differs little from the other, only that it grows to be rather a larger ſhrub, and the leaves are longer. The flowers are about the ſame colour as the Dwarf ſort; but neither of theſe ſcarcely ever produce berries: This makes them much leſs valuable than our Common Buckthorn, which will exhibit its black berries in plenty in the autumn, either for ſhow or uſe.

Common Buckthorn is well known in England. Where it does not grow common, about a habitation, a few of theſe ſhrubs ſhould be admitted; for it is a well looking tree, either in winter or ſummer, and its black berries in the autumn are no ſmall ornament. The Common Buckthorn will grow to be near ſixteen feet high, and will ſend forth numerous branches on all ſides. Theſe are ſmooth, and the bark is of a blueiſh colour. Many ſtrong ſharp ſpines come out from the ſides and ends of the branches. The leaves are oval, ſpear-ſhaped, about two inches long, and one broad. Their under ſurface is of a lighter green than the upper. They have ſerrated edges, and ſtand, ſometimes by pairs, ſometimes ſingly, on longiſh footſtalks on the branches. The flowers are produced in cluſters from the ſides of the

the branches, in June. Their colour is green: and they are ſucceeded by black berries, each containing four ſeeds. Syrup of Buckthorn is made of theſe berries, and is well known as a cathartic. From the juice of theſe berries alſo an admirable green colour is prepared, which is in great requeſt with miniature painters.

All the ſorts of Buckthorn are eaſily PROPAGATED, either by ſeeds or cuttings. The ſeeds of the Purging Buckthorn may be gathered in plenty in moſt parts of England; but the ſeeds of the Dwarf ſorts muſt be procured from abroad, where they grow naturally, for they produce no ſeeds with us. They ſhould be ſown as ſoon as poſſible after they are ripe, in almoſt any kind of garden mould made fine. They will not always come up the firſt ſpring; ſo that the beds muſt remain undiſturbed and weeded during the ſummer. After they are come up, and have ſtood in the ſeed bed a year or two, they may be planted out in the nurſery way, at ſmall diſtances. Theſe plants are alſo to be raiſed by cuttings, which ſhould be planted in the autumn; and if they are not planted very cloſe, they will want no removing until they are finally ſet out. If a large quantity of theſe plants is wanted, and little ground is prepared for the cuttings, they may be ſet very cloſe, and in the winter following taken up, and planted in the nurſery way, like the ſeedlings. In two or three years they may be planted out to ſtand.

2. FRANGULA, or BERRY-BEARING ALDER. This ſpecies affords us the following *Varieties* Common Black Berry-bearing Alder, Dwarf Berry-bearing Alder, and the American Smooth-leaved Berry-bearing Alder.

The *Common Black Berry-bearing Alder* will grow to the height of about ten feet. It will aſpire with an upright ſtem, and produce numerous branches on all ſides. The bark is ſmooth, of a blueiſh colour, and is all over ſpotted with white ſpots, which make it reſemble a blueiſh gray. The leaves are oval, ſpear-ſhaped, and grow irregularly on the branches. They are about two inches long and one broad. Their upper ſurface is ſmooth and of a ſhining green, and their under ſurface is poſſeſſed of many ſtrong veins that run from the midrib to the edges. The flowers are produced in bunches

bunches in June, each having a feparate footftalk, They are of a greenifh colour, and make no fhow; but they are fucceeded by berries, which are firft red, afterwards (when ripe) black, and are a great ornament to the tree. This plant will bear a moift fituation.

Dwarf Berry-bearing Alder is of very low growth. It feldom rifes higher than two feet. The branches are of a blueifh brown, and the leaves are nearly round. They are placed on fhort footftalks, and many ftrong veins run from the midrib to the border. It makes no fhow, either in the flowers or fruit; the firft being fmall, and the latter rarely happening.

American Smooth-leaved Berry-bearing Alder will arrive at the height of our common fort; and hardly in any refpect differs from it, either in leaves, flowers, or fruit.

3. ROUGH-LEAVED ALPINE FRANGULA, or Berry-bearing Alder, differs in no refpect alfo from the common fort, only that it is unarmed with thorns, will grow to be rather taller, and the leaves are tough, larger, and doubly laciniated.

There is a *Variety* of this fpecies, with fmooth leaves and of rather lower growth, called the *Smooth-leaved Alpine Frangula*.

The method of PROPAGATING thefe forts of the Berry bearing Alder is exactly the fame as that laid down for the Buckthorn; and if thofe rules are obferved, any defired quantity may be raifed.

4. PALIURUS, or *Chrifti Thorn*. The PALIURUS will grow to be a tree of near fourteen feet high, and may be trained to an upright ftem, which will fend forth numerous flender branches on all fides. Thefe are armed with fharp thorns, two of which are at each joint. One of thefe thorns is about half an inch long, ftraight, and upright; the other is fcarcely half that length, and bent backward. Between thefe is the bud for the next year's fhoot. The bark on thefe twigs is fmooth, and of a purplifh colour, and the fpines themfelves are of a reddifh caft. The joints alternately go in and out, forming at each bud an obtufe angle. The leaves are nearly of an oval figure, of a pale green colour, and ftand on very fhort footftalks. They are fmall, being fcarcely an inch in length, have three longi-

longitudinal veins, and are placed alternately on the branches. The flowers are produced in clusters from the sides of the young shoots. They are of a yellow colour; and though each single flower is small, yet they will be produced in such plenty all over the plant, that they may make a very good show. June is the time of flowering; and they are succeeded by a small fruit, that is surrounded by a membrane.

The plant under consideration, says HANBURY, "is undoubtedly the sort of which the crown of thorns for Our Blessed Saviour was composed. The branches are very pliant, and the spines of it are at every joint strong and sharp. It grows naturally about Jerusalem, as well as in many parts of Judæa; and there is no doubt that the barbarous Jews would make choice of it for their cruel purpose. But what farther confirms the truth of these thorns being then used, are the antient pictures of Our Blessed Saviour's crucifixion. The thorns of the crown on his head exactly answer to those of this tree; and there is great reason to suppose these were taken from the earliest paintings of the Lord of Life; and even now our modern painters copy from them, and represent the crown as composed of these thorns. These plants, therefore, should principally have a share in those parts of the plantation that are more peculiarly designed for religious retirement; for they will prove excellent monitors, and conduce to due reflection on and gratitude to *Him who hath loved us, and has washed us from our sins*," &c.

These deciduous sorts may be PROPAGATED by seeds and layers. The soil for the seed should be that taken from a fresh pasture, with the sward; and having lain a year to rot, and been turned three or four times, to this a fourth part of drift sand should be added; the whole being well mixed, the seeds should be sown half an inch deep. They rarely come up before the spring twelvemonth after sowing; so that the beds must be undisturbed all the summer, and kept free from weeds. After the plants are come up, they may stand a year or two in the seed bed, and be then planted out in the nursery, at the usual distance: In about three years they will be fit to be finally planted out. These plants may also be propagated by layers; but this is not always a very

very eaſy taſk, and it is ſeldom that plants can be obtained under two years. Nicking them like carnations is a very uncertain method to be practiſed on theſe twigs; for the end of the nick where the root is expected to ſtrike will ſwell, and be covered with a cloſe watery ſubſtance, without ſending out any fibres; and the branch growing in the ground will in two or three years grow this out, and thus all hopes of a root will be loſt. By twiſting them, alſo, is an uncertain method (though many plants may be raiſed this way); for if the twiſting be too great, you kill the twig deſigned for the layer; and if it is too little, you may look at the end of two or three years, and find no roots at your layers. However, by a gentle twiſt, juſt breaking the bark, plants may be raiſed. HANBURY continues, "Finding theſe methods precarious and uncertain, I had recourſe to another, by which I obtained numbers of plants. With a ſharp knife I made a gentle nick or two the depth of the bark, about the bud and thorns which are at a joint. Having done this in two or three places in every ſhoot, and having laid them in the ground, every twig had ſtruck root, and were become good plants by that time two years; many of which were fit to plant out for good, and the ſmaller proper for the nurſery ground to gain ſtrength."

5. The ALATERNUS. The *Varieties* of this Species are,

The Common Alaternus.
The Broad-leaved Alaternus.
The Jagged-leaved Alaternus.

The *Common Alaternus* is again *variegated:* There are of it, the Gold-ſtriped, the Silver-ſtriped, the Blotch-leaved, the large and the ſmaller growing *Alaternus*; and whoever is for having them in plantations of the preſent kind, will ſtill increaſe the variety. This is indeed objected to by ſome, as, they ſay, they cannot be Evergreens; others again think they are moſt proper, as they retain their leaves, and appear amongſt others, of different colours, like flowers in ſummer. The branches of theſe ſorts of *Alaternus* are numerous; and the younger branches are covered with a ſmooth green bark. In winter, indeed, they will be brown, and ſome of a reddiſh colour; others will have their ſides

next

next the ſun red, and the oppoſite green. The leaves are oval, of a lucid green in the common ſorts, and look very beautiful. Their edges are crenated, and they grow alternately on the branches. The flowers are produced in April, from the wings of the leaves, in little cluſters: They are of a greeniſh colour, but make no ſhow: and are ſucceeded by berries, which are very grateful to blackbirds, thruſhes, and the like kinds of birds.

The *Broad-leaved Alaternus* is the grandeſt looking tree of all the ſorts: It will grow to the greateſt height, if permitted to ſhoot freely, though it may be kept down to any height wanted. The leaves are the longeſt of any of the ſorts, and their edges are lightly crenated. They differ a little in figure from the preceding ſort, being more heart-ſhaped. They are of a fine ſhining ſtrong green colour, both in winter and ſummer; and this tree produces flowers and ſeeds like the other.

The *Jagged-leaved Alaternus* has as different a look from the other as any two Evergreens whatever. It is a well looking upright tree, and the branches are covered with a ſmooth fine bark, which in winter is of a reddiſh colour. The leaves, like thoſe of all the ſorts, grow alternately. They are long and narrow, and are ſo jagged as to cauſe them to have a particular look. Their ſurface is ſmooth and ſhining, and their figure lanceolate; and this, together with the nature of their ſerratures, cauſes in the tree a beautiful as well as ſingular look. The flowers are produced in the ſame manner as the others; and are ſucceeded by berries, which are uſed by painters in compoſing ſome of their yellows. There are variegated ſorts of the Jagged-leaved *Alaternus* in both ſilver and gold ſtripes, which are indeed very beautiful; but they are very apt to turn green, if planted in a rich ſoil; ſo that to continue the ſtripes in perfection, the worſt ſort of hungry land ſhould be allotted them.

There are more Varieties of the *Alaternus*, but their differences are ſo inconſiderable as ſcarcely to be worth enumerating. All the ſorts have been confounded by the unſkilful with thoſe of *Phillyrea*, which have indiſcriminately paſſed one for the other: That the Gardener. therefore, may be guarded from running again

into

into these errors, he must observe, that the leaves of all the sorts of *Phillyrea* grow always opposite by pairs, whereas those of the *Alaternus* grow singly and alternately on the branches, which first gave occasion to the shrub's being so called. The Botanist will see a more material difference, when, upon examining the flowers, he finds they belong to distinct classes.

6. The NARROW-LEAVED BUCKTHORN grows to be a shrub of ten or twelve feet high, sending forth several branches from the sides from the bottom to the top. They are covered with a blackish or dark coloured bark, and each of them is terminated by a long sharp thorn. The leaves are very narrow, fleshy, astringent, of a strong green colour, and grow together in bunches on the sides of the branches. The flowers come out from the sides of the branches in small bunches: They are of an herbaceous colour, appear early in the spring, and are succeeded by large round berries, like those of the Sloe Bush, which are harsh and sour to the taste, and of a fine black colour when ripe. The fruit of this sort continues on the trees all winter, making a beautiful appearance among the narrow clustered leaves at that season.

7. OLIVE-LEAVED BUCKTHORN will grow to be eight or ten feet high, sending forth numerous branches, each of which is terminated by a long sharp spine. The leaves are small, oblong, obtuse, undivided, veined, smooth, of a thickish consistence, and grow two or three together on their own separate footstalks. The flowers come out from the sides of the branches in the spring. They are small, of a whitish green colour; and are succeeded by round black berries, about the size and colour of those of the Common Purging Buckthorn.

These Evergreen sorts are to be PROPAGATED, 1. By layers. This business must be done in the autumn, when the last summer's shoots should be laid in the ground. These will often strike root at almost every joint; though they have been found in some strong soils, upon examining them in the autumn, after being layered a whole year, without any roots; so that it would be proper to give the layer a slit at the joint,

and

and bend it ſo in the ground as to keep it open; and it will have plenty of root by the autumn. Another thing to be obſerved is, that in order to obtain good layers, the plants deſigned to be increaſed ſhould be headed the year before, and this will cauſe them to ſhoot vigorouſly; and from theſe ſhoots the ſtrongeſt and beſt layers may be expected; many of which will be good plants to ſet out where they are to remain, while the weakeſt may be planted in the uſual nurſery way, to gain ſtrength. 2. Theſe plants may be raiſed by ſeeds, the variegated ones excepted, for they muſt always be encreaſed by layers. The ſeeds will be ripe in September, or the beginning of October, when they ſhould be guarded from the birds, or they will ſoon eat them all. Soon after they are ripe they ſhould be ſown, for even then they will often remain two years before they come up. The beds ſhould be compoſed of fine light mould, and they ſhould be ſown an inch deep. If few or no plants appear in the ſpring, you muſt wait, and weed the beds with patience, until the ſpring following, when you may expect a plentiful crop. Let them ſtand two years in the ſeed bed, with conſtant weeding, and frequent watering in dry weather; and in March let them be planted out in the nurſery, where they will be afterwards ready for removing when wanted. As theſe trees produce plenty of good ſeeds, by this means a prodigious quantity of plants may be ſoon raiſed; and thoſe from ſeeds are always obſerved to grow ſtraighter and to a greater height than thoſe raiſed from layers; ſo that where many of theſe trees are wanted for large plantations, the raiſing them from ſeeds is the moſt eligible method.

All the ſorts of *Alaternus* are very hardy, and may be planted in almoſt any ſoil or ſituation; but the Narrow and Olive-leaved Buckthorn ſhould be ſtationed in a dry, warm, well ſheltered place.

RHODODENDRON.

LINNEAN Class and Order, *Decandria Monogynia*: Each flower contains ten males and one female. There are ſeven SPECIES; ſix of which are here treated of:

1. RHODODE'NDRON *Ferrugi'neum*: The FERRUGINEOUS DWARF ROSE BAY; *a low deciduous shrub;* native of the Alps, Apennines, and other mountains of Europe.

2. RHODODE'NDRON *Hirſu'tum*: The HAIRY DWARF ROSE BAY; *a low deciduous ſhrub;* native of the Alps and many mountains of Switzerland and Austria.

3. RHODODE'NDRON *Chamæci'ſtus*: The CHAMÆCISTUS, or CILIATED-LEAVED DWARF ROSE BAY; *a low deciduous ſhrub;* native of Mount Baldus, and near Saltzburg in Germany.

4. RHODODE'NDRON *Dau'ricum*: The DAURIAN DWARF ROSE BAY; *a low deciduous ſhrub;* native of Dauria.

5. RHODODE'NDRON *Ma'ximum*: The AMERICAN MOUNTAIN LAUREL; *an evergreen ſhrub*, native of Virginia.

6. RHODODE'NDRON *Po'nticum*: The PONTIC DWARF ROSE BAY; *an evergreen ſhrub;* native of the East, and of moist ſhady places near Gibralter.

1. The FERRUGINEOUS DWARF ROSE BAY is a ſhrub of about two or three feet in growth. The branches are numerous, irregular, and covered with a dark brown bark, having a tinge of purple. The leaves are of two very different colours; the upper ſurface is of a fine green, but the under is of an iron colour. There will be numbers of these on every twig; and they grow in a pleasing irregular manner: They are of a lanceolated figure, have their ſurfaces ſmooth, and are little more than an inch long. Their edges are reflexed; but they have no ſerratures, and, on the whole constitute a great beauty when in leaf only. The flowers grow at the ends of the branches, in round

bunches. Their petals are funnel-ſhaped, of a pale roſe colour, appear in June, and are rarely ſucceeded by ſeeds in England.

2. Hairy Dwarf Rose Bay is a ſhrub of about the ſame, or rather of a lower growth. The branches of this ſpecies alſo are numerous, and the bark with which they are covered is of a lightiſh brown colour. They are ornamented with plenty of leaves, in an irregular manner. They are not ſo large as thoſe of the former ſort; but are of the ſame figure, only a little more inclined to an oval. They ſit cloſe to the branches, and have no ſerratures, but hairs on their edges like the eyelaſhes. Their under ſurface alſo is poſſeſſed of the ſame ſort of hairs, which are all of an iron colour. The flowers will be produced at the ends of the branches, in bunches, in May. Theſe are alſo funnel-ſhaped, of a light red colour, make a good ſhow, and are ſucceeded by oval capſules, containing ripe ſeeds, in Auguſt.

3. Chamæcistus, or Ciliated Dwarf Rose Bay, will grow to be about a yard high. The branches are numerous, produced irregularly, and covered with a purpliſh bark. The leaves are produced in great plenty, and without order, on the branches. They are oval, ſpear-ſhaped, ſmall, and their under ſurface is of the colour of iron. The edges alſo are poſſeſſed of many iron-coloured hairs, which are placed like thoſe on the eyelids. The flowers are produced at the ends of the branches, in bunches. They are of a wheel-ſhaped figure, pretty large, of a fine crimſon colour, and make a handſome ſhow. They appear in June, and are ſucceeded by oval capſules, containing ripe ſeeds, in September.

4. Daurian Dwarf Rose Bay is a low ſhrub, ſending forth many branches covered with a browniſh bark. The leaves are broad, naked, ſmooth, and come out without order on ſhort footſtalks. The flowers are wheel-ſhaped, large, and of a beautiful roſe colour: They appear in May; and are ſucceeded by oval capſules full of ſeeds, which do not always ripen in England.

All theſe deciduous ſorts are PROPAGATED beſt by the ſeeds, and as they grow naturally on the Alps,

 Apennines,

Apennines, and other ſnowy and cold mountains, and are ſeldom made to grow and flouriſh fair in gardens, it will be the beſt way for a Gentleman who has extended his plantation, and has any part of it mountainous, hilly, or rocky, on the north ſide, to get ſome ſpots well cleared of all roots and weeds; and theſe being made fine and level, let the ſeeds be ſown therein. They will want no covering; a gentle patting down with the ſpade will be ſufficient; for the ſeeds are ſo exceedingly ſmall, that they will be waſhed into the ground deep enough by the firſt ſhower of rain that follows. Whoever is not content with ſowing ſeeds, and covering them no more than what they will get by being patted down, muſt only lightly duſt ſome earth over them; for if they are covered half an inch, the general depth for moſt ſeeds, you muſt expect no crop. After the young plants come up, they muſt be watered in dry weather, weeded, and in the winter protected from the froſts, which will deſtroy them. And here one thing is to be obſerved, that though the north ſide, at the foot of or on a hill, is thought moſt proper for their growth, as being moſt ſuitable to their nature, yet a place muſt be choſen for them that has trees and hedges to ſhelter them from the northern black froſts; for theſe trees, hardy as they are, will be liable to be deſtroyed by them, for want of ſnow, as in other places, to cover them and keep them warm in the winter ſeaſon. After theſe plants are come up, they ſhould be thinned; and leaving only a proper number in each reſpective place, and being protected for the firſt two or three winters, either by mats or hand glaſſes, in the ſevereſt weather, they will be afterwards ſtrong enough to be left to themſelves, eſpecially if the places are tolerably ſheltered. If a Gardener has no other ground than his ſeminary for raiſing plants, his beſt method will be to prepare a compoſt for theſe ſeeds in the following manner: Take four buſhels of earth from ſome neighbouring hill, which if rocky, that neareſt the ſurface, on which the ſheep have been uſed to lie and dung, will be the beſt; but if it be of any other nature, the mould neareſt the ſurface, mixed with the following, will do very well: Take ſix buſhels of maiden earth from a rich loamy paſture, that has been dug up with the

ſward,

ſward, and by frequent turning is well rotted and mixed, and four buſhels of drift or ſea ſand. Let theſe be well mixed together, and of this let the bed be made. The bed being made level and fine, the ſeeds ſown, and gently patted down with the ſpade, or at fartheſt no other covering than being gently duſted over with the fineſt mould, may be left to nature. This bed ſhould be in a ſhady well ſheltered place; and the plants after they are come up ſhould be weeded and watered in the ſummer, and protected from froſts by mats in the winter. In the ſpring they may be pricked out in beds in the nurſery ground, at a very ſmall diſtance, that they may be hooped and matted if the following winter ſhould prove very ſevere. The ſecond winter they will require no other trouble than pricking furze buſhes round the bed for their defence; and after that they may be ſet out to ſtand.

5. The AMERICAN MOUNTAIN LAUREL is a plant ſo diſtinguiſhed becauſe, in America, it grows naturally upon the higheſt mountains, and on the edges of cliffs, precipices, &c. There it will grow to be a moderate ſized tree; with us it ſeldom riſes higher than ſix feet. The branches are not numerous, neither are they produced in any order. The leaves are large and beautiful, of an oval ſpear-ſhaped figure, and a little reſemble thoſe of our Common Laurel. They are of a ſhining ſtrong green on their upper ſurface, though paler underneath; but they loſe this delicacy as they grow older, altering to a kind of iron colour. Their edges are acutely refl xed, and they grow irregularly on ſhort fooiſtalks on the branches. The flowers are produced at the ends of the branches about Midſummer, though ſometimes ſooner; before which time the buds will be large and turgid; and indeed, as they begin to ſwell early in the autumn before, theſe have a good effect, and look well all winter. When the ſhrub is in blow, the flowers appear cloſe to the branches, in roundiſh bunches. Each is compoſed of one petal, which is divided at the rim into five parts, one of which is dotted in a pretty manner. They are very beautiful, and alter their colour as they grow older; for at firſt the petal is of a very pale bluſh colour, which dies away to a white; but the outſide, which is a peach colour, is

not ſubject in ſo high a degree to this alteration. They will continue, by ſucceſſion, ſometimes more than two months; and are ſucceeded by oval capſules, full of ſeeds.

6. Pontic Rose Bay grows to about four or five feet high, ſending forth ſeveral branches without order from the ſides. The leaves are ſpear-ſhaped, gloſſy on both ſides, acute, and placed on ſhort footſtalks on the branches. The flowers are produced in cluſters from the ends of the branches; each of them is bell-ſhaped, and of a fine purple colour. They appear in July, and are ſucceeded by oval capſules containing the ſeeds, which ſeldom ripen in England.

The propagation of theſe Evergreen ſorts muſt be from ſeeds, which we receive from the places where they grow naturally. The beſt way is to ſow them very thin in the places where they are deſigned to remain; and if theſe places be naturally rocky, ſandy, and ſhady, it will be ſo much the better (eſpecially for the firſt ſort; the ſecond requires a moiſtiſh ſoil, in a warm ſhady place); if not, a quantity of drift ſand muſt be added to the natural ſoil, and all made fine and level. Some ſpots for the reception of the ſeeds are to be pitched on. A few ſeeds ſhould be put in each, and covered about half an inch deep, and then ſome ſticks ſtuck round them to direct to the true places, that they may not be diſturbed by hoeing the weeds, but that theſe may be all carefully plucked up by the hand as often as they appear; for it will be a whole year, and ſometimes two or more, before the plants come up. This careful weeding muſt always be repeated; and after the plants come up, thoſe that grow too cloſe may be drawn the ſpring following, and each ſet in a ſeparate pot, and then plunged into a hotbed, to ſet them growing. The plants that remain without removing will be the ſtrongeſt and beſt, and will be more likely to produce flowers than any other; though this ſeems to be a plant that will bear tranſplanting very well, eſpecially if it is not to be carried at too great a diſtance for the roots to dry, and a ball of earth be preſerved to them. Whenever they are not to be raiſed and remain in the places, the beſt way is to ſow them in pots filled with ſandy earth, or ſuch as is made ſo by at leaſt a third

part

part of ſand being added. After the plants come up, they may be planted in ſeparate pots the ſpring following, and then ſet forward by a plunge in the bed; and afterwards they may be any time turned out into the places where they are to remain, which ought to be in a naturally ſandy ſituation, otherwiſe there will be little hopes of ſeeing them in any degree of perfection.

R H U S.

LINNEAN Claſs and Order, *Pentandria Trigynia*: Each flower contains five males and three females. There are twenty-four SPECIES; eight of which are ſufficiently hardy to ſtand this climate:

1. RHUS *Coria'ria*: The TANNER'S SUMACH, or the ELM-LEAVED SUMACH; *a tall deciduous ſhrub*; native of Turkey, Paleſtine, Syria, Italy, and Spain.

2. RHUS *Ty'phynum*: The VIRGINIA SUMACH; *a deciduous ſhrub*; native of Virginia.

3. RHUS *Gla'brum*: The SMOOTH SUMACH; *a tall deciduous ſhrub*; native of North America.

4. RHUS *Coppa'llinum*: The LENTISCUS-LEAVED SUMACH; *a deciduous ſhrub*; native of North America.

5. RHUS *Ve'rnix*: The VARNISH TREE, or POISON ASH TREE; *a deciduous ſhrub*; native of North America, alſo of Japan.

6. RHUS *Toxicode'ndron*: The TOXICODENDRON, or POISON OAK; *a low deciduous ſhrub*; native of North America.

7. RHUS *Ra'dicans*: The RADICANT TOXICODENDRON; *a deciduous ſhrub*; native of Virginia and Canada.

8. RHUS *Co'tinus*: The VENETIAN SUMACH, or COCCY'GRIA; *a deciduous ſhrub*; native of Italy, Spain, and many parts of Europe.

1. The TANNER'S SUMACH will grow to be about twelve feet high; and the branches are covered with a browniſh hairy bark. It is ſaid that this bark is equal

to that of the Engliſh Oak for tanning of leather, and that the leather from Turkey is chiefly tanned with it. The leaves of this ſhrub, which are placed alternately on the branches, have a grand look. They are pinnated, and each ends with an odd foliole. The midrib of each is garniſhed with about eight pairs of folioles, which all terminate with an odd one. The folioles of which the compound leaf is compoſed are oval, and not large, being ſcarcely two inches long, and three fourths of an inch broad; but the whole leaf makes a fine ſhow. Their colour is a light green; their under ſurface is hairy, and they are ſawed at their edges. The flowers, which are produced in large bunches at the ends of the branches, are of a whitiſh colour, with a tinge of green. Each is compoſed of many ſpikes, on which the flowers ſit cloſe. They come out in July; but are not ſucceeded by ripe ſeeds in England, like ſome of the ſubſequent ſorts. The leaves and ſeeds are poſſeſſed of many excellent virtues.

2. VIRGINIA SUMACH. Of this ſpecies there are ſeveral *Varieties*; ſuch as, the Common Stag's Horn, Large Virginian, and Dwarf Sumach.

The *Stag's Horn Sumach* is ſo called from the younger branches much reſembling a ſtag's horn, called the Velvet Horn. It will grow to be about ten feet high, and the older branches are covered with a ſmooth browniſh bark, in ſome places of a grayiſh colour, whilſt the younger ones are covered with a hairy down, which much reſembles the velvet horn of a ſtag. The leaves have a noble look; for they are large and pinnated. The folioles are oblong, and larger than thoſe of the preceding ſort: about ſeven pairs are ſtationed along the midrib, which are terminated by an odd one. Their under ſurface is hairy, and they die to a purpliſh ſcarlet in the autumn. The flowers are produced in June, at the ends of the branches: they will be in large tufts, but make no ſhow; though ſome admire them when ſucceeded by ſeeds in the autumn; for at the end of that ſeaſon, even after the leaves are fallen, there will be large tufts of ſeeds, of a ſcarlet colour, left at the ends of the branches, which have an uncommon appearance.

The *Large Virginian Sumach* differs in no reſpect from the

the preceding ſpecies, only that it ſhoots ſtronger, and grows to be larger, even ſixteen or eighteen feet high, and is a more regular tree. The young ſhoots alſo are of a more reddiſh colour; and though poſſeſſed of the like hairy down, on the whole do not ſo much reſemble thoſe of the velvet ſtag's horn as the other.

Dwarf Sumach differs in no reſpect from the Common Stag's Horn, except that it is of a very low growth, ſeldom riſing higher than three feet.

3. SMOOTH SUMACH. This includes many notable *Varieties*, commonly called New England, Smooth Carolina, and Canada Sumach.

New England Sumach will grow to about ſixteen feet high, ſending forth many ſtrong ſhoots from the root and the ſides, covered with a ſmooth downy bark. The radical ſhoots will often be near an inch in diameter in one ſummer's growth. The young branches alſo from the ſides will be large: they are ſmooth, though a little downy in the ſummer; and the bark in the winter is of a light brown colour. The leaves of this ſort are the largeſt of any, being compoſed of ten or more pairs of folioles, proportionally large, and which are terminated by an odd one. The flowers are produced at the ends of the branches, in large looſe panicles: They are of a greeniſh yellow colour, and come out in June, but are not ſucceeded by ſeeds with us.

The *Carolina Sumach* ſeldom riſes to more than ten feet high. The branches are ſmooth, of a fine purpliſh colour, and duſted over with a whitiſh powder. The leaves are pinnated like the other, and the flowers are produced in panicles at the ends of the branches. They are of a fine ſcarlet colour, appear in July, and are ſucceeded by bunches of ſeeds, which in autumn are of a very beautiful red, though they never ripen in England.

The *Canada Sumach* grows to about ten feet in height, and the branches, which are ſmooth and of a purpliſh colour, are duſted over, like the former, with a kind of whitiſh powder. The leaves are pinnated like the other, and the folioles are on both ſides ſmooth; but their ſurfaces are of two colours, the upper being of a ſhining green, whilſt the under is hoary. The flowers are red, and produced in July, in large panicles, at the ends of the branches. They appear as if a whitiſh powder had been duſted in among them, which attracts notice; but their ſeeds do not ripen in England.

4. Lentiscus-leaved Sumach. The chief *Varieties* of this ſpecies are, the True Lentiſcus-leaved, and the Canada Lentiſcus-leaved Sumach.

The *True Lentiſcus-leaved Sumach* ſeldom riſes to more than four feet in height, and the branches are covered with a ſmooth brown bark. The leaves alſo are pinnated, and are the moſt beautiful of all the ſorts; for the folioles, though ſmall, are of a ſhining green. There are about four or five pairs on the midrib, which are beautifully arranged, having a membrane or wing on each ſide running from pair to pair: they are terminated by an odd one, reſemble in appearance thoſe of the Lentiſcus, and are the greateſt ornaments of this ſhrub. The flowers are produced in July, at the ends of the branches. They are of a greeniſh colour; and though produced in large looſe panicles, make no great figure; neither do the ſeeds ripen with us.

Canada Lentiſcus-leaved Sumach grows to be ten feet high. The leaves have chiefly the properties of the former, but are larger, leſs delicate, and duſted or pounced over with a whitiſh matter. The flowers are produced in the ſame manner as the other: they are greeniſh, and ſucceeded by ſeeds in England.

5. The Poison Ash. This is called the Poiſon Tree becauſe it abounds with a milky poiſonous juice, and is diſtinguiſhed by the title Poiſon Aſh, becauſe the leaves ſomewhat reſemble thoſe of the Aſh tree. It is called alſo by ſome the Varniſh tree, being the ſhrub from which the true varniſh is collected. The Poiſon Aſh, with us, will grow to the height of about eight feet; and the branches, which are not very numerous, are covered with a ſmooth light brown bark, tinged with red. The leaves are pinnated, and the folioles of which each is compoſed conſiſt of about three or four pairs, with an odd one. Theſe are of an oblong pointed figure, of a fine green colour, and have their edges entire. In the autumn, they die to a red or purple colour, and at that time their leaves, juſt before they fall, make a charming appearance, ſome being red, others

others purple, others between both; the colours of the footſtalks and midribs will alſo be various, thereby in the ſame tree affording a variety of ſhades. The flowers are ſmall, and make no ſhow: they are whitiſh, and produced in May, from the wings of the branches. There will be male and female flowers on different plants; and the females are ſucceeded by ſmall roundiſh fruit, which ſeldom ripens in England.

6. The POISON OAK is a lower ſhrub, ſeldom growing to be more than four or five feet high. The branches are ſmooth, and of a light brown colour. It will coſt the Gardener ſome trouble to keep theſe plants properly, as upright ſhrubs; for they will ſend out ſhoots from the bottom, which will naturally trail on the ground, and ſtrike root. But theſe muſt be conſtantly taken off; for were they to be neglected a few years, a ſingle plant would have ſpread itſelf to ſuch a diſtance as to occupy a great ſpace of ground, in a manner not becoming a well ordered ſhrubery or wilderneſs. The leaves of this ſhrub are trifoliate. Each foliole has a ſhort pedicle to itſelf, and the common footſtalk of the whole three is very long. They are of a ſhining green, ſmooth, and have their edges ſometimes ſinuated, though generally entire. They are roundiſh, angular, large, and on the whole make a good ſhow. The flowers are of a whitiſh colour, are produced from the ſides of the branches, in July, and are ſucceeded by cream-coloured berries, which growing in the autumn, and even in the winter, after the leaves are fallen, in a kind of panicles, are by many taken notice of. It bears a moiſt ſituation.

There are ſeveral *Varieties* of this ſpecies; ſome with hairy leaves, ſome with leaves very downy, others of fine upright growth. In other reſpects their difference is inconſiderable.

7. RADICANT TOXICODENDRON. Of this ſpecies there are ſeveral *Varieties*; ſome of which are of upright growth, though the ſtalks of all have, more or leſs, a tendency to lie on the ground, and ſtrike root at the joints. The leaves of all the ſorts are trifoliate, of an oval figure, ſmooth, and entire. The flowers are greeniſh, appear in June and July, and are ſucceeded

ſucceeded by roundiſh yellow berries, which rarely ripen in England.

8. The VENETIAN SUMACH is a ſhrub of about ten feet growth, and has many valuable properties to recommend it. The bark on the older branches is of a light brown colour, whilſt that on the young ſhoots is ſmooth, and of a purple hue. The leaves are nearly of an oval figure, and ſtand ſingly upon long footſtalks on the branches. From theſe the tree receives great beauty: they are of a delightful green, are ſmooth, and when bruiſed emit a ſtrong ſcent, which by many is thought very grateful; and on that account only makes this ſhrub deſirable. The flowers are produced at the ends of the branches, in July, in a ſingular manner: The end of the laſt year's ſhoot about that time will divide itſelf, and produce hair-like bunches of purpliſh flowers, ſo as to cover the tree; and in the autumn, though they do not perfect their ſeeds with us, theſe tufts will ſtill remain, be of a darker colour, and almoſt cover it; on account of which ſingular oddneſs this ſhrub is valued by ſome perſons. The bark is uſed by the tanners; whilſt the wood and leaves are ſought after by the dyers; the former being ſaid to dye a yellow, and the latter, together with the young branches, to dye a good black.

The PROPAGATION of the Sumach is not very difficult; for the ſecond, third, and fourth ſorts, with their Varieties, produce ſuckers in ſuch plenty as to over-run, if not taken off, all that is near them. Theſe ſuckers when taken up will be each a good plant; nay, their very roots will grow; and though they be thrown upon a bed, and dug careleſsly in, even then many young plants will ſpring from them.

The POISON OAK and RADICANT TOXICODENDRON alſo PROPAGATE themſelves very faſt by their trailing branches, which ſtrike root as they go, and each of which will be a plant.

The VENETIAN SUMACH is eaſily encreaſed by layers; for the young ſhoots being ſlit and layered in the autumn, by the autumn following will be good plants, either for the nurſery ground, or where they are to be planted out to ſtand.

The

The ELM-LEAVED SUMACH and the POISON ASH, however, do not throw out ſuckers in this manner, and theſe are to be PROPAGATED from the ſeeds, which we receive from the places where they naturally grow. An eaſt border of garden mould made fine) ſhould be prepared; and in this the ſeeds ſhould be ſown as ſoon as poſſible after we receive them. The depth they will require will be about half an inch. After being ſown, and the border dreſſed up, nothing more need be done till the weeds begin to come up, which will be before the plants: as often as theſe appear, they muſt be plucked up; and when the hot parching weather comes on, the border muſt be ſhaded in the heat of the day, and, every evening, ſhould be gently ſprinkled over with water. In the beginning of June many of the plants will come up; though they frequently remain, at leaſt the greateſt part of them, until the ſecond ſpring before they make their appearance. After the plants are come up, they will want no other care than ſhading, weeding, and now and then a watering during the firſt ſummer, and if the winter ſhould be ſevere, they ſhould be matted, eſpecially the Elm-leaved ſort, which is rather the moſt tender whilſt young. After this they will require no other care than weeding until they are two-years-old ſeedlings; when, in the ſpring, they ſhould be taken up and planted in the nurſery ground, and in two or three years more will be fit to ſet out for good. And here it muſt not be omitted to obſerve, that the other ſorts before mentioned, which propagate themſelves ſo faſt by ſuckers, may be raiſed this way if the ſeeds can be obtained; and, indeed, whoever has not the conveniency of procuring a few plants of each, and can have the ſeeds, muſt practiſe this method with them, by which he will ſoon procure plenty.

ROBINIA.

LINNEAN Clafs and Order, *Diadelphia Decandria*. Each flower contains ten males and one female; the males being divided into two fets at the bafe. There are nine SPECIES; five of which will bear the open air of this country.

1. ROBI'NIA *Pfeud-aca'cia:* The ACACIA, or TWO-THORNED ACACIA; *a deciduous tree*; native of moft parts of North America.

2. ROBI'NIA *Hi'fpida:* The THORNLESS ACACIA; *a tall deciduous fhrub*; native of America.

3. ROBI'NIA *Caragá'na:* The CARAGANA: *a deciduous fhrub*; native of Siberia.

4. ROBI'NIA *Frute'fcens:* The SHRUBBY ASPALATHUS; *a deciduous fhrub*; native of Siberia and Tartary.

5. ROBI'NIA *Py'gmæa:* The DWARF ASPALATHUS; *a low deciduous fhrub*; native of Siberia.

1. The COMMON ACACIA, or TWO-THORNED ACACIA, will grow to the height of thirty-five or more feet. The branches are covered with a fmooth purplifh-coloured bark, and armed with ftrong fpines, which are placed at the buds. Each bud, efpecially of the young vigorous fhoots, will be generally guarded by two of thefe fpines, one of which will be on one fide, while the other will occupy the oppofite place. The branches are very brittle, and in fummer, when the leaves are on, are often broke by the high winds. The leaves come out late in the fpring; but for this they make ample amends by the beautiful foliage they will difplay foon after. They are pinnated leaves the moft beautiful of all the compound forts. The folioles of which each is compofed are of a fine green; and as there are no lefs than nine or ten pair of them placed along the midrib, with an odd one, the whole leaf appears very large; and all the tree being thus ornamented has a noble look, even at that time. But this fhrub will be in its greateft beauty when in flower; for thefe

theſe will be produced in long pendulous bunches, in June. They are of the papilionaceous kind; their colour is white; and when the tree blows freely, its head will be enchantingly covered with them; for they will hang all over it in a free and eaſy manner; ſome bunches appearing wholly in view, others again half hid by the waving leaves, that will ſometimes alternately hide and ſhew them; at which time alſo, when there is a current of air, the flowers themſelves receive freſh beauty from being thus agitated. But this is not all: nature has granted them a ſmell, which is very grateful; ſo that in an evening, or after a ſhower, they will perfume the circumambient air to ſome diſtance: Thus they will prove a feaſt to all thoſe who will attend at thoſe times, as they will never fail of regaling one of the ſenſes by their grateful and profuſe fragrance. Theſe flowers, it is to be lamented, are of ſhort duration; and are ſucceeded by pods, which in ſome ſeaſons will perfect their ſeeds with us.

The principal *Varieties* of this ſpecies are, the *Scentleſs*, *Prickly-podded*, *Roſe-coloured*, *Scarlet*, *Smooth-podded*, *&c. Acacia*.

2. The THORNLESS ACACIA, or HISPID ACACIA, is of lower growth; the young branches, and the footſtalks and very cups of the flowers, are covered with prickly hairs. The flowers are produced rather earlier than thoſe of the other ſorts; they are large, and of a moſt beautiful roſe colour. They have no odour like the others; but have a moſt beautiful appearance when in blow.

3. CARAGANA riſes, with a ſhrubby ſtalk, to the height of about eight or ten feet, ſending forth ſeveral branches, which are covered with a greeniſh yellow bark. The leaves are abruptly pinnated; the folioles are oval, ſpear-ſhaped, pointed, and conſiſt of about five or ſix pair arranged along the midrib. The flowers come out from the ſides of the branches, on ſingle footſtalks: they are ſmall, of a yellowiſh colour, appear in May, and are ſucceeded by ſmooth compreſſed pods containing the ſeeds, which will be ripe in September.

4. The SHRUBBY ASPALATHUS is a beautiful flowering ſhrub. Its growth will be ſeven or eight feet; and the branches naturally grow upright, The bark is

ſmooth,

ſmooth, and of a yellowiſh colour; but that of the youngeſt twigs partakes more of a purpliſh colour on one ſide, and is on the other often of a light green with a yellow tinge. The leaves are each compoſed of about four folioles, which are oval and pointed. The flowers are produced in May, from the joints of the branches, upon ſingle footſtalks: they are of a fine yellow colour, and of the butterfly make; and ſo adorn the tree when in blow, as to render it inferior to few of the flowering ſhrubs. Theſe flowers are ſucceeded by pods, containing ripe ſeeds, in the autumn.

5. DWARF ASPALATHUS is a pretty little ſhrub, ſending forth ſeveral ſlender branches, which are covered with a golden bark. The leaves are quaternate, wedge-ſhaped, obtuſe, have no footſtalks, and, unleſs very ſevere weather happens, continue on the plant the greateſt part of the winter. The flowers come out from the ſides of the branches, on ſingle footſtalks: they are ſmall, of a yellow colour, appear in May, and are ſucceeded by ripe ſeeds in the autumn.

The PROPAGATION of all theſe ſorts is very eaſy, and may be done, 1. By ſeeds. If theſe are ſown the beginning of March, half an inch deep, in a bed of any common garden mould, plants will come up in May, which will want no other care than weeding all the firſt ſummer, and no protection of any kind in the winter; for they are all hardy enough. In the following ſpring they ſhould be planted out in the nurſery ground, a foot aſunder, and two feet diſtant in the rows; and here (the firſt three ſorts) they ſhould not ſtand longer than two or three years before they are ſet out to remain, as they will grow exceedingly faſt, and by that time will be perhaps ſix feet in height. The fourth ſort being of lower growth, the plants may be pricked in beds, a foot aſunder, which will be room enough for them to grow in, before they be finally ſet out. It may not be amiſs to obſerve alſo, that the ſeeds of this ſort often remain until the ſecond ſpring before they come up; ſo that when they do not appear the firſt after ſowing, the beds muſt be kept weeded all ſummer; and, if the ſeeds were good, there will be no fear of a crop the following ſpring. 2. Theſe ſorts are eaſily propagated by cuttings, which if planted in October, in a

moiſtiſh

moiſtiſh ſhady border, many of them will grow. Here they ſhould ſtand two years, when they will be proper plants to be planted out; though we muſt obſerve, that the fourth ſort may remain longer before they are ſet out; and as the cuttings of that ſort have often failed growing, the moſt certain method, and what is generally practiſed when there are no ſeeds, is to encreaſe it by layers. 3. The firſt ſorts will encreaſe themſelves by ſuckers, in ſufficient plenty; for the old plants will ſpawn at a conſiderable diſtance, and afford ſuch a quantity of free-ſhooting ſuckers, that they will be all good plants, fit to be ſet out for continuance.

ROSA.

LINNEAN Claſs and Order, *Icoſandria Polygynia*: Each flower contains about twenty males and many females. There are eighteen SPECIES; thirteen of which we here enumerate:

1. RO'SA *Cani'na*: The DOG ROSE, or HEP TREE; *a deciduous ſhrub*; common in our hedges, and moſt parts of Europe.

2. RO'SA *Pimpinellifo'lia*: The BURNET ROSE, or CAT WHIN; *a deciduous ſhrub*; natural to England and moſt parts of Europe.

3. RO'SA *Spinoſi'ſſima*: The Scotch ROSE; *a deciduous ſhrub*; native of Scotland, England, and moſt parts of Europe.

4. RO'SA *Alpi'na*: The ALPINE ROSE: *a deciduous ſhrub*; native of the Alps of Switzerland.

5. RO'SA *Eglante'ria*: The EGLANTINE, or the SWEETBRIAR; *a deciduous ſhrub*; native of England and Switzerland.

6. RO'SA *Cinnamo'mia*: The CINNAMON ROSE; *a deciduous ſhrub*; grows in the ſouthern parts of Europe.

7. RO'SA *Caroli'na*: CAROLINA ROSE; *a deciduous ſhrub*; native of North America.

8. RO'SA

8. Ro'sa *Villo'sa:* The Apple Rose; *a deciduous shrub*; native of most parts of Europe.

9. Ro'sa *Centifo'lia:* The Hundred-leaved Rose; *a deciduous shrub:* it is not known where this Rose grows naturally.

10. Ro'sa *Ga'llica:* The Gallican Rose; *a deciduous shrub*; grows naturally in most parts of Europe.

11. Ro'sa *Sempervi'rens:* The Evergreen Rose, or Musk Rose; *an evergreen shrub*; native of Germany.

12. Ro'sa *Penduli'na:* The Long-fruited Rose; *a deciduous shrub*; native of Europe.

13. Ro'sa *A'lba:* The White Rose; *a deciduous shrub*; native of Europe.

1. The Dog Rose grows all over England, and is seldom cultivated in gardens. It is, nevertheless, possessed of many beauties, if observed with due attention; and, if it was not so very common, would deserve a place in the choicest Collection.

The *Varieties* of this species are, the Hep Tree with Red Flowers, the White-flowered Hep Tree.

2. Burnet Rose is a small growing shrub, seldom rising higher than one yard. The flowers are single, and make no great figure; but what renders this Rose valuable is, that the leaves are pinnated in such a manner as to resemble those of the Burnet, which occasions its being so called, and by which it constitutes an agreeable variety among the leafy tribe.

The *Varieties* of it are, Red-flowered, Burnet-leaved Rose, Black Burnet-leaved Rose, White Burnet-leaved Rose.

3. Scotch Rose. The *Varieties* of this species are all of low growth, and known by the respective names of,

Dwarf Scotch with a White Flower,
Dwarf Scotch with a Red Flower,
Dwarf Scotch with a Striped Flower,
Dwarf Scotch with a Marbled Flower.

They are all beautiful flowering shrubs. The White-flowering sort will grow to the highest size, as it will commonly grow to be three feet, whilst the others seldom rise to above two feet in height. The branches are

are upright and numerous, and ſmartly ſet off by their beautiful pinnated leaves; for the leaves of theſe ſorts excel thoſe of all other Roſes in delicacy, the folioles being ſmall, of a good green colour, and arranged along the midrib in the manner of thoſe of the Burnet. The flowers will be produced from the branches in vaſt profuſion; and though they are all ſingle, they make a ſhow inferior to few ſhrubs. In winter they will be full of heps that have the appearance of blackberries; and if the weather be mild, the young buds will ſwell early, and appear like ſo many little red eyes all over the ſhrub, which is a promiſe of the reviving ſeaſon. The young branches of all theſe ſorts are exceedingly full of prickles.

4. Alpine Rose. This is uſually called the Roſe without Thorns, the branches being perfectly free from all kinds of prickles. They are exceedingly ſmooth, of a reddiſh colour, and look well in winter. The flowers are ſingle, and of a deep red colour. They come out in May, before any of the other ſorts; and the plant is valued by ſome people on that account. They are ſucceeded by long narrow heps, which look ſingular, and, together with the early appearance of their flowers, and their beautiful twigs, that are wholly free from the armature of the other ſorts, cauſe this ſpecies to be much admired.

5. Eglantine, or Sweetbriar. The *Varieties* of this ſpecies are, Common Sweetbriar, Semi-double Sweetbriar, Double Red Sweetbriar, Maiden Bluſh. Double Sweetbriar, Sweetbriar with Yellow Flowers.

The *Common Sweetbriar* is well known all over England. The branches, which are of a reddiſh caſt, are all over cloſely armed with prickles; the flowers are ſingle, and of a pale red colour, like thoſe of the Common Wild Briar. The leaves conſtitute the value of this plant; for they are poſſeſſed of ſo grateful an odour, as to claim admittance for this ſort into the firſt claſs of aromatic plants: the odoriferous particles they emit are ſweet and inoffenſive; and they beſtow them in ſuch profuſion, eſpecially in evenings or after a ſhower, as to perfume the circumambient air to a conſiderable diſtance. For this reaſon, plenty of Sweetbriars ſhould be planted near much-frequented walks;

or if the borders of theſe are deſigned for more elegant flowering ſhrubs or plants, they may be ſtationed at a diſtance, out of view, and then they will ſecretly liberally beſtow their ſweets, to the refreſhment of all. For noſegays, alſo, there is nothing more proper than ſprigs of the Sweetbriar, when diveſted of its prickles; for they will not only have a good look as a fine green in the center of a poſy, but will improve its odour, let the other flowers of which it is compoſed be what they will.

Semi-double Sweetbriar differs in no reſpect from the Common, only that the flowers conſiſt of a double ſeries of petals that ſurround the ſtamina. The leaves are poſſeſſed of the ſame fragrance; but this ſort is thought more valuable on account of the flowers, which, being poſſeſſed of more petals, make a better figure.

Double Sweetbriar. The number of petals are ſo multiplied in this ſort as to form a full flower; and it ſeems to differ in no other reſpect from the other Sweetbriars. The flowers are red, and ſo large and double as to be equal in beauty to many of the other ſorts of Roſes. As by the fragrance of their leaves they afford us a continual treat during the ſummer months, as well as by their fair flowers at the time of blowing, all who pretend to make a Collection are careful of procuring plenty of this ſort.

Double Bluſh Sweetbriar is a moſt valuable, and at preſent a very ſcarce, plant. It ſeems to have a tendency not to grow ſo high as the other ſorts of Sweetbriars. The branches are green, and cloſely armed with ſtrong prickles. The flowers are of a pale red or bluſh colour, and every whit as double as the Cabbage Provence Roſe: it cabbages in the ſame manner, and is very fragant. No one need be told the value of a Roſe which has every perfection and charm, to the higheſt degree, both in the leaves and flowers, to recommend it.

Sweetbriar with Yellow Flowers. The flowers of this ſort are ſingle; the petals are of a bright yellow colour; but it differs in no other reſpect from the Common Sweetbriar.

6. CINNAMON ROSE. The *Varieties* of this ſp cies are.

are, Single Cinnamon Rofe, Double Cinnamon Rofe,

The *Single Cinnamon Rofe* is a much ftronger fhooter than the Double fort, which is better known. It will grow to be ten or twelve feet in height. The young branches are of a reddifh colour. The flowers are fingle, and have the fame hue as thofe of the Double. It is rather a fcarce plant at prefent; on which account chiefly it is thought valuable.

The *Double Cinnamon Rofe* will grow to about fix or feven feet high, and the branches are many and flender. The prickles are pretty numerous, and the young fhoots in winter are of a red colour, with a purplifh tinge. This fort, which ufhers in the flowery tribe of Double Rofes, will be in blow fometimes pretty early in May. The flowers are fmall, but very double: they are of a purplifh red, very fweet, and have a little of the fmell of cinnamon, which occafions this Rofe to be fo called; and on that account only, not to mention their early appearance, this fort is defirable.

7. CAROLINA ROSE. The *Varieties* of this fpecies are ufually called, Wild Virginian Rofe, Pennfylvania Rofe, Pale Red American Rofe.

The *Wild Virginian Rofe* will grow to be nine or ten feet high. The branches are covered with a fmooth red bark, and guarded by a very few prickles. It produces its flowers in Auguft, when moft of the other forts are out of blow, and is by many valued for that reafon. The flowers are fingle, of a red colour, are produced in clufters, and will continue blowing from the beginning of Auguft until October. Neither is this the fole beauty this fort affords us; for the flowers will be fucceeded by heps, which in winter appear like fo many red berries all over the fhrub. Thefe heps ferve as food for birds, and are therefore much frequented by thrufhes and others of the whiftling tribe, who will be ready to ufher in, by their fweet warbles, the earlieft dawn of fpring. This tree grows wild in Virginia, and many parts of North America, from whence we receive the feeds, and propagate it not only on fome of the above accounts, but becaufe it is naturally an upright well growing tree, and makes a good figure in winter by its red and beautiful fhoots.

The *Pennſylvania Roſe* ſeems to differ in nothing from the former, except its ſize, it ſeeming to be a plant of lower growth; and the Pale Red ſort occaſions variety only from the lobes of the flowers.

8. Apple Rose. This ſpecies is a curioſity, not ſo much from the ſingularity of the ſhoots, leaves, or flowers, as fruit. The ſhoots, indeed, will be ſtrong and bold, and in winter diſtinguiſh the tree from others by a degree of eminence. They are then covered with a ſmooth reddiſh bark; and the prickles which guard them are thinly placed, though thoſe are very ſtrong and ſharp. Many think this tree has a good look in winter, and value it much on that account. As to the leaves, they are nearly the ſame as the other ſorts of Roſes; but are large, and very hairy, and downy underneath. The flowers are ſingle, of a red colour, and are ſucceeded by heps as large as little apples. To their account the value chiefly of this ſort is to be placed; for being thus large, they occaſion a ſingular look; and this is heightened by being all over beſet with ſoft prickles. For uſe as well as beauty this ſort is propagated by ſome; for theſe heps or fruit, when preſerved, make a ſweetmeat greatly eſteemed.

9. Hundred-leaved Rose. This is a very extenſive ſpecies, and includes all *Varieties* whoſe ſtalks are hiſpid, prickly, and have leaves growing on footſtalks which are not armed with prickles; and whoſe flowers have oval, hiſpid germina and footſtalks. Of this kind are, the Deep Red Provence, the Pale Red Provence, the Large Cabbage Provence, the Dutch Provence, the Childing Provence, the Moſs Provence, the Great Royal Roſe, the Bluſh Hundred-leaved Roſe, the Dutch Hundred-leaved Roſe.

The *Provence Roſes* are all well known. The Red and the Pale Provence ſorts differ, in that one is a deep, the other a pale red; the petals are larger and looſer than the Cabbage Provence, and make varieties. The Cabbage Provence is the beſt of all the ſorts; and if its commonneſs does not detract from its value, is inferior to no Roſe. The Dutch Provence has a tendency to cabbage, and is of a deeper red than the Common Provence. The Childing is of lower growth than any of the other ſorts, ſeldom growing to be more than

than four feet it is naturally of upright growth, and the bark is brown and prickly. The flowers at firſt are globular, though they will afterwards open at top, and diſplay their petals folded a little like thoſe of the Belgic. All theſe are beautiful roſes, and greatly ornamental either to ſhruberies or gardens.

The *Moſs Provence* is a ſort that has been ſought after of late more than any of the others Its branches are of a duſky brown, and they are all over cloſely beſet with prickles. The flowers are like thoſe of the Common Provence; though they have a ſtronger footſtalk, and grow more upright. About the calyx of the flower grows a kind of moſs, which is of a yellowiſh green colour, and by which it will be wholly ſurrounded. This Roſe has not been many years known in England, and from whence it was firſt brought is uncertain. It ſeems to owe its excellence to the moſſy ſubſtance growing about the footſtalk and calyx of the flower; but were this as common as the other ſorts of Provence Roſes, that would be looked upon as an imperfection; for though this flower naturally is poſſeſſed of the ſame agreeable fragrance as the other Provence Roſes, yet this moſſy ſubſtance has a ſtrong diſagreeable ſcent, and is poſſeſſed of a clammy matter.

Great Royal Roſe is one of the largeſt, though not the compacteſt, Roſes we have. It will grow to be eight or nine feet high. The branches are brown, and have a number of prickles. The flowers are red, and poſſeſſed of a very grateful odour, and the petals very large. Upon the whole, this is a ſort very much coveted, and is one of the beſt Roſes in England.

The *Bluſh and Dutch Hundred-leaved Roſes* differ in no reſpect, only that the flowers of one are of a paler red than thoſe of the other; and both theſe ſorts may contend for the prize of beauty with any of the Roſe tribe. They ſeldom grow more than four feet high. The branches are green and upright, and have very few ſpines. The flowers are large, and exceedingly double: Each is compoſed of numerous ſhort petals, which are arranged in ſo regular a manner as to form a complete flower; and it is on account of the extraordinary number of theſe petals that this Roſe takes the name of Hundred-leaved Roſe. We ſeem to do injuſtice to this Rofe, when we do not pronounce it the faireft of the whole lift; but when we reflect on the furpaffing delicacy and beauty of many other forts, we are obliged to give the preference to none.

10. Gallican Rose. Under this title are arranged all thofe Rofes whofe branches and footftalks of the leaves are hifpid and prickly, and whofe flowers have oval, hifpid germina, and grow on hifpid footftalks. Of this kind are,

The Semi-double Red Rofe,
The Old Double Red Rofe,
The *Rofa Mundi*, or Variegated Rofe,
The York and Lancafter Rofe,
The Semi double Velvet Rofe,
The Full-double Velvet Rofe,
The Blufh Belgic Rofe,
The Red Belgic,
The Blufh Monthly,
The Red Monthly,
The White Monthly,
The Striped Monthly,
The Red Damafk,
The White Damafk,
The Blufh Damafk,
The Doubled Virgin,
The Marbled,
The Great Spanifh,
The Yellow Auftrian Rofe,
The Copper-coloured Rofe,
The Double Yellow,
The Franckfort Rofe.

11. The Musk Rose. The Evergreen fort is naturally a climbing plant, but if planted fingly will form itfelf into a bufh of five or fix feet high: its flowers are fingle, white, and fragrant.

Befides the Evergreen, there are two deciduous *Varieties* of this fpecies, called, the Single Mufk Rofe, and the Double or Semi double Mufk Rofe.

Single Mufk, or White Clufter, is a fcarce and valuable Rofe. The young fhoots are covered with a fmooth green bark, and are not poffeffed of many fpines; thofe few they have are very ftrong, and of a dark brown colour. This fort produces its flowers in Auguft,

Auguſt, in very large cluſters; they are of a pure white; and the tree will continue to exhibit its ſucceſſion of flowers until the froſt puts a period to the blowing. The ends of the branches are frequently killed by the froſts in the winter; ſo that early in the ſpring they ſhould be gone over with the knife, and all dead wood taken off, which would have an ill look, amongſt the healthy leaves and young ſhoots.

The *Semi double and Double Muſk*, or White Cluſter Roſes are late flowering ſorts They will begin blowing in Auguſt, and continue ſo till the froſt puts an end to the glories of that ſeaſon. The ſtalks are covered with a ſmooth green bark, which will be armed with a few very ſtrong, brown, crooked ſpines. The flowers are of a pure white, and produced in large cluſters, at the ends of the branches. Theſe at preſent are not common, and are much coveted by the curious.

12. PENDULOUS-FRUITED ROSE grows only to about five or ſix feet high, ſending forth ſeveral hiſpid branches from the bottom to the top. The leaves are compoſed of many oval folioles, arranged along the midrib, and their footſtalks have few or no prickles. The flowers have oval, ſmooth germina, grow on hiſpid footſtalks, and are ſucceeded by long pendulent fruit, full of ſeeds.

13. The WHITE ROSE. The characteriſtics of this ſpecies are, the ſtalks and footſtalks of the leaves are prickly, the flowers have oval, ſmooth germina, and grow on hiſpid footſtalks. Of this kind are,

The Double White Roſe,
The Semi-double White,
The Dwarf White,
The Maiden's Bluſh Roſe.

All the ſorts of Roſes are to be PROPAGATED, 1. By layers. For this purpoſe, in order to obtain plenty of them, a ſufficient number ſhould be planted for ſtools; and after theſe have been planted a year or two, they ſhould be headed near the ground, which will make them throw out plenty of young ſhoots. In the autumn, theſe ſhould be layered in the ground. The beſt way to do it is by a ſlit at the joint, though a gentle twiſt will often do as well, particularly for all the ſorts

 of

of Monthly Roſes, Damaſk Roſes, and Sweetbriar, which will readily take if the bark be juſt broke, and will often ſend forth roots at every joint by the autumn following. Moſt of the other ſorts do not ſtrike root ſo freely ; ſo that amongſt them, by the autumn after layering, few will be found ſtrong enough, and with root ſufficient to be planted out to continue. However, in general, they will have roots, and oftentimes very good ones. In the autumn every layer muſt be taken up, the ſtools neated up, and a freſh operation performed on the young ſhoots that may have ſhot the preceding ſummer. The layers that have been taken up ſhould be planted in the nurſery, at no very great diſtance, and the ſorts ſhould be kept ſeparate and booked, number ſticks being made to the ſeparate ſorts, that they may be diſtinctly known. The Moſs Provence and the Muſk Roſes do not ſtrike root ſo freely by layers ; neither does the Apple-bearing Roſe ; ſo that for all theſe ſorts you muſt often wait two years before you take off the layers from the ſtools, and ſometimes longer ; which is the reaſon of theſe plants being rather ſcarce, they not being to be expeditiouſly propagated in plenty. 2. Theſe trees may be propagated by ſuckers, which moſt of the ſorts have a natural tendency to throw out ; and theſe may be taken up, and the ſtrongeſt and beſt rooted ſet out to ſtand, whilſt the weakeſt may be planted in the nurſery for a year or two, to gain ſtrength. But here we muſt obſerve that the Moſs Provence, Muſk, and Apple-bearing Roſes ſeldom throw out ſuckers ; ſo that we muſt not wait for them from theſe ſorts, but muſt get forward with our layering. 3. The Common Sweetbriar is to be propagated by ſeeds. Theſe ſhould be ſown as ſoon as they are ripe, in a bed of common garden mould made fine. They generally remain until the ſecond ſpring before they come up, and afterwards will require no other care than weeding until the ſpring following, when they may be taken up, and planted in the nurſery at ſmall diſtances ; and in two or three years time they will be good plants for the ſhrubery, wilderneſs, or hedges. And indeed as great quantities of theſe odoriferous plants are often wanted, this is the eaſieſt and moſt expeditious way of raiſing them in plenty.

By

By ſeeds alſo the Burnet-leaved, Apple-bearing, and Red or White Scotch Roſes may be raiſed; which are doubtleſs diſtinct ſpecies, and will preſerve the ſorts by ſeeds.

RUBUS.

LINNEAN Claſs and Order, *Icoſandria Polygynia*: Each flower contains about twenty males and many females. There are eighteen SPECIES; four of which are applicable to our purpoſe:

1. RU'BUS *Frutico'ſus*: The COMMON BRAMBLE; *a well known trailing plant*; common in moſt countries in Europe.

2. RU'BUS *Hiſpidus*: The CANADA BRAMBLE; *a trailing plant*; native of Canada.

3. RU'BUS *Cæ'ſius*: The DEWBERRY or CÆSIUS; *a trailer*; native of moiſt places in moſt parts of England and Europe in general.

4. RU'BUS *Odora'tus*: The VIRGINIA RASPBERRY; *a deciduous ſhrub*; native of Virginia and Canada.

1. The COMMON BRAMBLE admits of the following *Varieties*:

The Double-bloſſomed Bramble, the Bramble without Thorns, the Bramble with White Fruit, the Cut-leaved Bramble, the Variegated Bramble.

The *Double-bloſſomed Bramble* differs in no reſpect from the Common Bramble, only that the flowers are very double. The ſtalks, like that, are cloſely armed on all ſides by ſtrong crooked prickles, that turn backwards. They are, like that, channelled; and in the winter have ſome of a reddiſh purple colour, others green, ſome red on one ſide and green on the other. The leaves alſo are ſhaped like the hands, and are compoſed ſometimes of three, ſometimes of five lobes. They have their upper ſurface ſmooth, and of a fine green colour, whilſt their under is of a whitiſh colour. The footſtalks that ſupport them are prickly, and a ſeries

series of prickles are arranged all along the midrib of each lobe. They continue on the plants most part of the winter, at the beginning of which they are green; but after Christmas they turn brown, and seldom look well after. This is the description of the Common Bramble, and of the Double sort also, which differs in no other respect than in the doubleness of the flower. They are produced in the same manner at the ends of the shoots, each of which is exceedingly double. The petals are whiter and as a profusion of these ornament the ends of most of the shoots in the same manner as the flowers of the Common sort, they make a show, and are beautiful beyond expression. It may be kept down and confined, to have the appearance of a flowering shrub. The flowers are succeeded by no fruit. It will thrive and flower exceedingly well under the drip of trees; so that for old plantations, this is an useful plant for the under shrubs,. as it will flourish where hardly anything else will grow.

Bramble without Thorns is not near so strong a shooter as the Common Bramble, the shoots being more trailing and slender, perfectly smooth, and of a blueish colour; and on this account it is that this plant is held as a curiosity. A curiosity, indeed, it is; and many have expressed their agreeable surprize to find a Bramble that they could familiarly handle without hurt The leaves of this sort have a blueish tinge, and the footstalks and midrib are intirely free from prickles. It flowers in the same manner as the Common Bramble, though the flowers are rather smaller; and are succeeded by black berries, on which the insects do not seem to swarm in such plenty as they do on the other sort.

Bramble with White Fruit is deemed curious only on that account, and has (says HANBURY) often given occasion to a hearty laugh, by a bull which has been made by many on their first seeing this fruit, who have cried out with surprize, "Here is a Bramble that bears white blackberries." It is, therefore, the colour of the fruit that makes this sort coveted, though the leaves are of a lighter green than any of the other sorts, and on that account make a variety among the leafy tribe.

Bramble with Cut Leaves differs from the Common only in that the leaves are cut in an elegant and beautiful

tiful manner. It affords a variety in no other reſpect; and thoſe that are fond of ſuch, are ſure of meeting one in this, whoſe leaves being thin and elegantly cut, make the plant have a different look from the other ſorts.

Variegated Bramble differs in no reſpect from the Common Bramble, only it is a weaker plant. The leaves are ſtriped; and it is valuable only to thoſe who are fond of variegated ſhrubs.

2. AMERICAN BRAMBLE. The ſhoots of this ſpecies are long, ligneous, procumbent, rough, and hairy. The leaves are trifoliate. naked, cut at the edges, ſerrated, and grow on hiſpid footſtalks. The footſtalks of the flowers alſo are hiſpid. They come out from the ends and ſides of the branches, in July and Auguſt; and are ſucceeded by round reddiſh fruit in the autumn.

3. CÆSIUS, Small Bramble, or DEWBERRY BUSH. The ſtalks of this ſort are weak, ſlender, prickly, and trailing. The leaves are trifoliate, large, and uſually of a duſky green colour. The flowers are whitiſh, come out from the ends and ſides of the branches, in July and Auguſt, and are ſucceeded by large blue fruit, which will be ripe in the autumn, and of which an excellent wine is made.

All theſe ſorts may be PROPAGATED by cuttings. They ſhould be planted in the autumn, in a ſhady border, and by the autumn following they will be fit to remove. But as a crop from cuttings often fails, the beſt way will be to throw ſome mould over the ſhoots, as they ſtrike in the ſpring; and when they have ſhot two or three feet farther, cover them afreſh, and ſo on all ſummer. By this means, thoſe parts that were firſt covered will have either ſtruck root, or they, together with all the others, will be preparing to ſtrike root; ſo that, being cut into lengths, and the parts before covered planted again in earth, and about three or four inches of the uncovered part being above ground, almoſt every one of the cuttings of this nature being thus prepared will grow, and thus plenty of plants may be ſoon obtained.

4. The VIRGINIA RASPBERRY. All the ſorts of Raſpberries are ſpecies of *Rubus*, and are propagated for their fruit; but this ſort is cultivated ſolely to mix with

with our flowering ſhrubs. It riſes from the ground like the Common Raſpberries, though it will naturally grow higher; but its growth is either higher or lower in proportion to the nature of the land or ſituation, as it will grow higher by two or three feet in a deep, rich, moiſt ſoil, than it will in a ſoil of the oppoſite nature. The ſtalks are of a brown colour, and wholly without prickles; and the ſtrongeſt will divide into ſeveral ſmaller branches. The leaves are exceedingly large for a ſhrub of that height; from whence the plant derives no ſmall beauty. They are broader than they are long, and of a fine green on both ſides, the upper being of a dark, the under of a lighter colour. Each is divided into an uncertain number of lobes, which are ſerrated, and end in acute points. Theſe leaves grow alternately on footſtalks that are of a proportionable length and ſtrength to the ſize of the leaves, they being often eight or nine inches broad, and ſeven or eight in length. The flowers are produced in July, in plenty, at the end of the ſtalks; and the ſucceſſion will be continued for often more than two months; though they are always the moſt beautiful on their firſt appearance. They are of a purpliſh red, a colour which is very deſirable at that time, when moſt of the other ſhrubs that are in blow will have yellow flowers. Each ſtands on a long footſtalk; and many of them being collected into a kind of looſe bunch, they make a tolerable figure. They are ſeldom ſucceeded by any fruit with us; and when this happens, it is of no flavour, and on that account of no value.

It is eaſily PROPAGATED from the ſuckers, which it ſends forth in ſuch abundance, that from a few plants, in a few years, almoſt any deſired quantity may be obtained: nay, ſo faſt do they creep and ſend forth ſtalks on all ſides, that, unleſs they are conſtantly taken up as they grow, they will ſoon overſpread and choke all ſmaller plants that grow near them. The beſt time for taking off the ſuckers is the autumn; though they will grow very well if planted either in the winter or ſpring.

R U S C U S.

LINNEAN Claſs and Order, *Dioccia Syngeneſia:* Male flowers containing three ſtamina, and female flowers containing one piſtil; upon diſtinct plants. There are four SPECIES:

1. RUSCUS *Aculea'tus:* The COMMON BUTCHER'S BROOM; *an evergreen ſhrub*; native of England, Italy, and France.

2. RU'SCUS *Hypophy'llum:* The BROAD-LEAVED BUTCHER'S BROOM; *an evergreen ſhrub*; native of Italy.

3. RU'SCUS *Hypoglo'ſſum:* The HYPOGLOSSUM; *an evergreen ſhrub*; native of Italy and Hungary.

4. RU'SCUS *Racemo'ſus:* The ALEXANDRIAN LAUREL; *an evergreen ſhrub*; native place not known.

1. The COMMON BUTCHER'S BROOM will riſe with tough, ligneous, ſtreaked, green, ſpreading ſtalks, to about a yard in height. Theſe proceed from a large, white, tender, creeping root, which will, if the plant has remained long, be found very deep in the ground. The leaves are of an oblong figure, of a dark duſky green colour, and grow alternately on the ſtalks. Their edges are intire; they are of a thick ſtiff conſiſtence; and their points are prickly, and as ſharp as needles. The flowers grow on the middle of the upper ſurface of the leaves, and will be ripe in June. They are ſmall and greeniſh; and the females are ſucceeded by large beautiful red berries, of a ſweetiſh taſte. This plant is of great uſe to the butchers, who gather it to make different beſoms, both for ſweeping of their ſhops and cleaning of their blocks; from whence it has the appellation of Butcher's Broom. The young tender ſhoots of this ſhrub, in the ſpring, may be eaten like hop-tops, or aſparagus, and ſome people are very fond of them. The ſeeds and roots are much uſed in medicine.

2. The BROAD-LEAVED BUTCHER'S BROOM has large white roots, with long thick fibres, and from theſe

riſe

rise pliable stalks, which will grow to be near a yard high. These stalks are of a very fine green colour, and are very tough and numerous. They produce their leaves in an alternate manner, are of a very fine shining green colour, and of a thick consistence They are longer and broader than the other sort; their figure is oval, and they end in acute points. The flowers of this sort grow on the under surface of the leaves, near the middle. These are small, and of a greenish white. They are produced in July; and the seeds that succeed them are small and red, and will be ripe in winter.

3. The HYPOGLOSSUM is the lowest of all the sorts, as the stalks seldom get to above a foot high, and has very few pretensions, indeed, to be called a shrub; nevertheless, it may justly claim a place at the edge at least of all evergreen shruberies. The roots are nearly of the same nature with the other sorts, and the stalks are numerous and pithy. They are of a dull green colour, and striated; and they produce their leaves in an irregular manner, being sometimes alternate, whilst others again may be seen standing opposite by pairs. These leaves are of a lanceolated figure, and are of the same dull green colour with those of the stalks. They are from three to four inches long, and about one broad. They grow without any footstalks, being narrow at both ends, and their edges naturally turn towards the center of the upper surface. They are free from serratures; and from the stalk or base of the leaves run several veins the whole length, which gradually diverge from the middle, but approach again in the same manner until they all end in the point of the leaf. Each of these leaves produces another small leaf of the same shape, from the middle of its upper surface; and from the bottom of these small leaves are produced the flowers. These will be ripe in July, are small and yellowish, and the fruit that succeeds them is large and red, and will be ripe in winter.

4. The ALEXANDRIAN LAUREL has the same kind of white scaly roots with long thick fibres as the others, and the branches are very numerous and pliable. They are smooth and round, of a shining green colour, and produce others smaller, alternately from the bottom to the top. They will grow to be four or five feet high, and

and their pliable branches are neverthelefs brittle near the bottom. The leaves grow chiefly on the fmaller fide fhoots, and on thefe they are placed alternately. They fit clofe to the branches, are fmooth, of a delightful fhining green colour, and have feveral fmall veins running the whole length, diverging from the middle, but approaching again to end at the point. They are from two to three inches long, and about one broad, are of an oblong lanceolated figure, and end in very acute points. The flowers are produced in long bunches, at the ends of the branches. Each of them is fmall, and of a yellowifh colour; and they are fucceeded by large red berries, which will be ripe in winter.

There is a *Variety* of this fort with red flowers. "This fpecies of *Rufcus*," fays HANBURY, "is fuppofed to be the Laurel which compofed the wreaths worn by the antient victors and poets; and indeed with good reafon, not only on account of its pliablenefs, by which it might be eafily wrought for fuch purpofes, but the wreaths on the antient bufts, &c. feem to figure to us the leaves and flender branches of the plant we are treating of."

There is another fort of *Rufcus*, which has oval acute-pointed leaves, growing by threes round the ftalks, and which produce the flowers and fruit from the midrib, on the under furface; alfo another fort, with oval acute-pointed leaves, which produces the flowers from the midrib, on the upper furface. But as thefe are only Varieties of the above forts, have the fame kind of roots, produce the fame kind of flender pliable branches, and have their flowers fucceeded by nearly the like kind of berries, nothing more need be faid of them.

All thefe forts may be eafily PROPAGATED. 1. After having obtained a plant or two of each, their roots will increafe fo faft, and will proportionally fend forth fuch a quantity of ftalks, that each of them will foon form itfelf into a little thicket: thefe, then, are to be taken up and divided; and from one original root or off-fet many will be foon produced. The beft time for this work is early in the autumn; though they will grow very well if divided and removed in the fpring, or any

time

time in the winter. 2. These plants are also to be encreased by seeds. This, however, is a slow way; but must, nevertheless, be practised, when the plants cannot be obtained. The beds for their reception must be made fine, and cleared of the roots of all weeds. They will require no other compost than that of good common garden mould. They should be sown an inch and a half or two inches deep, and the beds should be neated up to lie undisturbed, for they will not come up before the second, and sometimes the main crop the third, spring after sowing. All the summer they should be kept clean from weeds; and if the beds wear away so as to endanger the seeds being laid bare, a little fine mould should be riddled over them, to supply what may be lost by wear in weeding, settling, &c. After they are come up, they will require no other care than weeding, for they are very hardy; and when they come too thick in the spring after the frosts are over, the strongest should be drawn out and planted in beds six inches asunder. This will make room for the others to flourish; and though mention is made of removing these plants after the frosts are over, it is not because they are tender and subject to be destroyed by it, but if they are removed in the autumn, or early in the winter, being then small, the frosts generally throw them out of the ground, to the great danger, if not intire loss, of the whole stock of the new-removed seedlings. This, however, is considered by few Gardeners who have not paid dear for their experience, and is what is chiefly recommended by our modern authors, to transplant seedlings of most sorts from the beds in October; which, indeed, would be an excellent month, were no frosts to ensue. But good thought and experience, by fatal practice, have taught the Gardener now, to defer the removing his small seedlings until the spring, when they will not be liable to be turned out of their warm beds when they should least like it, by the rigours of the winter. But to return: After the seedlings are two or three years old, whether they have been removed or not, they will by that time be good strong plants, fit for removing, and may be then taken up and planted out.

S A L I X.

LINNEAN Clafs and Order, *Dioecia Decandria:* Male flowers containing two ftamina, and female flowers containing one piftil, upon diftinct plants. There are feveral SPECIES; fifteen of which are cultivated in this country.

1. SA'LIX *A'lba:* The COMMON WHITE WILLOW; *a deciduous tree*; common about towns and villages in moft parts of Europe.

2. SA'LIX *Vitelli'na:* The GOLDEN WILLOW; *a low deciduous tree*; native of England and moft parts of Europe.

3. SA'LIX *Purpu'rea:* The PURPLE WILLOW; *a deciduous tree*; native of England and the South of Europe.

4. SA'LIX *Penta'ndria:* The SWEET WILLOW; *a deciduous tree*; native of mountainous and marfhy fwampy grounds in moft parts of Europe.

5. SA'LIX *Babylo'nica:* The WEEPING WILLOW; *a deciduous tree*; native of the Eaft.

6. SA'LIX *Hermaphrodi'tica:* The SHINING WILLOW; *a deciduous tree*; grows about Afton in Cumberland, and alfo Upfal in Sweden.

7. SA'LIX *Tria'ndria:* The TRIANDROUS WILLOW; *a deciduous tree*; native of Switzerland and Siberia.

8. SA'LIX *Phylicifo'lia:* The PHYLICA-LEAVED WILLOW; *a low deciduous tree*; native of the North of Sweden.

9. SA'LIX *Amygdali'na:* The ALMOND-LEAVED WILLOW; *a deciduous tree*; native of England and moft parts of Europe.

10. SA'LIX *Hafta'ta:* The HASTATED WILLOW; *a deciduous tree*; native of Lapland and Switzerland.

11. SA'LIX *Fra'gilis:* The CRACK WILLOW; *a deciduous tree*; native of England and the North of Europe.

12. SA'LIX *He'lix:* The ROSE WILLOW; *a low de-*

ciduous tree; native (though not common) of England and the Southern parts of Europe.

13. SA'LIX *Ca'prea:* The SALLOW; a well known *low deciduous tree*; native of England and most parts of Europe.

14. SA'LIX *Vimina'lis:* The OZIER; *a low deciduous tree*; native of England and most parts of Europe.

15. SA'LIX *Glau'ca:* The GLAUCOUS WILLOW, or ALPINE SALLOW; *a deciduous shrub or tree*; native of the Alps of Lapland and the Pyrenees.

1. The WHITE WILLOW. This is a tall growing tree, and being universally known needs no description. The silvery elegance of its leaves would render it very *ornamental*, were it not for its too great commonness: it is a quick grower, and its wood is *useful* when lightness and a cleanness of grain is required *.

2. The

* HANBURY, speaking of Aquatic Forest Trees, says, " The sorts used for plantations of these trees have hitherto been our Common White and Red Willow. These, however, seem now to give place to more sorts, which have been lately introduced. A few years ago I saw in the public papers an advertisement of a Willow which would grow large enough for masts of ships, &c. in twenty or thirty years; and in another paper there was an account, that these trees might be seen in full maturity at one Squire Angel's, about three miles from Westminster Bridge. I went to examine them, but when I came found them the Common White Willows, which, having liked the situation, had grown to a great size and beauty. I enquired out the author of the advertisement, but found he knew nothing of the nature of these Willows, and that he had his account from a basket maker near Westminster Bridge. Upon applying to the basket maker, he disavowed knowing anything of the trees growing by Mr. Angel's, but said he had two sorts of Willows, which would answer in every respect to the first advertisement; that they were of all others the freest shooters; that they were not so subject to rot in the sides as the large White Willow Tree; but that they would grow sound to timber, fit for masts of ships, &c. in less than thirty years. He added, that he had cuttings many years ago brought him from the coast of France, by a Captain whose name I have forgot. I immediately procured some cuttings of these sorts, which grow to a miracle, and seem as if they would answer the promised expectation; so that these now are the trees of which our future timber plantations should consist: nay, whether they are designed for the basket makers or for hurdles, they ought to have their share; and should always be preferred to be planted out for standards for lopping, by the sides of rivers, rills, ditches, &c. The cuttings of these two sorts have been dispersed into

almost

2. The GOLDEN WILLOW may be admitted into ornamental plantations, not for any extraordinary figure these trees will make in summer, but from the show they make in winter; for their bark is smooth, and of a clear yellow; and in that season they have a singular and striking effect among other trees. This will not grow to near the size of the other sort.

3. The PURPLE WILLOW or RED WILLOW is a free shooter, and will grow to a size almost as large as the Common White Willow. A few of these only should be admitted into our plantations; for they have no singular look in summer; but in winter their bark appears of a red colour, which makes a pretty variety among other trees at that season; but it is, nevertheless, not near so striking as the yellow sort.

4. The SWEET-SCENTED WILLOW. This will grow to be a large timber tree, and the branches are covered with a smooth brown bark. The leaves of this sort resemble those of the Bay Tree, and are by far the broadest of any of the sorts of Willows. They are smooth, and have their upper surface of shining green; but their under surface is paler, and they are serrated at their edges. They emit, especially when bruised, a grateful odour; so that as an aromatic it claims a place in these plantations among others of its own growth. Indeed it deserves it; for air will frequently be perfumed by the fragrance of its leaves after a shower to a considerable distance; so that it will readily join with other aromatics in perfuming the air with their spicy odours. It delights in a very wet situation *.

5. The WEEPING WILLOW of Babylon will grow to be a large tree; and no tree is more proper to be planted by rivers, ponds, over springs, &c. than this;

almost every quarter of England; so that there is no doubt but that in a few years the planting of them alone for timber will become general, as they may be encreased at pleasure, by every slip or twig.

In March 1786 we measured a Willow growing near the Cathedral in Lichfield, whose girt was near fourteen feet; then in full growth.

* On the 4th of December 1793, the Sweet Willow was throwing off its seed! the tops of the trees white with down, bearing full-grown seeds, blowing about as snow! Q. Was this owing to the mildness of the season?

for its slender branches are very long and pendulous; the leaves, also, are long and narrow; and when any mist or dew falls, a drop of water will hang at the end of each of these leaves, which, together with the pendulous branches and leaves, cause a most pleasing appearance. Lovers garlands are said to have been made of the wreaths of this Willow, the branches of which are very slender and pliable; and the plant itself has always been sought after for ornamental plantations, either to mix with others of the like growth in the largest quarters, or to be planted out singly over springs, or in large opens, for the peculiar variety they will occasion by the elegance of their outline.

6. Shining Willow is a large growing tree, sending forth several slender branches, which hang down, and are covered with a pale brown bark. The leaves are smooth, glandulous, serrated, and of a yellowish green colour. The flowers are numerous hairy catkins, and the male flowers have two stamina only. They appear early in the spring; and the females are succeeded by downy seeds, like the Common Willow.

7. Triandrous Willow is a large growing tree, sending forth numerous erect, flexible branches, which are covered with a grayish bark. The leaves are oval, smooth, spear-shaped, acute-pointed, serrated, green on both sides, and eared at their base. The catkins are long, narrow, loose, and appear early in the spring. This sort is planted by the basket makers, to mix with other kinds for their different sorts of work.

8. Phylica-leaved Willow. This is a tree of rather lower growth than the former. The branches are numerous, flexible, tough, and serviceable for several articles in the basket way. The leaves are spear-shaped, smooth, serrated, and waved on their edges. The flowers are long catkins, which come out early in the spring from the sides of the branches; and they soon afford a large quantity of down, which is wafted about with the winds to a considerable distance.

There is a *Variety* of this with broad leaves.

9. Almond-leaved Willow. This is a Willow of the middle size, sending forth numerous flexible tough branches, covered with a light green bark. The leaves are spear-shaped, smooth, serrated, acute, eared at their

their baſe, and of a light green colour on both ſides. The flowers are oblong catkins, which turn to a light down in the ſummer.

There are ſeveral ſorts of this ſpecies, that are of inferior value to this, which is generally diſtinguiſhed from the others by the name of the *Old Almond-leaved Willow*. The branches are very tough and flexible, and when planted in the Ozier way, and grown to be one year's ſhoots from the ſtools, are very ſtrong, and highly ſerviceable for the different purpoſes of baſket making.

10. HASTATED WILLOW. This is a middle-ſized tree for the Willow kind, ſending forth ſeveral long, green ſhoots from the ſtools, which are full of pith, but nevertheleſs tough, and ſerviceable to the baſket maker. The leaves are nearly oval, acute, ſmooth, ſerrated, ſit cloſe to the branches, and have broad appendices at their baſe. The flowers are an oblong, yellow catkin, and come out in the ſpring from the ſides of the young ſhoots, almoſt their whole length.

11. CRACK WILLOW is another middle-ſized tree for the Willow kind. The branches are very brittle, and covered with a browniſh bark. The leaves are oval, ſpear-ſhaped, long, ſmooth, ſerrated, green on both ſides, and have glandulous footſtalks. The catkins are long, ſlender, and the ſcales are looſely diſpoſed.

There is a *Variety* of this ſpecies with a yellow bark, which it caſts every year, called the *Almond-leaved Crack Willow*. Both ſorts are unfit for the baſket makers uſe, being very brittle; on which account this ſpecies gained the appellation of Crack Willow.

12. ROSE WILLOW. This is of much lower growth than the former. The body of the tree is covered with a rough, yellow bark. The branches are upright, tough, and of a reddiſh colour. The leaves are ſpear-ſhaped, narrow, ſmooth, of a blueiſh green colour, and, towards the upper part of the branches, are nearly oppoſite to each other. The flowers come out from the ſides of the branches, and numbers of them are joined together in a roſe-like manner. They are of a greeniſh white colour, and have a ſingular and beautiful look.

There are two or three *Varieties* of this ſpecies. The leaves of one are downy underneath; the ſtalks of another are brittle, and the leaves green on both ſides; whilſt another has its leaves of a light green on the upper ſurface, and glaucous underneath. They are all low growing plants, and ſeldom cultivated for uſe.

13. SALLOW. The Sallow is well known all over England, and delights in a dry rather than a moiſt ſoil. It is a tree rather below the middle growth. The branches are numerous, ſmooth, of a dark green colour, and their chief uſe is for hurdle wood and the fire; though the trunk, or old wood, is admirable for ſeveral uſes in the turnery way. The leaves are oval, rough, waved, indented at the top, and woolly underneath. The catkins are very large, yellow, appear early in the ſpring, and are much reſorted to by the bees, on their firſt coming out of their hives at that early ſeaſon.

There is a *Variety* of this ſpecies with long leaves, which end in acute points; and another with ſmooth leaves, beautifully ſtriped with white, called the *Striped Sallow.*

14. OZIER is a tree of rather low growth, though the ſhoots grow amazingly long and ſtrong in one year from the ſtools. The leaves are ſpear-ſhaped, narrow, long, acute, almoſt intire, of a blueiſh green on their upper ſide and hoary underneath, and grow on very ſhort footſtalks. This is the moſt propagated of all the kinds for baſket making: it admits of ſeveral ſorts of different value, but all are nevertheleſs uſeful to the baſket maker.

The *Varieties* uſually go by the names of the *Green Ozier*, the *Old Baſket Ozier*, *Welſh Wicker*, &c. &c. &c. *

15. GLAUCOUS WILLOW. This is a low Alpine Willow, of little uſe for ornament or profit. The leaves are oval, oblong, intire, of a glaucous colour, and poſſeſſed of fine hairs on their under ſide. The catkins are large, oval, of a white colour, and appear about the time of thoſe of the Common Sallow.

* For the management of OZIER BEDS, ſee Vol. I. Article WOODLANDS.

All

All the SALICES may be PROPAGATED by planting the cuttings, which may be done at all times of the year, for they will grow if it is in ſummer; though the beſt ſeaſon is the winter, or early in the ſpring, juſt before they begin to ſhoot. The cuttings ſhould be of the laſt year's wood, ſhould be in height in proportion to their thickneſs, and always ought to be planted in an upright poſition.

SALSOLA,

LINNEAN Claſs and Order, *Pentandria Digynia:* Each flower contains five males and two females. There are ſixteen SPECIES; one only of which is adapted to our Collection.

SA'LSOLA *Frutico'ſa:* The SHRUBBY GLASSWORT, or the STONECROP TREE; *an evergreen ſhrub*; native of the ſea coaſts of England, France, Spain, and Perſia.

The STONECROP TREE is a ſhrub of about four or five feet growth. It will ſhoot rather higher, if permitted; but is never more beautiful than when about a yard high. The branches are numerous, naturally grow upright, are covered with a gray bark, and are very brittle. As to the leaves, they are very much like the Common Stonecrop of our walls, which is well known, being narrow, taper, and fleſhy like them. They are of the ſame light pleaſant green, and the branches are ſtored with them in plenty. The flowers make no ſhow; neither is there anything that is deſirable to the Gardener that ſucceeds them. This is a very hardy ſhrub; but, as we have introduced it as an Evergreen ſhrub, it may not be improper to give a hint or two for its being properly ſtationed. It ſhould be et in a well ſheltered place; for although the leaves remain on all winter, yet our ſevere black froſts ſuddenly coming on them, when in an open expoſed place, deſtroy them, and cauſe them to turn black; and although

 the

the ſhrub will ſhoot out again early in the ſpring, yet the black deſtroyed leaves will look very diſagreeable all winter, and be as blots among others that are leſs ſubject to theſe diſaſters. One hint more may be neceſſary; and that is, whenever this ſhrub is planted, either in ſmall or large gardens, among deciduous or evergreen trees, not to circumſcribe the tree, with ſtrings or baſs mattings, in order to confine the branches and keep them cloſer: this will effectually deſtroy all the branches and leaves, if not the whole plant; for being thus cloſely confined, the free admiſſion of the air will be excluded, which will cauſe theſe ſucculent leaves to rot and decay. This precaution is the more neceſſary, as their upright branches being heavy laden with ſuch plenty of ſucculent leaves, are ſubject to be blown down from the bottom by the high winds: and as they then muſt of courſe look irregular, and may probably overſpread ſome little plant that grows near them, it is a common thing to tie them up again to the other branches. This cuſtom, however, ought never to be practiſed; but when any of them happen to be blown down in that manner, they ſhould be taken off and thrown away.

Nothing is more eaſy than the PROPAGATION of the Stonecrop Tree; for it is increaſed by layers, cuttings, and ſuckers. In ſhort, if ſome of theſe ſhrubs are planted, they will ſoon ſend forth many ſtalks from the roots; and if the whole be then taken up, theſe, without any other trouble, may be divided, and will each of them be a good plant; and thus, in a few years, from a plant or two of this ſhrub, numbers may be obtained.

SAMBUCUS.

LINNEAN Claſs and Order, *Pentandria Trigynia*: Each flower contains five males and three females. There are four SPECIES; three of which are here treated of; the fourth, *Sambucus Ebulus*, or Dwarf Elder, is an herbaceous plant.

1. SAM-

1. SAMBU'CUS *Ni'gra:* The COMMON ELDER; *a deciduous ſhrub or tree*; common in moſt parts of England, but is ſaid to be originally a native of Germany.

2. SAMBU'CUS *Canadénſis:* The AMERICAN ELDER; *a deciduous ſhrub*; native of Canada, Pennſylvania, and Virginia.

3. SAMBU'CUS *Racemóſa:* The MOUNTAIN ELDER, or the MOUNTAIN RED-BERRIED ELDER; *a tall deciduous ſhrub*; native of the mountainous parts of the South of Europe.

1. The COMMON ELDER admits of many *Varieties:*

The Black Elder.
The White-berried Elder,
The Green-berried Elder,
The Parſley-leaved Elder,
The Gold-ſtriped Elder,
The Silver-ſtriped Elder,
The Silver-duſted Elder.

The *Common Black Elder* is too well known to require any deſcription. It will grow to thirty feet high, with a large trunk; and in this caſe its wood is very valuable. The leaves and flowers have a ſtrong and diſagreeable ſmell, which renders it improper to be planted near buildings or walks which are much frequented; but if they could be planted ſingly, or a ſmall clump of them, at a diſtance from any place of reſort, there is no tree in the world will make a grander figure, or be more ſtriking when in blow; for at that time they will be covered all over with large bunches of white flowers, which will aſſume an air of majeſty at that diſtance, equal to any of the flowery tribe. "Neither may a few of them only be ſtationed in this manner; but any acute corner of the plantation, that ſhews itſelf at a diſtance, may end with one of theſe trees; for there it will diſplay its gaudy pride when in blow, and the eyes of all be feaſted by its delicious appearance, whilſt the ſenſe of ſmelling is no way incommoded by its ſtrong diſagreeable ſcent."

The *White-berried Elder* differs from the former in that the berries are whiter; the bark, alſo, of the young ſhoots, is whiter; the buds, likewiſe, at their firſt appearance, are inclined to a whiter colour; the leaves, too, are of a paler green; and the plant in general has not

not ſuch a ſtrong diſagreeable ſcent, though it nevertheleſs has a proportionable ſhare. A plant or two only of this ſort is to be admitted, merely for variety; though where they are required for the ſake of the berries to make wine, a hedge of them may be planted, in a place that is little frequented, and they will plentifully furniſh the owner with berries for his purpoſe.

Green-berried Elder differs, in that the berries are green, the bark, alſo, of the young ſhoots, is of a darker gray than that of the White; and the buds, at their firſt appearance, have nearly as dark a colour as that of the Common Elder. We muſt have only a plant or two of this ſort for variety; and where the berries are wanted for wine, a hedge of them may be planted in ſome diſtant place, in the ſame manner as thoſe of the White ſort.

The *Parſley-leaved Elder* varies in no reſpect from the Common ſort, except in the nature of the leaves; which are laciniated in ſuch a manner as to reſemble the leaves of ſome ſorts of parſley. Theſe leaves occaſion a wonderful variety in ſhrubery quarters among the leafy tribe, and on their account the plant is deemed worthy of a place in any Collection; though the flowers poſſeſs the ſame nature with the Common ſort, and emit the ſame diſagreeable ſcent.

The *ſtriped* ſorts are diſtinguiſhed by their different coloured ſtripes; whilſt the Silver-duſted kind is remarkable for leaves finely powdered or duſted over, in a pounce-like manner, cauſing thereby a very beautiful and ſtriking appearance.

2. The AMERICAN ELDER is of a lower growth than any of the above ſorts, ſeldom riſing higher than eight or ten feet. The young ſhoots are of a reddiſh colour. The leaves on the lower part of the plant are trifoliate; others are compoſed of about two or three pairs of folioles, terminated by an odd one. Theſe folioles are ſerrated, and of a pleaſant green colour; neither do they emit ſo ſtrong a ſcent as any of the other ſorts. The flowers are produced in the ſame manner as the folioles; and are ſucceeded by berries of a reddiſh colour. Though theſe berries have not quite ſuch a ſtrong diſagreeable taſte as the Common Elder berries, yet they have a kind of phyſical flavour: nevertheleſs,

thelefs, they are liked by fome perfons, who are as fond of them as they are of fome forts of fruit. What was faid of the firft fort, recommending its being planted fingly, or in fmall clumps at a diftance, will hold good in all thefe forts, which when in blow will equally have the fame noble appearance as that, except the American, which is of lower growth, and confequently of lefs figure than the others, and as fuch lefs proper for the purpofe.

3. The MOUNTAIN ELDER will grow to about ten or twelve feet high, and is a tree that is with great juftice univerfally admired. The bark of the young fhoots is of a reddifh colour, and the buds in winter will be very large and turgid, and of a ftill deeper red. The leaves are pinnated with an odd one; their folioles are ferrated; they are placed at a good diftance on the midrib, which is pretty long; and they die to a reddifh colour in the autumn. The reddifh coloured branches, with their large turgid buds, have a fingular and noble look in winter amongft other trees; and in the fpring, as flowering fhrubs, thefe trees feem to attempt to vie with any of the flowering tribe; for in April, and the beginning of May, they will produce their bunches of flowers at the ends of every joint of the laft year's fhoots. Thefe bunches of flowers are of an oval figure; a figure in which compound flowers are not commonly produced. They are not, however, of fo clear a white as any of the other forts, being tinged with green; and although the tree will be covered with them, they have not the fame ftriking appearance; but this defect is made amends for by the peculiar form which they affume, and the fcarcity of the plant itfelf. Were there nothing but the above-recited properties to recommend this fhrub, it might juftly claim admiffion in plenty into our choiceft plantations: but thefe are not all its beauties; what remains is much more ftriking and engaging; for thefe oval bunches of flowers are fucceeded by oval bunches of berries, that are of a deep fcarlet colour. A crop, indeed, does not always enfue; but when it does, no tree is more fingularly beautiful than this is rendered by them, chiefly occafioned by their colour and form, which any one muft conceive to be delightful.

All

All the sorts of Elder are PROPAGATED by cuttings. These should be of the last year's shoot, and each cutting should consist of three joints; two of which must be in the ground, whilst the third is left above, to make the shoot. October is the best month for this business; and almost any soil will do, though the moister it is the better. These cuttings may be either planted very close, and removed the autumn following into the nursery ground, or they may be planted a foot or more asunder, and then they will be of a sufficient distance until they are finally taken up, which may be any time after two years. Thus easy is the culture of these plants when known.

SMILAX.

LINNEAN Class and Order, *Dioecia Hexandria*: Male flowers containing six stamina, and female flowers containing three pistils, upon distinct plants. There are fourteen SPECIES; eight of which are as follow:

1. SMI'LAX *A'spera*: The ITALIAN SMILAX, or COMMON ROUGH BINDWEED, or PRICKLY BINDWEED; *a low climber*; a native of Italy, Spain, Sicily, and France.

2. SMI'LAX *Exce'lsa*: The ORIENTAL SMILAX; *a lofty climber*; native of many parts of the East.

3. SMI'LAX *Sarsapari'lla*: The PERUVIAN SMILAX, or SARSAPARILLA; *a climber*; native of Peru, Mexico, and Virginia,

4. SMI'LAX *Rotundifo'lia*: The CANADA SMILAX; *a climber*; a native of Canada.

5. SMI'LAX *Laurifo'lia*: The LAUREL-LEAVED SMILAX; *a climber*; native of Virginia and Carolina.

6. SMI'LAX *Tamo'ides*: The BRIONY-LEAVED SMILAX; *a climber*; native of Carolina, Virginia, and Pennsylvania.

7. SMI'LAX *Lanceola'ta*: The LANCE LEAVED SMILAX; *a climber*; native of Virginia.

8. SMI-

8. SMI'LAX *Herba'cea*: The IVY-LEAVED SMILAX; *a climber*; native of Virginia and Maryland.

1. The ITALIAN SMILAX is poſſeſſed of a long, creeping, white, fleſhy root, which ſends forth many ſlender, angular ſtalks, armed with ſtrong, ſhort, crooked ſpines, and having claſpers. If any thing is near for it to climb on, it will, by ſuch aſſiſtance, arrive at the height of ten or twelve feet. The leaves are cordated, end in acute points, are of a fine dark green colour, indented, have nine longitudinal veins, have their edges beſet with ſome ſhort ſpines, and are placed on tolerably long tough footſtalks. The flowers make no figure: They are white, and are produced from the wings of the ſtalks, in ſmall bunches, in June or July; and the female flowers will be ſucceeded by round red berries.

There is a *Variety* of this ſpecies which produces black berries; and from which it differs in no other reſpect; and which occaſions its being called by Gardeners the *Black-fruited Rough Bindweed.* There is alſo another ſort with brown fruit.

2. ORIENTAL SMILAX is a lofty climber; for being planted near pretty tall growing trees, it will aſcend to their very tops, and proudly, by ſuch aſſiſtance, ſhew itſelf to a great diſtance. The roots are thick, white, and fleſhy; and the ſtalks are angular, and armed with ſpines. The leaves are of a pleaſant green colour, and are nearly of a ſagittated figure. They are poſſeſſed of no ſpines, have longitudinal veins, and their footſtalks are tolerably long and tough. Their flowers are white, and are produced in ſmall bunches, in June and July; and the females are ſucceeded by round red fruit in their own coúntries, but not with us.

3. PERUVIAN SMILAX, or SARSAPARILLA, has alſo white, thick, fleſhy roots. Theſe ſend out angular ſtalks, that are armed with ſharp ſpines; but they will not climb up trees to near the height of the former. The leaves are ſmooth, being unarmed with ſpines. They are retuſe, oval, cordated, of a ſtrong green colour, have three nerves, and grow on ſtrong rough footſtalks. The flowers are produced in ſmall bunches, from the ſides of the branches. They are of little

figure,

figure, and the females are fucceeded by a fmall, round, red fruit, where they grow naturally.

4. CANADA SMILAX has long creeping roots, which fend forth round flender ftalks, that are thinly guarded with fharp ftraight fpines. The leaves are reniform, cordated, and have no fpines. They are broader than they are long, have five ftrong nerves, and fhort footftalks, from each of which grow two flender clafpers. The flowers are produced in fmall bunches, in June and July. They will be fucceeded by a fmall berry, which will not come to perfection here.

5. LAUREL-LEAVED SMILAX has round taper ftalks, that are befet with fpines. The leaves are of a ftrong green colour, and a thick confiftence. They have no fpines, have three nerves, are of an oval lanceolate figure, and are about the fize of thofe of our Common Bay Tree. The flowers are produced in fmall round bunches, in June and July, from the wings of the ftalks, and thefe are fucceeded by fmall black berries in the autumn. This fort is rather of a tender nature, and unlefs the foil be naturally dry and warm, and the fituation well fheltered, they will be pretty fure of being killed in the winter.

6. BRIONY-LEAVED SMILAX has large, flefhy, white roots, which fend forth round, taper, prickly ftalks. The leaves are oblong, heart-fhaped, have no fpines, but have many veins running lengthways. Their upper furface is of a fine ftrong green colour, and, being tolerably large, they make a goodly fhow. The flowers are produced in July, in fmall loofe bunches, and are fucceeded by black berries.

7. LANCE-LEAVED SMILAX. The ftalks are flender, taper, and free from prickles. The leaves are fpear-fhaped, pointed, and unarmed with fpines. The flowers come out in fmall clufters, and are fucceeded by red berries.

8. The IVY-LEAVED SMILAX. The ftalks are angular, herbaceous, unarmed with fpines, but poffeffed of clafpers, by which they lay hold of anything near them for fupport. The leaves are oval, free from fpines, feven-nerved, and grow on footftalks. The flowers of this genus make no fhow, being poffeffed of no ornament

ment except the ſegments of the calyx. Thoſe of this ſpecies are very ſmall, and are collected in ſmall umbels. They appear in June; and are ſucceeded by roundiſh berries, which ſeldom ripen in England.

Theſe ſorts are all eaſily PROPAGATED; indeed they will propagate themſelves, if a plant or two of each ſort can be obtained; for they are poſſeſſed of long creeping roots, which run under the ſurface of the ground, and will, both near the main plant and far off, ſend up young ones; which being taken up in the autumn or ſpring, or in any time of the winter, will be good plants for uſe. Thus will theſe plants by nature furniſh you ſoon with plants enough for your purpoſe, if one or two of each can be firſt procured, and planted in a light good ſoil, in proper beds prepared for the purpoſe, under warm hedges, or amongſt trees in well ſheltered places.

S O L A N U M.

LINNEAN Claſs and Order, *Pentandria Monogynia* Each flower contains five males and one female. There are forty SPECIES; one of which, though common, claims our attention.

1. SOLA'NUM *Dulcama'ra:* The WOODY NIGHTSHADE, or the BITTERSWEET; *a ligneous climber*; native of England and moſt parts of Europe.

The WOODY NIGHTSHADE. Were it not for the commonneſs of this plant, it would deſervedly claim a principal place in our eſteem, as one of thoſe ſorts that require ſupports to ſet them off; for beſides the flowers, which are of an exquiſite fine purple, and grow in bunches, it has many beauties to recommend it to our obſervation and care. The leaves ſtand on large footſtalks, and the upper ones are of a haſtated figure. Their beautiful purple flowers will be produced in ſmall cluſters, in June and July; and they are ſucceeded by oblong red berries, which will be ripe in autumn. This is the Common ſort, which is of all the moſt beautiful,

beautiful, though hardly ever propagated. The *Varieties* of it, however, are in great esteem with most people, and of these there are, 1. A Variety with *white flowers*, which is much coveted on that account; and although these flowers are not so beautiful as the purple ones, yet the sort being a rare plant, makes it desirable; and this is the sort that is cultivated, and which differs in no respect from the purple, only in its white flowers, thereby pleasing the spectator by the variety it affords. 2. The next remarkable Variety of the Woody Nightshade is that with beautifully *variegated leaves*. These plants are sedulously propagated for the sake of their finely striped leaves; so that there is scarcely a Nurseryman who does not raise plenty of them for sale amongst other shrubs; and they are so generally liked, that his disposing of them will be pretty certain. This plant, as has been observed, is only the Common Woody Nightshade with the leaves delightfully variegated, its flowers being of the same fine purple, and the fruit that succeeds them exactly the same. 3. Another Variety has *thick leaves*, which are *very hairy*. This sort grows chiefly in Africa, and must have a warm situation to live through our winters. It is, however, a very fine plant, and where such a situation is not found, ought to be treated as a greenhouse plant.

All these sorts are easily PROPAGATED by cuttings; for they will grow, if planted in any of the winter months, in almost any soil or situation, and will be good plants for removing by the autumn following. If the owner has only a plant or two of these, which he is desirous of multiplying with certainty, let him lay the young stalks upon the ground, and draw over them a little soil, and they will effectually be good plants by the next autumn: and this will be the surest way, as cuttings of most sorts, though they will for the most part take very well, are often attended with much hazard. The Variegated sort must be planted upon a poor soil, or it will be in danger of running away from its colours.

SORBUS.

SORBUS.

LINNEAN Claſs and Order, *Icoſandria Trigynia*: Each flower contains about twenty males and three females. There are three SPECIES:

1. SO'RBUS *Aucupa'ria*: The WILD SORB, or MOUNTAIN ASH, or QUICKBEAM, or QUICKEN TREE, or ROAN TREE; *a low deciduous tree*; native of England and moſt of the Northern parts of Europe.

2. SO'RBUS *Domeſtica*: The CULTIVATED SORB, or SWEET SERVICE; *a low deciduous tree*; native of the South of Europe.

3. SO'RBUS *Hy'brida*: The MONGREL SORB, or SEMI-PINNATED SERVICE; *a low deciduous tree*; native of Gottlandia.

1. The WILD SORB, or MOUNTAIN ASH. Although we generally ſee this ſpecies in a ſhrubby underwood ſtate, it will nevertheleſs, if properly trained, grow to a tree of the middle ſize. It has no claim to the appellation of Mountain Aſh, except ſome diſtant reſemblance of the common Aſh in the formation of its leaves. The flowers of this tree have a pleaſing effect in the ſpring, and its berries, in autumn and winter, render it highly *ornamental*. EVELYN enumerates its *uſes*: he ſays, " beſides the uſe of it for the huſbandman's tools, goads, &c. the wheelwright commends it for being all heart. If the tree be large, and ſo well grown as ſome there are, it will ſaw out into planks, boards, and timber. Our fletchers commend it for bows next to Yew, which we ought not to paſs over, for the glory of our once Engliſh anceſtors: In a ſtatute of Henry VIII. you have it mentioned. It is excellent fuel; but I have not yet obſerved any other uſe." HANBURY follows him, and places the Mountain Aſh among his Foreſt Trees. Its wood is undoubtedly pliable and tough. Its ſhoots, from the ſtool, are generally numerous ſtraight, and long. In the North of England the huſbandmen uſe them for

 whip-

whip-ſtocks to drive their teams with. For ſtakes, edders, and perhaps for hoops, no wood is better adapted than the Wild Sorb; and as an UNDERWOOD it ſeems well worth the planter's notice. It braves the bleakeſt ſituation and aſpect; but cannot bear much wet: it is truly a Mountain plant.

The PROPAGATION of the NATIVE SORB is from ſeeds or by layering. Having procured a ſufficient quantity of berries, they ſhould be ſowed, ſoon after they are ripe, in the ſeminary, about half an inch deep. They frequently lie till the ſecond ſpring before they make their appearance; and, in the ſpring following, may be planted out in the nurſery. We need not repeat that the ſeminary ſhould be kept clear of weeds, and that the young plants in dry weather now and then ought to be refreſhed with water; neither need the Gardener be reminded, that after they are planted in the nurſery way, digging the ground in the rows muſt be obſerved every winter, taking off all ſhoots alſo which would make the tree forked, and keeping the weeds hoed in the rows, till they are of ſufficient ſize to plant out where they are intended to remain.

This ſpecies will take very well from *layers*; ſo that whoever cannot procure the berries, and has a few of theſe trees, may cut them down cloſe to the ground, when they will throw out many ſtools; and if the year following theſe are laid in the ground in the ſame manner as carnations, they will have taken good root in one year. But trees cultivated this way will not grow ſo ſtraight and handſome, neither will they arrive at ſo great a magnitude as thoſe raiſed from the ſeeds.

The Quicken Tree will grow upon almoſt any ſoil, either ſtrong or light, moiſt or dry. It flouriſhes both on the mountains and in the woods; it is never affected by the ſeverity of the weather, being extremely hardy; and if even planted on bleak and expoſed places, it grows exceedingly well.

2. The CULTIVATED SORB, OR SWEET SERVICE, is ſo diſtinguiſhed from the other, becauſe it produces eatable fruit, which in France, Italy, and other parts, is ſerved up in deſſerts; and the tree is cultivated there

ſolely

ſolely on that account. It will grow to be larger than the Quicken Tree; and in many reſpects is ſuperior in beauty to moſt trees. It will grow with an upright ſtem; and the young ſhoots in the ſummer are ſo downy as to appear covered with meal. In the winter they are inclined to a purpliſh colour, and are ſpotted all over with whitiſh ſpots; the buds at the ends of them will be turgid, preparing for the next year's ſhoot. The leaves reſemble thoſe of the Quicken Tree; they are finely pinnated, and compoſed of ſeven or eight pair of lobes, which are terminated by an odd one. They are broader than thoſe of the Quicken Tree, ſerrated in a deeper and more irregular manner, and their under ſurface is of a much more downy nature. The flowers are white, grow in umbels, come out in May, and are ſucceeded by an agreeable fruit, which is large, fleſhy, and of various ſhapes in the different varieties.

3. The Mongrel Sorb. This ſeems to be a mongrel, between the Quicken Tree and *Aria*. It is an upright growing tree, and the young branches are of a whitiſh colour. The leaves are very downy, and pinnated at the baſe; but the upper lobes join together, thereby forming a half-pinnated leaf. The flowers are white, grow in umbels, and are ſucceeded by bunches of roundiſh berries, which will be ripe in the autumn.

The PROPAGATION of the Native Sorb has already been given; and that culture will ſerve for all the ſorts: but in order to have good fruit of the Sweet Service Tree, the beſt ſorts ſhould be grafted or budded upon pear or quince ſtocks.

Theſe trees are very hardy, for they will grow in almoſt any ſoil; though they make the ſwifteſt progreſs, and arrive at the greateſt height, in a moiſt ſituation.

SPARTIUM.

LINNEAN Clafs and Order, *Diadelphia Decandria:* Each flower contains ten males and one female; the males being divided into two fets at the bafe. There are fixteen SPECIES; feven of which are proper for our Collection:

1. SPARTIUM *Scopa'rium:* The COMMON ENGLISH BROOM; *a deciduous fhrub*; native of England, and the Southern parts of Europe.

2. SPA'RTIUM *Ju'nceum:* The SPANISH BROOM; *a tall deciduous fhrub*; native of Spain, Portugal, and fome parts of Italy and Sicily.

3. SPA'RTIUM *Radia'tum:* The STARRY BROOM; *a low deciduous fhrub*; native of Italy.

4. SPA'RTIUM *Monofpe'rmum:* The SINGLE-SEEDED BROOM; *a deciduous fhrub*; native of the barren parts of Spain.

5. SPA'RTIUM *Angula'tum:* The EASTERN BROOM; *a deciduous fhrub*; native of the Eaft.

6. SPA'RTIUM *Spino'fum:* The THORNY BROOM, or PRICKLY CYTISUS; *a tender deciduous fhrub*; native of the fea-coafts of the Southern parts of Europe.

7. SPA'RTIUM *Sco'rpius:* The PRICKLY BROOM; *a deciduous fhrub*; native of Spain and the South of France *.

1. The ENGLISH BROOM will grow to be about fix feet high. The branches are very flexible and numerous; they are angular; and the bark with which they are covered is of a delightful green. The leaves are both trifoliate and fingle, the lower part of the branches producing the former, the upper part the latter. The flowers are large, and produced in May, all along the fides of the laft year's fhoots, from the bottom to the top. They ftand upon fhort footftalks, and fo ornament each twig of which the whole fhrub is compofed, that they have a look grand beyond moft of thofe

* For another Clafs of Brooms fee GENISTA.

of

of the flowery tribe. Thefe flowers are fucceeded by compreffed pods, containing kidney-fhaped feeds, all of which are very well known.

2. SPANISH BROOM. There are two notable *Varieties* of this fpecies: Common Spanifh Broom, Double Spanifh Broom.

Common Spanifh Broom is a fine plant, and has been much fought after as a flowering fhrub. It will grow to be ten feet high. The branches are taper, placed oppofite, and covered with a fmooth green bark. The leaves, which are not very numerous, are of a fpear-fhaped figure, and, like the twigs, of a fine green colour. The flowers are produced at the ends of the branches, in loofe fpikes, in July; and there will be a fucceffion ftill kept up, at the end of each fpike, often until the froft puts a period to their blowing. The flowers of this fort, alfo, are fucceeded by compreffed pods, which contain kidney-fhaped feeds, that often ripen in the autumn.

The *Double Spanifh Broom* differs in no refpect from the other, except that the leaves are very double. The manner of growing, colour of the fhoot, and nature of the leaves, are exactly the fame; and it produces very full double flowers; but thefe flowers do not come out fo early as the fingle fort, it being often September before any of them will be in blow; and the fucceffion will be continued fo flowly, that fometimes not more than two or three flowers on a fpike will be fully out before the frofts nip them from any further blow. This fort is fucceeded by no feeds.

3. STARRY BROOM is a low plant, feldom growing more than a yard high, even when it has the advantage of culture; in the places of its natural growth, two feet it feldom afpires to. Notwithftanding the low growth of this fhrub, however, it will occupy a large fpace of ground in proportion to its fize, for it extends its flexible branches all around to fome diftance. The branches of which it is compofed are very narrow, angular, and grow oppofite by pairs. The leaves are trifoliate, grow oppofite to each other, and the folioles are awl-fhaped, placed oppofite, and fpread out in fuch a manner as to refemble the rays of a ftar, which occafions its being fo called. The flowers are produced

 in

in June and July, at the ends of the branches: they will be in a kind of ſmall cluſters or ſpikes, are of a bright yellow colour, and of the ſame figure with the former, but proportionally ſmaller. They are ſucceeded by ſhort hairy pods, in which are a few kidney-ſhaped ſeeds, which will be ripe in Auguſt or September.

4. SINGLE-SEEDED BROOM. The *Varieties* of this ſpecies are, the Common Yellow, and the White-flowered.

The *Common Single-ſeeded Broom* is a plant about ſix feet in growth. Its branches, which are very numerous and tough, are angular; and the leaves, which are not very many, are of a lanceolated figure. The flowers are produced in bunches from the ſides of the branches. Theſe bunches are ſmall; but being of a fine deep yellow colour, and alſo being in plenty all over the ſhrub, give it a beautiful look. This ſhrub blows in July; and the flowers are ſucceeded by ſhort pods, each of which contains a ſingle ſeed only, which ſeldom ripens in England.

White-flowered Single-ſeeded Broom, which is alſo called *The White Spaniſh Broom*, is of a more tender nature than the former ſort; yet not ſo tender but that it will thrive abroad with us, in any dry ſoil and well ſheltered ſituation, if the winters are not too ſevere. After this precaution, we would adviſe the Gardener not to venture his whole ſtock of theſe plants abroad, leſt a ſevere winter ſhould take them all off; but to have a few planted in pots, and ſet under ſhelter, that, in caſe the others ſhould be killed, a ſhare of theſe may ſupply their places. The White Spaniſh Broom, then, will grow to about eight feet high; and the branches are numerous, ſlender, and tough. Their bark is of a whitiſh colour, and they are taper, almoſt like a ruſh. The leaves, which are not many, are of a lanceolated figure. The flowers are white, come out in cluſters from the ſides of the branches in July, and are ſucceeded by ſhort pods, each of which contains one ſingle ſeed only.

5. The EASTERN BROOM will grow to about ſix or eight feet high. The branches of this, alſo, are numerous, ſlender, and tough. They are rather of a ſingular ſtructure, each of them affording ſix angles.

The

The leaves, which are few, are of different figures, ſome being found ſingle only, whilſt others are trifoliate. The flowers are produced in July, at the ends of the branches, in a kind of ſpikes. They are of a paler yellow than moſt of the other ſorts, and are rarely ſucceeded by ſeeds with us.

6. Prickly Cytisus, or Thorny Broom, has ſcarcely any buſineſs in this place, being generally reared as a greenhouſe plant; but as it will bear our moderately mild winters in a warm ſoil and ſituation, with this caution it may be introduced. It is about ſix feet in growth; and the branches are numerous, ſlender, tough, angular, and armed with long ſpines. The leaves are trifoliate; and the flowers are produced in cluſters, in June, at the ends of the branches. They ſtand on long footſtalks, are of a bright yellow, and make a good figure. They are ſucceeded by ſhort hard pods, which contain a few ſeeds of the ſame figure with the others.

7. Prickly Broom. The ſtalk of this ſpecies is woody, and ſends forth ſeveral ſlender, prickly branches, which ſpread themſelves every way. The leaves are oval, ſmooth, and in ſome Varieties hairy. The flowers are moderately large; and ſome are of a deep yellow colour, whilſt others are pale. They appear in July; and are ſucceeded by ſhort pods, containing the ſeeds, which ſeldom ripen in England.

All theſe ſorts of Broom, the Double-bloſſomed excepted, are to be propagated from ſeeds; and one method may be obſerved for all the ſorts. The ſorts that ripen their ſeeds in England are ſuppoſed to be ready at hand; the ſeeds of the others muſt be procured from the places where they grow naturally. The firſt week in April is the beſt time for ſowing the ſeeds; and this ſhould be either in drills, or on beds, half an inch deep. It will not be long before the plants appear; and as the hot weather comes on, they ſhould be ſhaded from nine o'clock in the morning till within an hour of ſunſet. Watering and conſtant weeding muſt be given them; and this is all the trouble they will require in ſummer. The reader will perceive our Common Broom to want none of this care; neither will the Common Spaniſh Broom need much of it; it is to be afforded those only which are less common, that we may be more certain of a plentiful strong crop. In the spring all these seedlings are to be taken up, and pricked out in the nursery ground, a foot asunder, and two feet distant in the rows. This work must be done when they are one-year-old seedlings; because they naturally send down a strong tap root, which, if deferred longer, will be grown so big as to endanger the growth of the plant. After they have stood in the nursery ground two years, they will be good plants for setting out where they are to remain. Thus may all the sorts of *Spartium* be raised by seeds; though it will be highly proper to have particular regard to the situation of the tenderer sorts; such as the White Spanish Broom, the Oriental, and the sorts called the Prickly *Cytisus* and Prickly Broom. These may be raised the same way; but the soil and situation must be naturally warm and well sheltered, and the beds should be hooped to be covered with mats in frosty weather, otherwise the whole crop will be in danger of being lost the first winter. In the spring they may be planted, some in pots (to preserve the sorts), others in the warmest places of the shrubery. Another method will not be improper to be followed in raising the tenderer sorts; namely, by sowing them in pots in April, and plunging them in a shady border up to the rim. At the approach of the first frost, they may be removed into the greenhouse, or placed under some shelter, when they will be effectually preserved until the spring; then they should be turned out, and planted in separate pots, which should be plunged in a shady border, and removed under cover in the winter. By thus protecting them for a winter or two, they will get stronger, and be able to resist the cold; and then a share may be planted out in the warmest situation, whilst the others may be removed into larger pots, to be kept, if wanted, as greenhouse plants.

SPIRÆA.

LINNEAN Clafs and Order, *Icofandria Pentagynia:* Each flower contains about twenty males and five females. There are nineteen SPECIES; fix of the ligneous kind are here treated of:

1. SPIRÆ'A *Salicifo'lia:* The COMMON SPIRÆA FRUTEX, or THE WILLOW-LEAVED SPIRÆA; *a low deciduous fhrub or herb*; native of Siberia and Tartary.

2. SPIRÆ'A *Tomento'fa:* The RED-FLOWERING SPIRÆA; *a low deciduous fhrub*; native of Philadelphia.

3. SPIRÆ'A *Hypericifo'lia:* The HYPE'RICUM-LEAVED SPIRÆA, or HYPERICUM FRUTEX; *a deciduous fhrub*; native of Canada.

4. SPIRÆ'A *Crena'ta:* The SPANISH SPIRÆA; *a low deciduous fhrub*; native of Spain and Siberia.

5. SPIRÆ'A *Opulifo'lia:* The GELDER-ROSE SPIRÆA; *a deciduous fhrub*; native of Virginia, Canada, and Carolina.

6. SPIRÆ'A *Sorbifo'lia:* The SERVICE-LEAVED SPIRÆA; *a low deciduous fhrub*; native of moift land in Siberia.

1. The COMMON SPIRÆA FRUTEX rifes to about four feet high. The root is fpreading; fo that befides the common ftalks which fend forth branches, others are produced from the roots called Suckers, which by the autumn will be as high or higher than any of the whole plant. The bark on all thefe is fmooth, and of different colours; that on the old ftalks is red, though for the moft part clouded with a dufky matter: The young fhoots that grow from thefe ftalks are lighter, though neverthelefs of a reddifh tinge; whilft the bark on the fummer fhoots, that fprung from the root, are nearly white. The leaves of this fpecies are of a fine green, and grow without order on the branches. They are fpear-fhaped, obtufe, naked, and their edges are ferrated. The flowers are produced in June, at the ends of the branches that grow from the main ftalk; and before thefe have done blowing, the fuckers that arife

arise from the roots will exhibit their flower buds at the ends. These are generally larger and fairer than those that were before in blow; and by these suckers a succession of flowers is often continued even until late in the autumn. The flowers are produced in double branching spikes, which are larger downwards, diminish gradually, and end with an obtuse spike at the top. They are of a pale red colour; and though separately each flower is small, yet being produced in these thick spikes, four or five inches long, they have a good look. These flowers, with us, are succeeded by no ornamental seeds. This plant cannot bear a wet situation.

2. The Red-flowering Spirea will grow to the height of about four feet; and the branches are covered with a purple bark. The leaves grow on these without order: they are of an oval, lanceolated figure, and unequally serrated. Their upper surface is of a fine green colour, but their under is downy: the stalks, also, are possessed of a good share of this meally kind of matter. The flowers are produced in July, at the ends of the branches, in double branching spikes, like the former; and being of a bright red colour, make a fine appearance.

There is a *Variety* of this species with white flowers.

3. The Hypericum-leaved Spirea, or Hypericum Frutex, will grow to the height of about five or six feet, and has beauty and elegance beyond description; not so much from its natural form of growth, or the colour of the bark or leaves, as from the flowers; for the branches are produced irregularly. The older shoots are covered with a dark brown bark; the younger shoots are smooth and lighter, and are tinged with red. The leaves are small, though of a pleasing dark green colour; they are produced irregularly on the shrub, and have their edges entire. The flowers are produced in May, almost the whole length of the branches: they are of a white colour; and though each flower is separately small, yet they are collected in umbels that sit close to the branches, which being thus ornamented their whole length, scarcely any thing but flowers, besides the main stalks, are to be seen; so that the shrub has the appearance of one continued flower, branched out into as many different divisions as there

are

are twigs; for every twig at a little diſtance will look like a long narrow ſpike of flowers; and theſe being all over the ſhrub, of a pure white, the ſhow they then make is delightful. It ſickens in a wet ſituation.

4. Spanish Spirea will grow to be about four feet high; and the branches, which are produced irregularly, are covered with a dark brown bark. The leaves are ſmall, of a pleaſant green colour, and ſerrated at their ends. The flowers are produced from the ſides of the branches, in May; they grow in roundiſh bunches, are of a whiter colour than, and being produced nearly the whole length of, the branches, make a charming ſhow, like the preceding ſort; from which this appears very little to differ, without being ſtrictly examined.

5. Gelder-rose Spirea. Of this ſpecies there are two *Varieties*, called, Virginian Gelder-roſe, and Carolina Gelder-roſe.

Virginian Gelder-roſe will grow to be ſeven or eight feet high. The branches are covered with a dark brown bark, which peels off in the winter, and diſcovers an inner, which is ſmooth, and of a lighter colour; ſo that in winter this ſhrub has a very ragged look. The leaves reſemble thoſe of the common currant buſh, which has occaſioned its being called by ſome the Currant-leaved Gelder-roſe. They are for the moſt part lobed like them; though all the leaves will not be alike, ſome being divided into more than three lobes, whilſt others are ſcarcely divided at all. They are ſerrated at their edges, are of a paliſh green colour, and placed irregularly on the branches, on long green footſtalks. The flowers are produced in June, at the ends of the branches: they are white at their firſt opening, and afterwards receive a reddiſh tinge, which is ſtill heightened before they die off. Each flower ſeparately is rather ſmall; but many of them grow together, each having its ſeparate footſtalks, in large umbels. The beauty of the Common Hawthorn is known to all; and it may not be amiſs here, as the ſimile is juſt, and that the Reader may have a true idea of the flowers, to mention, that each flower ſeparately has the appearance of a ſingle flower of the Hawthorn, and that they are produced in bunches. Theſe flowers are ſucceeded by the ſame kind of bunches of of reddiſh, cornered fruit, which cauſes a pretty variety in the autumn.

Carolina Gelder-roſe differs very little from the former ſort. The branches are covered with the ſame kind of falling bark; though the leaves are not lobated in the ſame manner; for theſe will be of different ſhapes; yet moſt of them are nearly oval, but end in points, and are all unequally ſerrated round their edges. The flowers of this ſort, alſo, are white, but grow in rounder and ſmaller bunches than the other. They are ſucceeded by the like kind of cornered fruit, which is of a reddiſh colour in the autumn.

6. SERVICE-LEAVED-SPIREA is a ſhrub of very low growth; a yard is the higheſt we ever yet knew it arrive to. The young branches are covered with a purpliſh bark. The leaves are beautifully pinnated, ſo as nearly to reſemble thoſe of the Service Tree. The folioles are oblong, and generally about four pair in number: they are uniformly ſerrated, and exceedingly ornamental to the ſhrub. The flowers are white, and produced at the ends of the branches, in July, in panicles. They are ſeldom ſucceeded by ſeeds in England.

The PROPAGATION of all the ſorts is very eaſy. It may be done by cuttings; for if the ſtrongeſt parts of the ſhoots of the laſt ſummer's growth be planted in October, in a ſhady border, moſt of them will grow, and become good plants by the autumn; ſo that by the autumn after that, they will be very proper plants to be ſet out to ſtand. But if a perſon has only a plant or two of a ſort, from which he can get but a very few cuttings, the beſt way is to layer them, and not hazard their growing this way; for although they will take freely, yet (ſays HANBURY) by ſome unſeaſonable weather, I have known whole crops of *cuttings of all ſorts* to fail. Thus, of the many thouſand cuttings of all ſorts I planted in the winter preceding the dry ſummer in 1762, very few grew; for although they were ſhaded and watered, and others planted in ſhady borders, yet ſuch large cracks and chaſms would open among them (as they did almoſt all over my plantations) as to cauſe watering to be of no ſervice; nay, the more I watered them, the harder the mould ſet, and the chaſms

chasms became greater; and notwithstanding many of the cutting were planted in parts that were possessed of a natural moisture, yet the crevices there were larger, and the ground harder; and all attempts to prevent it seemed to be in vain. Though this is the nature of the soil of few nurseries, I mention this to shew, that there is a hazard in planting of cuttings, unless the season should prove good; for this turn I had scarcely any grew: so that whatever trees will grow by cuttings, if a Gentleman has only a plant or two, and wants to have them encreased, the best way is to do it by layers; and hence, of all *the sorts before mentioned*, if the twigs be but laid in the ground in the autumn, they will have good roots by the autumn following, many of which will be plants strong enough to be planted in the shrubery, whilst the weaker may be set in the nursery ground for a year or two, to gain strength. Some of these sorts will throw out suckers, which will be good plants when taken up: nay, the first sort will propagate itself fast enough this way; for after it has stood a year or two, it will throw them out so vigorously, as has been before observed, that in one summer they will grow to be as high as the whole plant, and will have fair flowers at their ends in the autumn. And here the Gardener must observe, that after this sort is planted in the shrubery, the suckers must be constantly cleared off the old plants every winter, otherwise they will soon be so numerous and close, as to lose that beauty which always attends plants that arise with single or with few stems.

S T A P H Y L E A.

LINNEAN Class and Order, *Pentandria Trigynia*: Each flower contains five males and three females. There are two SPECIES:

1. STAPHYLE'A *Pinna'ta*: The COMMON STAPHYLEA, or BLADDER NUT; *a deciduous shrub*; native of many parts of Europe.

2. STA-

2. STAPHYLE'A *Trifólia:* The TRIFOLIATE STAPHYLEA, OR BLADDER NUT; *a deciduous ſhrub*; native of Virginia.

1. The COMMON STAPHYLEA will grow to be eight or ten feet high. The older branches are covered with a brown bark; that on the younger ſhoots is of a much lighter colour. The bark is exceedingly ſmooth; the twigs are very pithy, and when broken have a very ſtrong ſcent. The buds will be turgid and large early in winter, as if ready to burſt out of their ſtipulæ, and begin their ſhoots; this cauſes the plant at that ſeaſon to have an air of health and verdure, which of courſe muſt then be very pleaſing. The leaves are pinnated, of a light green colour, and, like all others of that nature, are very ornamental. They conſiſt of two pair of folioles, that are terminated with an odd one; which occaſions this ſort being frequently called the Five-leaved Bladder Nut. Theſe folioles are tolerably large, oblong, pointed, and ſtand on pretty long foot-ſtalks. The flowers are produced in long pendulous bunches, from the wings of the leaves; and are white. The buds appear in the ſpring, almoſt at the firſt dividing of the ſtipulæ, though they will not be in full blow until May. Theſe flowers are ſucceeded by large inflated bladders, in which the ſeeds are contained, and have a very ſtriking and ſingular look in the autumn. The nuts of this tree are ſmooth, and ſaid to be eaten as food by the poor people in ſome countries. They are alſo uſed by the Catholics, who compoſe ſome of their roſaries of them.

2. The TRIFOLIATE STAPHYLEA grows to about the ſame height with the former. The elder branches will be beſprinkled, as it were, all over with grayiſh ſpots. The bark on the younger branches is perfectly ſmooth, and of a yellowiſh colour. The buds will be ſwelled early in the winter, though they will not be ſo large and turgid as thoſe of the former ſort. The leaves are trifoliate, and grow by threes on a footſtalk; which has occaſioned this plant being diſtinguiſhed by the name of Three-leaved Bladder Nut. They are of a light green colour; and the folioles are generally pretty large, oval, pointed, and ſerrated at their edges. The flower buds appear at the firſt beginning of the buds to open

open in the ſpring; which has been known to be ſometimes ſo early as January; though the flowers will not be in full blow until May. Theſe flowers, like the former, are produced from the ſides of the branches, in long pendulous bunches: their colour is white; and they are ſucceeded by large inflated bladders, in which the ſeeds are contained. The ſeeds of both ſpecies ripen well in England.

Theſe ſpecies may be PROPAGATED by ſeeds, layers, or cuttings. 1. The ſeeds ſhould be ſown, ſoon after they are ripe, in the autumn, three quarters of an inch deep, in almoſt any ſort of common garden mould made fine. In the ſpring ſome ſhare of the plants will appear; though you muſt not expect the whole crop until the ſecond ſpring following: nay, if the ſowing of the ſeeds is deferred until the ſpring, ſcarcely any of them will come up until the ſpring after. All the ſummer the beds muſt be kept clear of weeds; and if it ſhould prove dry, a gentle watering ſhould be given the young plants, which will encreaſe their growth. The ſpring after the remainder of the crop will come up; and the buſineſs of weeding muſt be continued that ſummer. In the autumn the two-years-old plants ſhould be drawn out and planted in the nurſery, a foot aſunder, and two feet diſtant in the rows; and in the beginning of March the one-year-old ſeedlings ſhould be taken up, and planted in the ſame manner. The reaſon of deferring the planting out of the younger ſeedlings is, that, being ſmall when planted out in autumn, they are often thrown out of the ground by the froſt, and many of them loſt; whereas of larger plants there will be little danger. After they have ſtood two or three years in the nurſery, they will be good plants for any places where they are wanted. 2. Theſe ſhrubs may alſo be propagated by layers; and this muſt be performed in the autumn, on the ſhoots of the preceding ſummer, by ſlitting them at a joint, and laying them in the ground. The making of this ſlit will be neceſſary, or at leaſt the well breaking of the bark, otherwiſe they will not ſtrike root; and if this be done with judgment, they will have good roots by the autumn following, many of which will be good plants, and fit for the ſhrubery; whilſt the weaker may be planted in the

the nurſery ground for a year or two, to gain ſtrength. One caution is to be obſerved: If the layering is to be performed by twiſting the young ſhoots ſo as to break the bark, be careful not to over-do this; for being very pithy, it will kill them to be much twiſted; and if the bark is not well broke, they will not ſtrike root this way. 3. Theſe trees are to be encreaſed alſo by cuttings; from which they will grow very well. The cuttings muſt be the bottom part of the laſt ſummer's ſhoot, which ſhould be planted in October, in a ſhady border of light earth. If the ſpring ſhould prove dry, give them ſome watering, and there will be little fear but that moſt of them will grow.

STEWARTIA.

LINNEAN Claſs and Order, *Monadelphia Polyandria*: Each flower contains many males, and five females; the males being joined in one ſet at the baſe. There is only one SPECIES:

STEWA'RTIA *Malacode'ndron*: The STEWARTIA; *a deciduous ſhrub*; native of Virginia.

The STEWARTIA is a ſhrub of about eight or ten feet growth with us, and the branches, which are produced irregularly from the ſides of the main ſtem, are covered with a brown bark. The leaves are placed alternately on the branches, and are of much the ſize and make of thoſe of the Cherry Tree. Their upper ſurface is of a fine green, though they are lighter and hairy underneath, and have their edges moſt acutely ſerrated. In the beginning of June this tree will be in blow. The flowers are produced from the ſides of the branches: they are white, and ſeem to be compoſed of five large oval petals; but upon examining them to the bottom, we find them joined at the baſe. The flowers have a genteel look, are poſſeſſed of an air of delicacy; and this being at preſent a very ſcarce plant, makes it more valuable. It was named *Stewartia* in honour of the

Right

Right Hon. the Earl of Bute, as a compliment to his great ſkill in the ſcience of botany.

This plant is PROPAGATED by layers and ſeeds. 1. The young ſhoots ſhould be layered in autumn, by making a ſlit at the joint, as is practiſed for Carnations. In the ſpring, a tall hedge of ſome kind ſhould be made on the ſouth ſide of them, bending alſo a little towards the eaſt and weſt, that they may be ſhaded all the ſummer. In dry weather they ſhould be watered; and then they ſhould remain until the March following, when they ſhould be examined to ſee if they have ſtruck root; for ſometimes they will ſtrike root pretty freely, if ſo ſhaded and watered; and ſometimes they have diſappointed our expectations after waiting two years; though cuttings will ſometimes grow. In March, however, a ſufficient quantity of pots muſt be provided, filled with good garden mould, mixed with a ſhare of drift ſand; and the layers ſhould be taken up, whether they have ſtruck root or not, and planted in theſe pots, which muſt be plunged up to their rims in a bark bed. Thoſe layers that have no roots will have the parts ready for ſtriking, and this aſſiſtance will ſet them all forward; ſo that in a very little time they will become good plants. They muſt be hardened as ſoon as poſſible to the open air. For this purpoſe the pots ſhould be taken out of the beds, and plunged up to the rims in a ſhady place; and though theſe are hardy trees, it will be proper to take the pots up, and remove them into the greenhouſe, or under ſome ſhelter, for the firſt winter. At the latter end of March they may be turned out of the pots, with their mould, into the places where they are wanted to ſtand. 2. Another method of propagating theſe plants is from ſeeds, which we receive from abroad. Theſe ſhould be ſown in pots of light earth, about half an inch deep, and the pots ſhould be plunged up to the rims in a bark bed; where all the advantages of heat, water, and ſhade, muſt be afforded them; for without theſe requiſites, it is not often that they will grow.

STYRAX.

LINNEAN Clafs and Order, *Decandria Monogynia*: Each flower contains about ten or twelve males, and one female. There is only one SPECIES:

STY'RAX *Officina'lis*: The STORAX TREE; *a tall deciduous fhrub*; native of Italy, Paleftine, and Syria.

The STYRAX, in its native places of growth, will arrive to be more than twenty feet high; with us, twelve or fourteen feet is the height we may expect it to grow to. The branches are covered with a fmooth grayifh bark; and the younger fhoots are of a reddifh colour. The very wood of this tree is finely fcented; and in Turkey and other places where it naturally grows, that fragrant refin called Storax exfudes from its trunk, an incifion being firft made. The virtues of this refin are well known, and the tree is rendered valuable on that account. The leaves which ornament the flender branches, that are produced without order all around, are of a moderate fize, and of an oval, pointed figure. Their edges are a little waved, though free from ferratures. They grow on fhort footftalks, without any order, being fometimes by pairs, fometimes fingly, producing a pleafing irregularity. They a little refemble the leaves of the Quince Tree, and are of two colours; their upper furface is of a lucid green, but their under is hoary; and this difference of colours makes a good contraft efpecially when waving with the wind, on this charming fweet-fcented tree. The flowers are produced in June, from the fides of the branches, in bunches; feven or eight flowers will conftitute a tuft. Their form and colour fomewhat refemble thofe of the Orange Tree, and their odours are diffufed all around. Thefe flowers are fucceeded by no fruit with us; fo that the height of its beauty is when it is in full blow.

The PROPAGATION is from feeds, which we receive from

from abroad. These must be sown an inch deep, in pots of light sandy earth, which pots should be plunged in a shady well sheltered place, there to remain until the second spring after sowing. In March the seeds will be ready to sprout; and to assist them, it will be necessary to take up the pots, and set them up to the rims in a hotbed. When the plants come up, all convenient air must be given them; often water; and they should be hardened soon to the open air. They should be then set abroad in the shade, and in the winter should be removed into the greenhouse, and placed under shelter. In the spring it will not be necessary to force them a second time in the hotbed; for if the pots are set in a shady place up to the rims, and now and then a little watering afforded them, the plants will grow very well, and make good shoots that summer. Like greenhouse plants, at the approach of winter, they must be removed into shelter; and in spring they must be shook out of these larger pots, and each planted in a separate smaller pot; and being well watered, if they are plunged into a hotbed, it will set them growing finely. After they have had help this way, they must be soon hardened, and the pots taken up, and set up to the rims in mould in a shady place. In winter they should be placed in the greenhouse as before; and this method must be continued for six or eight years, treating them exactly as hardy greenhouse plants, and shifting them into fresh pots, as their encrease of size by growth requires. By this time they will be woody and strong; and may then, the beginning of April, be turned out of the pots, with the mould, into the places where they are designed to remain. If the soil be naturally dry and warm, and the place well sheltered, nothing but very severe frosts will injure them, especially after having stood a winter or two.

SYRINGA.

LINNEAN Claſs and Order, *Diandria Monogynia:* Each flower contains two males and one female. There are only two SPECIES:

1. SYRI'NGA *Vulga'ris:* The COMMON LILAC; *a tall deciduous ſhrub*; native of Egypt.

2. SYRI'NGA *Pe'rſica:* The PERSIAN LILAC; *a deciduous ſhrub*; native of Perſia.

1. The COMMON or EGYPTIAN LILAC. The *Varieties* of this Species are,

The Purple Lilac,
The Blue Lilac,
The White Lilac.

The *Purple Lilac* generally riſes to the higheſt ſize of any of the three ſorts, though the height of all of them is either greater or leſs, according to the ſoil in which they are planted. The Purple, in good light, rich earth, will grow to be ſixteen or twenty feet high; and the others, in the ſame ſort of mould, nearly as high. The Purple Lilac is naturally of an upright growth, though it ſoon divides into branches; and theſe alſo, as the tree grows older, into others, all of which are covered with a ſmooth browniſh bark. All winter the plant has a bold and healthy look, occaſioned by the large and turgid purpliſh buds, which will have begun to ſwell early the preceding ſummer, and which will burſt forth into leaf ſoon in the ſpring following. The leaves are large and ſmooth, and of a pleaſant dark green colour. They are of an oval, cordated figure, end in acute points, and grow oppoſite by pairs on the branches. The flowers will be produced in May, at the end of the ſame ſpring's ſhoot, in very large and almoſt conical bunches. They are of a purpliſh colour, are cloſely placed, and the number of which each bunch is compoſed is very great. "I have meaſured a bunch of them, ſays HANBURY, a foot long; and can any thing be thought to excel ſuch a profu-

profufion of flowers, in its aggregate ftate, of which each clufter is compofed! But many of thefe flowers appear all over the tree, mixed in an eafy manner among the delightful leaves; fome peeping as it were above them, and feveral reclining their tops, to make the appearance ftill more free and eafy. The value of thefe flowers is ftill heightened by their delightful fragrance; and when their blow is over, which it will be in a fortnight or three weeks, they have paid us their tribute, except what they afford from their leaves and manner of growth; for they are fucceeded by feed veffels, of fuch a colour and nature as none but the curious botanift can find any pleafure in obferving."

The *Blue Lilac* differs in no refpect from the Purple, except that the branches are rather more flender and lefs erect, and that it feldom rifes higher than twelve or fourteen feet. The branches are covered with a fmooth brownifh bark; and the buds in the winter will be turgid like the former, though fmaller; and they, as well as the young fhoots, will have a blueifh tinge. The leaves are exactly like the preceding fort, though they will have a caft of blue. The flowers are produced in May, in not quite fuch large bunches as the former fort; the bunches will be alfo loofe. They are of a fine blue colour, and admirably fcented; and the preference is to be given with juftice to neither of thefe trees.

The *White Lilac* feems rather a ftiffer plant than the Blue, and the branches grow more erect than any of the forts. The young branches are covered with a fmooth light coloured bark; and in winter the buds, which will be large and turgid, are of a herbaceous yellow colour, by which this fort at that feafon may be eafily diftinguifhed from the others. The leaves are of the fame figure and nature, though their colour is lighter, thereby making a variety. The flowers are of a fine white colour; and are produced in the fame kind of large clofe panicles as the others, which ftand upright. They are very fair, and, in the bunches, are fet very clofely together, which caufes them to be more erect than either of the two former forts. Thus may any perfon who has never feen thefe trees form an idea of their beauty when in blow; which will be very early,

 when

when the plants are ſmall; for they will begin flowering at the height of four or five feet, and will every year after afford greater plenty of flowers as they advance in growth. The bunches generally grow by pairs, two at the end of the ſame ſpring ſhoot, though of unequal ſize, the one being generally much larger than the other.

2. The PERSIAN LILAC. The *Varieties* of this ſpecies are:

The Common Perſian Lilac (or Perſian Jaſmine),
The White Perſian Lilac,
The Blue Perſian Lilac,
The Cut-leaved Perſian Lilac.

The *Common Perſian Lilac* ſeldom grows higher than five feet, and is deemed a moſt delightful flowering ſhrub. The branches are long, ſlender, flexible, and covered with a ſmooth browniſh bark, with a blueiſh tinge, on which are often ſeveral yellowiſh punctules. The buds will be large and turgid in winter; and the leaves and flower buds will come out early in ſpring. The leaves are of a lanceolated figure, of a fine green colour, and grow oppoſite by pairs on the branches. The flowers will be in full blow before the end of May. They are of a blueiſh colour, and are produced in the ſame kind of panicles as the other ſorts, though they will be ſmaller and looſer. Their odour is more heightened than that of the others; and the ſhrub, on the whole, is very valuable, though now pretty common. The long flexible branches have a natural tendency to hang downwards; and when in blow their bunches of flowers will greatly encreaſe this tendency; on which account it will be proper to place a few ſticks to ſupport them, which may be diſpoſed in ſuch a manner as to eſcape notice, unleſs by the niceſt examiner; and this will be proper, as the ſeeing the branches tied to ſticks in full view, would ſhew a degree of ſtiffneſs which would not look well.

White Perſian Lilac will grow to the ſame height with the former. The leaves, buds, and ſhoots are of a lighter colour. It produces its flowers at the end of May, in the ſame kind of panicles as the other (though theſe are of a white colour), and poſſeſſed of the ſame heightened odour.

Blue

Blue Persian Lilac differs from the preceding, in that the flowers are of a deep blue colour, thereby causing a pleasing variety on that account.

Cut-leaved Persian Lilac affords the greatest variety by its leaves; though the bark is rather darker, and the twigs seem slenderer, and are still more pendulous than the other sorts. The leaves of this sort are divided, almost to the midrib, into an uncertain number of segments; and as this occasions them to have a different, an unfrequent and a singular look, the value of the plant is much heightened on their account; particularly as it is in no respect diminished in the elegance and fragrance of its flowers.

The best way of PROPAGATING all these sorts is by layers; for if this work be performed in autumn, on the young shoots, they will be good plants by the autumn following. This method is particularly to be preferred in the first three sorts of Lilacs, as they naturally throw out such plenty of suckers as to weaken, unless constantly taken off, and diminish the beauty of the mother plants. Plants raised by layering will be less liable to throw out suckers, and consequently will be more valuable. The common way, indeed, is to take up the suckers, and plant them in the nursery for a year or two, and then set them out finally; but these plants will not be so valuable as the others, as they will be more liable to produce suckers, which to the Gardener, when he has got a sufficient stock of plants, become very troublesome.

The Persian sorts being less liable to put up suckers, may not only be encreased by layers; but when they do throw out any, the suckers may be taken up, and deemed good plants. Cuttings of these sorts, also, planted in August, in a shady moist place, will often grow.

The Persian Lilacs never produce seeds with us, but the first three sorts do; and by these the plants may be encreased; which also is a good method. The seeds ripen in the autumn; and in October they should be sown. They are rather small; and therefore the mould of the beds should be very fine, and they should be covered over lightly. In the spring they will come up, and will want no other care than weeding. In the spring

 following

following they may be planted in the nurſery, a foot aſunder, and two feet diſtant in the rows; and here they may ſtand two or three years, when they will be of a proper ſize to be finally planted out, and will flower in a year or two after. The differences of all theſe three ſorts are generally permanent from ſeeds; ſo that a perſon may ſow them with reaſonable hopes of obtaining the ſorts the ſeeds were gathered from.

T A M A R I X.

LINNEAN Claſs and Order, *Pentandria Trigynia*: Each flower contains five males and three females. There are only two SPECIES:

1. TA'MARIX *Ga'llica*: The FRENCH TAMARISK; *a tall deciduous ſhrub*; native of France, Italy, and Spain.

2. TA'MARIX *Germa'nica*: The GERMAN TAMARISK; *a deciduous ſhrub*; native of low overflowed places in Germany.

1. The FRENCH TAMARISK will grow to the height of about fourteen feet. The branches are few, and ſpread abroad in an irregular manner, ſome being upright, others horizontal, whilſt others decline with their ends towards the earth. The bark is ſmooth, and of a deep red or purpliſh colour next the ſun; but on the oppoſite ſide of the branch of a pale brown. The leaves are rather of a pale green, and very beautiful. They are very narrow; and upon examining them we find them ſcaly in a fine degree. The flowers will be produced in plenty at the ends of the branches: They grow in ſeemingly very large looſe panicles; but on examining them, we find that each is compoſed of numerous compleat flowers, which grow in ſpikes, and are produced near the extremities of the branches on the ſlender twigs all around. Each of theſe ſpikes ſeparately is but ſmall; and they are of a pale red colour. The flowers of each ſpike are exceedingly

ingly ſmall ; and the number of ſtamina is five, which differs from the other ſpecies by only having half the number. This ſort flowers in July, and we have known it in full blow in September, and ſometimes in October, and even November, when the weather has been all along mild. Nothing ornamental ſucceeds the blow.

2. The GERMAN TAMARISK is of lower growth, ſeldom aſpiring higher than eight or ten feet. It is a more regular tree than the former, as the branches all naturally grow in an upright poſition. They are very brittle, are ſcented, and covered with a ſmooth yellowiſh bark. The leaves have a ſcaly appearance, and ſtand much cloſer together than thoſe of the other ſort: They are of an exceedingly light green colour, and very ornamental. The flowers are produced in July, at the ends of the branches, in long looſe ſpikes. Each ſepa rate flower is ſmall, though much larger than the other ſort, and is poſſeſſed of ten ſtamina, which are alternately ſhorter. Theſe ſpikes attract the attention when in blow, and are acknowledged by all to have a fine look; neither is the noble appearance loſt when the flowers are faded; but it is continued in the ſpikes even until the ſeeds are ripe, which then ſeem to diſſolve into a ſhattered down and ſcales.

The PROPAGATION of theſe ſorts is very eaſy: Every cutting will grow that is ſet in winter, and will be a good plant by the autumn following. The encreaſing of theſe ſorts by layers has been recommended; but this is bad advice, not only as being unneceſſary trouble, when they will grow ſo freely by cuttings, but becauſe layers of this tree very often will not ſtrike root at all. We have layered them, and found them, after lying two years, without any roots; and the wound being grown up, differed from the other branches only in that the mould had a little altered the colour of the bark; which ſhould warn all perſons who want a ſtock of theſe plants to beware of layering: and this, no doubt, they will do when we aſſure them the cuttings will ſtrike root as freely as thoſe of the Common Willow. The beſt time for the work is October, though any time of the winter will do The cuttings ſhould be of the laſt ſummer's ſhoot; and a moiſt part of the garden

garden is moſt eligible for them to be planted in. In two years they will be good plants for the wilderneſs or ſhrubery, and may then be planted out in almoſt any ſoil, though they beſt like a light moiſt earth, eſpecially the German ſort; as in other countries, where it grows naturally, it is generally found in low watery grounds.

TAMUS.

LINNEAN Claſs and Order, *Dioecia Hexandria*: Male flowers containing ſix parts, and female flowers containing one part; upon diſtinct plants. There are only two SPECIES:

1. TA'MUS *Commu'nis*: The COMMON BLACK BRIONY; *a climber*; native of England, South of Europe, and the Eaſt.

2. TA'MUS *Cre'tica*: The CRETAN BLACK BRIONY; *a climber*; native of Crete.

1. The COMMON BLACK BRIONY. This has a very thick fleſhy root, full of a viſcous juice, blackiſh without, white within, and from which iſſue numerous ſlender twining ſtalks, which wind about themſelves, or any thing that is near them, and will mount, if ſupported, about twelve feet high. The leaves are heart-ſhaped, ſmooth, undivided, of a ſhining green colour, and grow alternately on the ſtalks. The flowers come out from the ſides of the ſtalks in long bunches. They are ſmall, of a whitiſh colour, appear in June and July, and the females are ſucceeded by round red berries, which ripen in the autumn.

There is a *Variety* of this with brown, and another with black berries.

2. CRETAN BLACK BRIONY. This has a large, fleſhy root, from which iſſue many ſlender twining branches, which, if ſupported, will riſe to about the height of the former. The leaves are trifid, or divided into three lobes. They are of a good green colour, ſmooth,

ſmooth, and grow alternately on the branches. The flowers come out in bunches, from the ſides of the branches. They appear about the ſame time as the former; and are ſucceeded by the like kind of red berries.

The PROPAGATION of both theſe ſorts is very eaſy: It is effected by parting the roots, or ſowing the ſeeds. 1. The beſt time of parting the roots is early in the autumn, that they may be eſtabliſhed in their new ſituation before the froſts come on. 2. The ſeeds alſo ſhould be ſown in the autumn, ſoon after they are ripe, otherwiſe they will often lie until the ſecond ſpring before they make their appearance. A very few of theſe plants in the ſhrubery quarters will be ſufficient. The beſt way is to well dig the ground under the trees or buſhes where you chuſe they ſhould grow; then put five or ſix berries in a place, covering them over about half an inch depth of mould. They will readily come up, will twiſt about the trees, and ſhew themſelves to greater advantage than when directed by art in their courſe.

TAXUS.

LINNEAN Claſs and Order, *Dioecia Monadelphia*: Male flowers containing many ſtamina joined in one ſet at the baſe, and female flowers containing one piſtillum; upon diſtinct plants. There are two SPECIES:

TA'XUS *Nucifera*: and

TA'XUS *Bacca'ta*. THE YEW; *an evergreen tree*; ſaid to be a native of Britain, and moſt parts of Europe. It grows alſo in Canada.

The YEW will grow to a great ſize. EVELYN mentions ſome very large ones in his time; and Mr. PENNANT, in a Tour in Scotland, took notice of a Yew in Fotheringall church-yard, the ruins of which girted fifty-ſix feet and a half. The leaves of this tree form perfect feathers: the young leaflets are of a pale yellowiſh hue; but

but the old leaves are of a darker green. Having been accuſtomed to ſee this tree, either as a ſubject of torture or a companion of the dead, and generally in an old declining and frequently diſeaſed ſtate, we are either wholly unacquainted with its natural beauties, or overlook them. If, however, the Yew, ſtanding ſingly, be ſuffered to form its own head, it becomes *ornamental* in a ſuperior degree; it throws out its lower branches to a great extent; and, ſhooting upwards, takes a ſtrikingly conical outline; putting on a looſe genteel appearance. The timber of the Yew is tough and durable. EVELYN enumerates its *uſes:* "Beſides the uſes of the wood for bows (for which the cloſe and more deeply dyed is beſt) the artiſts in box, cabinet makers, and inlayers, moſt gladly employ it; and in Germany they uſe to wainſcot their ſtoves with boards of this material: alſo for the cogs of mills, poſts to be ſet in moiſt grounds, and everlaſting axle-trees, there is none to be compared with it: likewiſe for the bodies of lutes, theorboes, bowls, wheels, and pins for pulleys; yea, and for tankards to drink out of." He mentions whole woods of theſe trees divers miles in circuit, growing in the neghbourhood of Box Hill, in Surrey. Theſe woods, or rather, we apprehend, *plantations*, have lately been taken down (a few ſtandards excepted); and the timber of ſuch trees as were found were ſold to the cabinet makers at very high prices, for inlaying: one tree in particular was valued at a hundred pounds, and half of it was actually ſold for fifty. The leaſt valuable were cut up into gate poſts; which are expected to laſt for ages: even ſtakes made from the tops of Yew have been known to ſtand for a number of years. We do not mention theſe circumſtances as an inducement for making plantations of Yew, ſo much as hints to thoſe who may have Yew Trees in their poſſeſſion. Indeed, ornamental and uſeful as the Yew Tree undoubtedly is, there is one great objection to planting it: we mean its poiſonous effects upon cattle. It is obſervable, however, that in the extenſive Yew plantations abovementioned cattle were admitted with impunity, and ſtill range amongſt the ſtragglers that are left, without any evil conſequence. They are browſed to the very bole: ſheep are particularly fond of the

the leaves, and, when the ground is covered with ſnow, will ſtand upon their hind legs, and devour them as high as they can reach. HANBURY, ſpeaking of this matter, ſays, "It has been thought dangerous to turn cattle into fields where Yew Trees grow; but I believe, no beaſts will touch them, unleſs compelled by extreme hunger. It is true, ſeveral have loſt both horſes and cows by their eating the leaves of this tree; but this accident muſt be attributed either to the abovementioned cauſe, or to the Gardener having thrown the clippings careleſsly in places where cattle come; who, particularly cows, will eat them when about half dry as greedily as new hay. By ſuch an accident, viz. a Gardener's having thrown the clippings of a Yew Tree over the wall, a neighbouring farmer of mine loſt ſeven or eight of his beſt cattle; which ought to be a caution to all Gardeners, whenever theſe trees are cut, to be careful that the clippings be either carried in for the fire or buried."

The PROPAGATION of this tree is from ſeeds. In autumn when the ſeeds are ripe, a ſufficient quantity ſhould be gathered; and being firſt cleared of their mucilage, let them be ſown neatly, in beds about half an inch deep. By being thus expeditious in planting them, many will come up the next ſpring; whereas, if the ſeeds are kept out of the ground till February, the plants will not appear until the ſpring after. During the ſummer the beds muſt be kept clean from weeds, and if the weather ſhould prove very dry, now and then watered. This will promote the growth of the hidden ſeeds, and at the ſpring may be expected a general crop. The plants being come up, no other care will be neceſſary, for two years, than keeping the beds weeded, and refreſhing them in dry weather with gentle watering. After they have gained ſtrength in theſe beds by ſtanding two years, a piece of ground muſt be prepared for them, in which they ſhould be planted at a foot aſunder. Here they may ſtand for three or four years, and may be then planted where they are deſigned to remain, or ſet out in the nurſery in rows two feet aſunder, and three feet diſtance in the rows, in order to be trained for hedges, or raiſed to a good ſize to be planted out for ſtandards.

There

There are two *Varieties:* one with very *ſhort leaves*, and another with *ſtriped leaves.* Theſe are encreaſed by *layers*, and the ſtriped ſort, HANBURY tells us, muſt be ſet in a very barren ſoil, or it will ſoon become plain.

THUYA.

LINNEAN Claſs and Order, *Monoecia Monadelphia:* Male flowers containing four parts joined at the baſe, and female flowers containing one part; upon the ſame plant. There are four SPECIES: Two of them are as follow:

1. THU'YA *Occidenta'lis:* The COMMON ARBOR VITÆ; *an evergreen tree or ſhrub*; native of the moiſt ſwampy parts of America and Siberia.

2. THU'YA *Orienta'lis:* The CHINESE ARBOR VITÆ; *an evergreen tree or ſhrub*; native of China.

1. The COMMON ARBOR VITÆ will riſe to thirty or forty feet high. The leaves of this tree are peculiarly formed, the leaflets being broad, and, in an advanced ſtate of the tree, thinly ſcattered: when bruiſed they emit a ſtrong, and, to moſt people, very diſagreeable ſcent. In a youthful ſhrub-like ſtate, the *Thuya* nevertheleſs gives no unpleaſing variety, and may be admitted amongſt *ornamentals.* EVELYN and HANBURY arrange it amongſt Foreſt Trees; and in Canada, the Indians, we are told, apply it to many *uſes.* HANBURY tells us "the wood is reddiſh, firm, and reſinous; ſo that we may eaſily judge of its value for curioſities of moſt ſorts when worked up by the reſpective artificers of turnery, joiners, cabinet makers, &c." He ſeems, however, to ſpeak from theory rather than from experience.

The Arbor Vitæ is to be PROPAGATED either from ſeeds, layers, or cuttings, the former of which produce the beſt trees, though the two latter methods of propagating are more generally practiſed. 1. In order to propagate this tree from ſeeds, theſe laſt ſhould be gathered

gathered as ſoon as they are quite ripe, which will be by the beginning of October. They muſt be ſown in pots or boxes of light fine earth, being covered about a quarter of an inch deep. The boxes ſhould immediately after be put in a well ſheltered place, ſo that the ſeeds in them, whilſt they are preparing to diſcloſe, may not be deſtroyed by violent froſts. Being thus protected till the month of February, they muſt be brought out, and ſet along a South wall, that the ſun warming the mould may ſet the powers of vegetation at work; and whenever ſevere weather is expected, they ſhould be removed into their ſhelter, but muſt be brought out again when the fine ſpringing weather returns. With this care, the plants will come up in the ſpring; whilſt, without it, they frequently lie until the ſecond ſpring before they make their appearance, by which neglect one year is loſt. When the young plants are up, and all danger of the froſt is ceaſed, they ſhould be ſet in the ſhade where they can have the free air; and in this place they may remain all ſummer. During that ſeaſon, little water ſhould be given them; keeping them clean from weeds is the principal trouble they will cauſe. By the autumn they will have made a poor ſhoot; for this reaſon they ſhould continue in their pots or boxes, which muſt be placed in the ſame ſheltered ſituation they had at firſt, where they may remain all winter. In the ſpring they may be brought out into the ſun again, to reap the benefit of his influence at that ſeaſon; and if they are ſet in the ſhade at the beginning of May to remain there all ſummer, it will forward their growth. The ſpring following, being then two-years-old ſeedlings, they ſhould be taken out of the boxes, and planted in beds nine inches aſunder. Here they may ſtand two years, before they are ſet in the nurſery. When they are taken from theſe beds with this intent, a moiſt ſeaſon ought always to be made choice of, and they ſhould be planted a foot and a half aſunder, and two feet and a half diſtance in the rows, where they may ſtand till they are ſet out to remain. 2. In order to propagate this tree from layers, the ground ſhould be dug, and made light round about the ſtools, and the branches laid down ſo deep as that the top eyes may but juſt peep above the ground, all being

being of the young wood. But if it ſhould ſo happen, that a few of the laſt year's ſhoots on the branches ſhould have ſhot out vigorouſly, and that there are many healthy twigs which would make good layers, that are not ſo long;. in order to have the greater plenty of layers, and that the ſhorter ſhoots may not be buried, it will be proper to ſhorten the longeſt, ſo that, being all laid in the ground, their noſes may juſt appear above the ſurface. This will be a means of preſerving every twig, and conſequently of propagating the greater number of plants from the ſame ſtool. When theſe plants are layered, the ſhoots ought to have a gentle twiſt or a ſmall nick, for without this they will not always ſtrike root: nay, if the land is ſtrong and heavy, it is great odds but you find them without root, as you laid them, only grown bigger. Thus will one year be loſt, which ſhews the neceſſity of obſerving theſe precautions. Being layered in this manner in the autumn, by the autumn following they will have taken root; and in the ſpring, when the ſevere froſts are paſt, they may be taken from the ſtools, and planted in the nurſery, at the diſtance directed for the ſeedlings. 3. In order to propagate theſe trees from cuttings, young ſhoots ſhould be taken from the trees in Auguſt, if rain has fallen; if not, the buſineſs muſt be deferred till it does: for work of this kind ſhould never be performed till the early autumnal rains have fallen upon the earth, and made it cool and moiſt. All theſe cuttings ought to be of the laſt year's ſhoot; and if a bit of the old wood be left at the end of each, it will encreaſe the certainty of ſucceſs. The ſituation theſe cuttings ſhould have ought to be ſhady and well ſheltered; and the ſoil in which they are planted, to enſure the greater ſucceſs, ſhould be a red loam. They may be planted almoſt as thick as you pleaſe: not more, however, than four or five inches aſunder, in rows; the rows may be a foot and a half diſtance from each other; and after they are planted, a little litter may be laid between the rows, to keep the froſt out of the ground in winter, and the ſun from over-drying them in the ſummer. This litter will not only keep down the weeds, but will ſave the trouble of watering, which will be much better for the plants; for theſe young plants,

plants, juſt ſtriking root, do not much like watering, at leaſt not in great plenty, as it often cauſes the tender fibres to rot at firſt ſtriking, and ſo deſtroys the young plant. In one year theſe cuttings will have good roots; ſo that the litter may be taken away, and the ſurface of the earth turned over in the ſpring, which will cheriſh the plants, and prepare them to ſhoot vigorouſly the ſucceeding ſummer. In the autumn, being then two years old, they may be taken up, and planted in the nurſery, at the diſtance directed for the ſeedlings and layers. After they are planted in the nurſery, they will require nothing more than the uſual care of keeping them clean from weeds, and digging between the rows in winter, till they are planted where they are to remain. This plant bears a very moiſt ſituation.

October is the beſt month for planting out theſe trees, though any of the winter or ſpring months will anſwer. When they are planted, they ſhould be ſet a yard aſunder, and thinned and managed as has been all along directed for others which are nearly of the ſame growth. The Arbor Vitæ gives great richneſs to tufts and maſſes of ſhrubs: it is full, yet flowing in feather-like tuftlets.

There is a *Variety* of this tree diſcovered by Mr. HANBUBY, which he has named The *American Sweet-ſcented Arbor Vitæ:* this ſeems to remove a principal objection to the Common ſort; namely, the diſagreeableneſs of its ſmell. He ſays, "It came up from ſome ſcattered ſeeds at the bottom of a box I had from Pennſylvania. It has the ſame duſky look in winter as the Common ſort, though it is better furniſhed with branches; neither are they produced ſo horizontally, or hang down in the manner of the Common ſort. What makes this ſort moſt valuable is the property of its leaves; for being bruiſed, they emit a moſt refreſhing odour, which is by many ſuppoſed to be as fine an aromatic as any we have; whereas the leaves of the othe ſorts being bruiſed, to moſt people are fœtid and diſagreeable. Whether this property will be continued by ſeeds, I have not yet experienced."

2. The CHINESE ARBOR VITÆ is a much more beautiful plant than the Common ſpecies; for its branches are more numerous, and grow in a more pic-

 tureſque

turesque erect manner, and the leaves are of a fine pleasant light green colour; whereas those of the other in winter are of a dark disagreeable green, inclined to a dusky brown, which is the worst property of this tree in the winter season. The branches of the Common Arbor Vitæ are of a dark brown colour, and the bark on the young branches is smooth; the bark of the Chinese is also smooth, and of a light brown. The leaves of this sort, like the others, are imbricated, that is, they grow over each other; but they are more numerous and smaller, and grow closer together; and being of so fine a green, which continues all winter, makes this sort the most valuable, though not to the rejection of the others, even in pleasurable plantations; for those cause good variety by their manner of growth, as well as the colour of their leaves. The flowers of none of the sorts have any beauty; they have males and females distinct; and the females of the Common Arbor Vitæ are succeeded by smooth cones, whereas the cones of the Chinese sort are rugged. They are larger than the Common sort, and are of a fine gray colour.

This species, as well as the Sweet-scented sort, may be PROPAGATED by layers and cuttings, as has been directed above for the Common sort.

TILIA.

LINNEAN Class and Order, *Polyandria Monogynia*: Each flower contains many males and one female. There are only two SPECIES:

1. TI'LIA *Europæ'a*: The EUROPEAN LIME, or the LINDEN TREE; *a tall deciduous tree*; native, it is said, of England, and most parts of Europe.

2. TI'LIA *America'na*: The AMERICAN LIME; *a deciduous tree*; native of Virginia and Canada.

1. The EUROPEAN LIME will grow to eighty or ninety feet high, and from twenty to thirty feet in cir-

circumference. The foliage is peculiarly ſoft and delicate, and its flowers ſweet in the extreme. It naturally forms a moſt perfectly elliptical head; and even in winter its general appearance is rendered pleaſing, by the elegance of its long ſlender twigs. As ſtandards, eſpecially in a rich deep ſoil, Limes are peculiarly eligible; they are, in ſuch ſituations, of very quick growth, and except the Oak and the Eſculus, few or no trees exceed them in point of *ornament*. The *wood* of the Lime is light, ſoft, and peculiarly fine grained: it ranks with that of the Sycamore and the Poplar, and may ſerve upon many occaſions as a ſubſtitute for the Beech: indeed, in one point of view, it ſeems to exceed any of thoſe woods, and ſtands upon its own baſis; namely, for the purpoſe of the carver: we cannot, however, upon the whole, recommend it in general terms to the planter as a timber tree: land, ſuch as this tree requires to render it of quick growth, ought rather to be applied to the more uſeful purpoſe of huſbandry, or, if convenient or neceſſary to be planted, ſhould be occupied by the more valuable Oak or Aſh; for which neceſſary woods, a certain and perpetual market may be expected.

The European ſpecies affords ſeveral *Varieties:* as,

The Narrow-leaved Mountain Lime,
The Broader-leaved Mountain Lime,
The Elm-leaved Lime,
The Green-twigged Lime,
The Red-twigged Lime.

All theſe are very inconſiderable differences; and though, if nicely obſerved, they cauſe ſome variety, yet that is ſo ſmall as not to deſerve much pains to procure them, except the Red-twigged ſort, which of all others is the moſt beautiful; becauſe, when diveſted of their leaves, its young branches exhibit their fine ſmooth red bark all winter, which has a pleaſing effect in all places; though in the younger plants this effect will be more ſtriking and delightful, as the bark only is red of the laſt year's ſhoots; and the ſmaller the plants are, the more of theſe and the leſs of older wood the compoſition of the tree will be; whereas, when the trees get older, the twigs will be ſhorter and leſs viſible; and though ſtill of a red colour, yet not of ſo delicate a red

 as

as the young plants wear on their bark at first. Sometimes these trees will run away from their colour, and grow with green branches; but as this is not common, the Red-twigged sort must be still allowed to be preferable to all others; and the seeds of this must always be sown for the raising of sorts.

The PROPAGATION of the EUROPEAN LIME is from seeds, cuttings, and layers. HANBURY, however, says, "That trees from layers or cuttings never grow so handsome nor so fast as those from seeds. These should be gathered from thriving healthy trees of the true Red-twigged kind; and then by far the greatest part of the young plants will be of that sort. The seeds will be ripe in October; and let a dry day be made choice of for gathering them. As the seeds grow at the extremity of the branches, and as it would be tedious to gather them with the hand, they may be beaten down by a long pole, having a large winnowing sheet, or some such thing, spread under the tree to receive them.

When you have got a sufficient quantity, spread them in a dry place, for a few days, and then sow them. The manner of sowing them is in beds of rich mould, about an inch deep, and about an inch asunder all over the bed. The plants will appear the first spring, and should stand in the seminary two years, when they should be removed to the nursery, planting them in rows, about two feet and a half asunder, and a foot and a half in the rows; and here they may remain until wanted for use.

2. The AMERICAN LIME. Of this species also there are a Variety or two, which indeed differ very little in appearance from any of the Common European sorts; for the leaves are heart-shaped like theirs. There are a larger and a smaller leaved sort. Their edges are finely serrated, and end in acute points. These beautifully cordated leaves, that thus run into acute points, have their under surface of a paler green than their upper. The larger leaved kind is by far the finest sort, and the branches vary from all others of this genus, in that they are covered with a dark brown bark. The flowers excite no attention in the Gardener; but the Botanist is delighted when he finds they are furnished with nectaria, whereas the flowers of our Common Lime Tree have none. The flowers are produced in bunches,

bunches, like our Common ſort, but make no better figure. They are very fragrant; and are ſucceeded by coriaceous capſules, containing the ſeeds.

The PROPAGATION of this ſpecies is the ſame as that of the European ſort, if ſeeds can be procured from abroad; if not, a few plants muſt be obtained. Theſe ſhould be planted in a light rich ſoil, if ſuch can be had, for in ſuch they ſhoot the ſtrongeſt; though almoſt any other will do. After theſe plants have ſtood a year or two, they ſhould be headed near the ground, for ſtools. They will then ſhoot out many young branches from theſe, which may be layered in the autumn; though, if they ſtand two years, there will be greater plenty of young twigs for layering; for every ſhoot of the firſt ſummer will the year following divide into ſeveral. When the layering of theſe is to be performed, which ought to be in the autumn, the ſtrong two-years ſhoots muſt be brought down; and if they are ſtiff and do not bend readily, they muſt have a gentle plaſh with the knife near the bottom; a ſlit ſhould be made at the joint for every one of the youngeſt twigs, and their ends bent backwards, that the ſlit may keep open. This being done, the mould muſt be levelled among the layers, and the ends of them taken off to within one eye of the ground. The buſineſs is then done; and the autumn following they will have all good roots, many of which will be ſtrong, and fit to plant out to ſtand, whilſt the weakeſt may be removed into the nurſery ground, in rows, to gain ſtrength. All the ſorts of Lime Trees will alſo grow from cuttings; but this is found to be an uncertain method; and if it was more certain, plants raiſed either by them or layers are not near ſo good as thoſe raiſed from ſeeds, which way ought always to be practiſed where they can be obtained. Where that is not to be done, any art muſt be uſed to obtain ſome few plants; and if the Gardener ſhould happen to procure a cutting or two of the American ſorts, ſet them in pots, and plunge them in the bark bed; let him water and ſhade them, and they will be ſure to grow; and theſe he may afterwards encreaſe at pleaſure.

VIBURNUM.

Linnean Claſs and Order, *Pentandria Trigynia:* Each flower contains five males and three females. There are eleven Species; eight of which are proper for our collection.

1. Vibu'rnum *La'ntana:* The Common Viburnum; or Wayfaring Tree, or pliant meally Tree; *a deciduous ſhrub or tree*; native of England, and moſt of the Northern parts of Europe.

2. Vibu'rnum *Denta'tum:* The Saw-leaved Viburnum; *a deciduous ſhrub*; native of Virginia.

3. Vibu'rnum *Nu'dum:* The Entire-leaved Viburnum, *a deciduous ſhrub*; native of Virginia.

4. Vibu'rnum *Prunifo'lium:* The Plum-leaved Viburnum, or the Black Haw; *a deciduous ſhrub*, native of Virginia and Canada.

5. Viburnum *O'pulus:* The Marsh Elder; *a tall deciduous ſhrub*; native of moiſt grounds in England, and moſt parts of Europe.

6. Vibu'rnum *Acerifo'lium:* The Maple-leaved Viburnum; *a deciduous ſhrub*; native of Virginia.

7. Vibu'rnum *Caſſinoi'des:* The Bastard Cassi'ne; or Cassioberry, or South Sea Thea; *a deciduous ſhrub*; native of Virginia.

8. Vibu'rnum *Ti'nus:* The Laurusti'nus; *an evergreen ſhrub*; native of Italy and Spain.

1. The Common Viburnum will grow to be twenty or more feet high, though it may be kept down to any height deſired; and in ſuch gardens as are at a diſtance from the places where it grows common, and in which it has not been before obſerved; in ſuch gardens it is enquired after, and attracts the attention of thoſe who walk therein, almoſt as much as any ſhrub in the whole Collection. The branches are not very numerous, and in winter they are covered with a ſmooth grayiſh bark, inclined to a brown colour, eſpecially near the bottom of the ſhoots. The younger, as they ſhoot,

ſhoot, are white and downy, and the ends, eſpecially in winter, feel ſoft and woolly. The branches are long, and exceeding tough. They will often ſhoot near ſix feet from the bottom in a year; and make the beſt bands for fagoting. The leaves are very large, heart-ſhaped, very full of large veins, and have their edges ſerrated Their upper ſurface is of a dark green colour, but their under is white, and like cotton; and they are placed oppoſite by pairs on the branches. The flowers are produced at the ends of the branches: the buds will be formed the preceding ſummer, which continue to get larger in the autumn; all winter they will be in a ſtate of increaſe, and at that ſeaſon they terminate the ends of the branches like ſo many rough buttons. The flowers, when out, will be in large umbels, to form which theſe buds encreaſe in ſize all ſpring, but ſhew little of what may be expected from them until about May, when they begin to divide, and ſhew that they are growing to be bunches of flowers. In June, they will be wholly out, and formed into large umbels; they are of a white colour; and have a good appearance. Theſe flowers are ſucceeded by berries, which are alſo ornamental, and cauſe variety; for they will be firſt of a fine red colour, and afterwards of a deep black. This plant likes a dry ſituation.

There is a *Variety* of this ſort with more oval leaves; but the differences are very inconſiderable in all reſpects. There is alſo the *Striped-leaved Viburnum*, which is coveted by thoſe who are fond of variegated plants.

2. The Saw-leaved Viburnum is ſo called, becauſe the leaves are more beautifully ſerrated than any of the ſorts. It is at preſent not very common. Its branches, leaves, and flowers, are not ſo large as the former, but they are of a more genteel growth. It will grow to the height of about ten feet. The bark is ſmooth, and of a light colour; and the leaves are of a fine light green. They are tolerably large, though nothing like thoſe of the other ſorts, and ſtand on longiſh footſtalks, which give them a fine air. They are ſtrongly veined, and have their edges finely ſerrated. They are of a roundiſh oval figure, and are placed oppoſite by pairs on the branches. The flowers are produced in June, at the ends of the branches, in very

large round bunches: Their colour is white; they appear in June; and are ſeldom ſucceeded by any berries in England.

3. ENTIRE-LEAVED VIBURNUM. The ſorts of Lauruſtinus are evergreens, and have all entire leaves; but this ſpecies of Viburnum agrees in every reſpect in deſcription with two ſorts, one of which ſheds its leaves in winter, whilſt the other retains its verdure during that ſeaſon. The deciduous kind grows to about ten feet high. The younger branches are covered with a ſmooth deep red bark; whilſt that of the older, though ſmooth, is of a dark brown colour. The leaves are pretty large, and of a delightful ſhining green on their upper ſurface; but their under is paler, and much veined: they are of a lanceolated, oval figure, though their ends are rounded; their edges are entire, and they ſtand oppoſite by pairs on the branches. The flowers are produced in July, at the ends of the branches, in large umbels; their colour is white; and they have much the reſemblance of thoſe of the Common Lauruſtinus, though they are rather ſmaller. They have a genteel look; and are ſucceeded by berries, which never ripen with us.

4. PLUM-LEAVED VIBURNUM, or BLACK HAW. This ſpecies, for the moſt part, goes by the name of Black Haw, becauſe the fruit a little reſembles that of the Haw, though of a black colour. It will grow to be about ten feet high; and the branches are covered with a ſmooth reddiſh bark. The leaves are oval, and not ſo large as any of the other ſorts, being ſeldom more than two inches long, and proportionally broad. They are of a light pleaſant green colour, and have their edges finely ſerrated. Their footſtalks are pretty ſhort, and they grow for the moſt part oppoſite by pairs on the branches. The flowers are produced in June, at the ends of the branches, in large umbels. Their colour is white; but they are ſeldom ſucceeded by berries in England.

5. MARSH ELDER. Of this ſpecies there are two notable Varieties: Marſh Elder with Flat Flowers. Gelder Roſe.

The *Marſh Elder with flat flowers* will grow to be a tree near twenty feet high. The young branches are covered

covered with a ſmooth and almoſt white bark. They are often produced oppoſite by pairs; though in general they are of an irregular growth. The young ſhoots will be cornered; and this is more perfect in the more vigorous ones, being compoſed of five or ſix flat ſides. The leaves are large and ornamental, of a fine green colour and a ſoft contexture, compoſed of three large lobes, which are jagged at their edges, and grow on glandulous footſtalks. In autumn theſe leaves have exquiſite beauty; for they die to ſo fine a red, as to have a ſtriking effect at that ſeaſon. The flowers are produced in large umbels, in the beginning of June, all over the tree, and have a grand look. Each umbel is compoſed of very many hermaphrodite flowers, which of themſelves make no great figure; but they are ſurrounded by a border of male flowers, which are white, and are ſo ornamental to each buſh as to throw a luſtre over the whole tree. Neither does this ſhrub ceaſe to exhibit its beauties when the flowers are over; for beſides what it affords by its leaves, which are inferior to few other trees, both in ſummer and autumn, the hermaphrodite flowers will be ſucceeded by fine ſcarlet berries, which will grow in ſuch large bunches, and be produced in ſuch plenty all over the ſhrub, as to give it an appearance ſuperior to almoſt any thing of the berry kind; and were it not for its commonneſs, this would, on their account only, be ranked amongſt trees of the firſt value.

The *Gelder Roſe*, or *Snowball Tree*, is a Variety only of the preceding ſort; its origin was accidental, and it is kept up and continued by culture in our gardens. The nature of the ſhoots and ſize of the tree, together with the colour of the bark, differ in no reſpect from the former. The leaves alſo are of the ſame form, are produced in the ſame manner, and die away to the ſame delightful red in the autumn. The Variety this ſort occaſions, then, is by the flowers; and by theſe this variety is ſo great, as to be exceeded by ſcarcely any two diſtinct ſpecies whatſoever. They are produced in the beginning of June, all over the tree, in large globular bunches. Each bunch is compoſed of numerous male flowers, of the ſame nature with thoſe that ſurround the hermaphrodite flowers of the former ſort.

fort. Their colour is white, like thofe; but being produced in large globular heads, and in great plenty, have a much finer appearance. HANBURY adds, "It is delightful to fee this tree ufher in the month of June, as it were, with its glorious flowers, which will then at a diftance have the appearance of balls of fnow, lodged in a pleafing manner all over its head."

6. MAPLE-LEAVED VIBURNUM. This is a middle fized fhrub, fending forth feveral branches, which are rough, and full of pith. The leaves are compofed of three principal lobes, like thofe of the Maple Tree, and grow on fmooth footftalks. The flowers come out from the fides of the branches, in umbels. Their colour is white; they appear in June; and are rarely fucceeded by feeds in England.

7. BASTARD CASSINE, Caffioberry Bufh, or South Sea Thea, is rather tender, will grow to about ten feet in height, forming itfelf into a bufh by rifing with three or four ftems, and fending forth numerous branches from the bottom to the top. The leaves are of an oblong, lanceolated figure, ferrated, grow oppofite by pairs, and continue on the trees until the nipping frofts come on; infomuch that in the early part of a mild winter, they have been taken for an Evergreen. Thefe leaves are of an exceedingly bitter nature, if chewed; and it is faid, that an infufion of them proves efficacious in removing pain, bracing a relaxed ftomach, and reftoring a loft appetite. The flowers are produced in bunches from the fides of the branches. Their colour is white; they appear at the end of July; and are fucceeded by red berries in the autumn. Whenever this plant is to have a fhare in a Collection, a naturally warm and dry foil, that is well fheltered, muft be fought for, otherwife there is a chance of lofing it by frofts; or if the plant is not wholly deftroyed, the young branches will be killed, and the tree fo haggled, as to have rather a bad appearance with others in the fpring.

The firft fix forts are very eafily PROPAGATED, either by feeds, layers, or cuttings. No particular art need be ufed for the feeds, whether they be of the forts of our own ripening, or of thofe we receive from abroad. A border of common garden mould, made fine, will be fufficicient; though it may be proper to obferve, that many

many of them will lie until the ſecond ſpring before they appear. The beds, before and after the plants are come up, will want nothing except weeding; and when they are a year or two old, they may be planted in the nurſery, at ſmall diſtances; and in two or three years more they will be fit to be finally planted out. 2. They are all eaſily propagated by layers alſo; for if branches are pegged down, and the mould anyhow thrown on them, they will have plenty of roots by the next autumn; and moſt of them will be good plants for almoſt any place. This freedom, however, ſhould be given to none but thoſe of our own country; for the American ſorts, as being ſtrangers, demand more care and neatneſs in the performance. 3. They are alſo eaſily propagated by cuttings; for the young ſhoots of theſe trees cut into lengths, and planted in a moiſt garden ſoil, in the autumn, will any of them grow; and this is our common method of propagating them. However, if a perſon has only a few plants of the American kinds, the beſt way is to make ſure of encreaſing them by layers.

If a large quantity is wanted, the beſt way to PROPAGATE the MARSH ELDER is by ſeeds. As the GELDER ROSE is a male flowering Variety, and never produces any ſeeds, it muſt always be propagated by layers or cuttings, by which the Variety will always be preſerved.

The BASTARD CASSINE is PROPAGATED by layers. The young ſhoots are fit for this purpoſe; and when they have taken root, if they are planted in pots, and protected for two or three winters, until they are grown ſtrong plants, either in a greenhouſe, or under a hotbed frame or ſome cover, there will be leſs danger of loſing them than by planting them immediately in the nurſery, or where they are to remain for continuance. However, a perſon who has not theſe conveniences, muſt fix on the warmeſt and beſt ſheltered ſpot he can find; and having prepared the ground, let the layers be taken from the old plants in the ſpring; if the weather be moiſt, it will be ſo much the better; and let him plant them in the nurſery, row by row, at two feet aſunder. In the ſummer, they ſhould be watered in dry weather, and when the winter froſts begin to come on, the ground ſhould be covered with peaſe ſtraw almoſt rotten, old thatch or tanners bark, to keep them from penetrating the roots. By this means many of the plants will be preſerved; and this care may be repeated every winter until they are planted out to ſtand. But this is not ſo good or ſo ſafe a method as potting them, and managing them as before directed; for they may be then turned out of their pots, when wanted, mould and all together, without feeling the effect of a removal.

8. The LAURUSTINUS is one of the greateſt ornaments of our gardens in the winter months, not only as it is a fine Evergreen, but becauſe, during that ſeaſon, it will either be in full blow, or elſe exhibit its flowers and buds in large bunches ready to burſt open, in ſpite of all weather that may happen; and the boldneſs of theſe buds, at a time when other flowers and trees ſhrink under oppreſſive cold, is matter of wonder and pleaſure. There are many VARIETIES of *Lauruſtinus*; but thoſe moſt remarkable are, The *Narrow-leaved Lauruſtinus*, The *Broad-leaved Lauruſtinus*, The *Hairy-leaved Lauruſtinus*, The *Shining-leaved Lauruſtinus*, The *Silver ſtriped Lauruſtinus*, The *Gold ſtriped Lauruſtinus*.

The *Narrow-leaved Lauruſtinus* is ſo called, becauſe, of all the ſorts, the leaves of this are ſmalleſt. It is generally planted among the low ſhrubs; though we have known it trained up againſt a wall to fourteen or ſixteen feet high. It produces its branches irregularly, which will grow ſo thick and cloſe as to form a buſh; for it hath that appearance when planted ſingly in open quarters. The bark in ſummer is green, and often a little hairy and glandulous; in winter it is frequently of a dark brown colour. The leaves grow by pairs, ſtanding oppoſite, on ſtrong and very tough footſtalks. They are of an oval figure, and their edges are entire. Their upper ſurface is ſmooth, and of a ſtrong green colour; but their under is lighter, and a little hairy; and they are at all ſeaſons very ornamental. The flowers are produced in large umbels, and are well known. It generally will be in full blow in January, February, March, and April; during which time it will be covered with bloom, cauſing a delightful effect.

The *Broad-leaved Lourustinus* differs from the former ſort,

ſort, in that the leaves are broader, and the roots proportionally ſtronger. It will arrive to a greater height than the other ſorts, and the umbels of the flowers are larger, though they will not be produced in ſuch plenty: it nevertheleſs makes an excellent figure.

The *Hairy-leaved Lauruſtinus* is as free a ſhooter as the other, and the leaves are frequently as large, and differs from that in ſcarcely anything but that the leaves are hairy; the young ſhoots alſo are hairy to a great degree. In this reſpect it makes a ſmall Variety. It flowers like the other ſorts; but blows rather later than thoſe.

The *Shining-leaved Lauruſtinus* is ſtill of about the ſame growth, and the leaves are large and fair. They are of an oval figure; and their upper and under ſurfaces are both ſhining, though their under is veined, and of a paler green. It differs only in that the leaves and young ſhoots are ſmooth, ſhining, and free from hairs; and being of this lucid green, force eſteem. It generally flowers later than the firſt two ſorts.

The *two variegated ſorts* are only one or other of the above ſorts, ſtriped with white or yellow; though the ſorts ſtriped with ſilver we have met with have been the Broad-leaved kinds; but the Gold-ſtriped ſorts have always been the firſt, or Narrow-leaved kind, with leaves ſtriped or blotched with yellow; and on theſe accounts, thoſe who are fond of variegated plants covet them in their Collection.

All theſe ſorts are eaſily PROPAGATED; for if in winter a little mould be anyhow thrown amongſt the young branches, they will ſtrike root, and be good plants by the next autumn. Notwithſtanding theſe plants, however careleſsly the mould be thrown, will grow, it is not here recommended to the Gardener to practiſe that cuſtom; it is expected that he be always neat in all his work; it is mentioned here only to ſhow what may be done; but let him gently lay the branches down, ſtrip off ſome of the lower leaves, and with his hand draw the mould amongſt the young ſhoots, and leave them neated up, as if a workman had been there; and theſe will be all good plants by the autumn, the ſtrongeſt of which may be ſet out to remain, whilſt the youngeſt

youngest may be planted out in the nursery, at small distances, to gain strength.

By the severe frost of 1794-5, the Laurustinus was very much hurt, especially in exposed situations.

VINCA.

LINNEAN Class and Order, *Pentandria Monogynia*: Each flower contains five males and one female. There are five SPECIES; three of which will bear our open air.

1. VI'NCA *Ma'jor*: The LARGE-LEAVED PERVINCA, or PERIWINKLE; *a shrub or creeper*; native of England, France, and Spain.

2. VI'NCA *Minor*: The SMALL-LEAVED PERVINCA, or PERIWINKLE, or The COMMON PERIWINKLE; *a creeper*; native of Germany, France, and England.

3. VI'NCA *Lu'tea*: The YELLOW PERVINCA, or PERIWINKLE; *a creeper*; native of Carolina.

1. The LARGE GREEN PERIWINKLE has smooth stalks of a pale green colour, which, if supported, will arise to about four or five feet high; but, unsupported, the tops turn again at about two feet high, and thus at a distance form the appearance of a round evergreen shrub of that low size; and when they are designed for this, the suckers must be always taken off, otherwise they will soon form themselves into a pretty large bed; for they will send out these at some distance from the rotten plant, and the very tops bending to the ground will often take root, which, unless taken away or prevented, will soon spread abroad, and take off the shrub-like appearance of the plant. The leaves are of a delightful evergreen, and stand opposite by pairs on strong footstalks. Their edges are entire, and they are of an oval heart-shaped figure. They are smooth and shining, and very ornamental in the winter months. The

The flowers are produced from the wings of the ſtalks, almoſt all the year round; are blue; but there will be ſometimes white ones ſeen amongſt them. They are compoſed of one petal, ſtanding ſingly on upright footſtalks. The tube is narrow, and nearly of a funnel ſhape; but their brim is large and ſpreading, ſo as to form a pretty large well looking flower.

2. The COMMON GREEN PERIWINKLE has ſmooth green ſtalks, like the former, though they are much more weak and ſlender, and will trail along the ground, and ſtrike root at almoſt every joint: ſo that they will ſoon run a great way, their general ſurface putting on a kind of rock-work-like appearance; though if they are planted near other ſhrubs, they will riſe to two or three feet high, and will cauſe a pretty look amongſt them this way. The leaves are ſmooth, and of a fine ſhining green colour. They are of an oval figure, their edges are entire, and they ſtand oppoſite by pairs on ſtrong ſhort footſtalks. The flowers are compoſed of one petal. They ſpread open at the rims, and grow from the wings of the ſtalks in the ſame manner as the former, though they are much ſmaller; and as they are not ſo ſubject to flower in winter, that is another reaſon for their being held leſs valuable.

The *Varieties* of this ſpecies are,

The Green Periwinkle with Blue Flowers,

The Green Periwinkle with White Flowers,

The Green Periwinkle with Double Blue Flowers,

The Green Periwinkle with Double White Flowers,

The Green Periwinkle with Double Purple Flowers,

The Gold-ſtriped Periwinkle with White, Blue, and Double Flowers,

The Silver-ſtriped Periwinkle with White, Blue, and Double Flowers.

All theſe ſorts are *Varieties* of the Common Periwinkle; though they may differ in the colour or properties of the flowers, or the variegation of the leaves. The White-flowering Periwinkle is this very ſort, only the flowers are white; the Double Periwinkle is the ſame ſort, only the flowers are double, and of a reddiſh colour; the Gold ſtriped Periwinkle is alſo this ſort, only the leaves are beautifully variegated with a gold colour;

colour; and the Silver-ſtriped with that of ſilver: The variegations are ſo completely done, and their ſtripes ſo little ſubject to vary or run away, that they are highly eſteemed amongſt the variegated tribe. There are Double Blue and Double White flowers belonging to both theſe ſorts; and theſe are all the hardy Varieties Nature affords us from this genus.

3. YELLOW PERIWINKLE has a twining ſlender ſtalk, which twiſts about whatever is near it. The leaves are oblong, and not much unlike thoſe of ſome of our Willows. The flowers are both ſingle and double; and thus continue in ſucceſſion from June to the end of ſummer. This ſpecies muſt have a warm light ſoil, and a well ſheltered ſituation.

The PROPAGATION of theſe ſorts may be eaſily ſeen to be not very difficult. With regard to the firſt ſort, the ſuckers it naturally ſends out may be taken up and multiplied at pleaſure; and the ends of the ſhoots that turn again, and ſtrike root into the ground, will be good plants when taken off: Nay, the very cuttings will grow; ſo that any deſired number of theſe plants, be it ever ſo great, may be ſoon obtained. With regard to the other ſorts, there is no end of their multiplying; for as they will ſtrike root, if permitted to lie on the ground, at every joint, one good plant of each ſort will produce a hundred of the like in a ſeaſon or two.

All theſe ſorts are very hardy, and will grow under the drip of trees, and flouriſh in all ſoils and ſituations. No plants are more proper to be ſet among low or larger ſhrubs, either in the evergreen or deciduous quarters, to form tufts or beds in the reſemblance of rock-work, or to be placed near other ſhrubs, by whoſe aſſiſtance their ſlender ſtalks may be ſupported to the height Nature will admit them to riſe.

V I S C U M.

LINNEAN Clafs and Order, *Diœcia Tetrandria*: Male flowers containing four parts, and female flowers containing one part; upon diftinct plants. There are nine SPECIES; one of them common in many parts of this ifland.

VISCUM *Album*: The MISLETOE; *a parafitical plant*; native of England and moft parts of Europe.

The MISLETOE is a fingular plant. It will grow upon trees only; more efpecially upon the Crab, the Hawthorn, and the Maple: It is not unfrequent upon the Afh; but feldom, very feldom indeed, is feen upon the Oak; and but rarely upon the Willow. It has a thick flefhy leaf ftanding ftiff upon the twigs, which are green and forked. The whole of the plant is of a green colour, and of the fhrubby, bufhy kind, rifing in numerous ftems; dividing into forked branches; and thefe again into forked twigs, thick fet with leaves. This thickens the general furface of the plant, and forces it into a fpherical or more generally a hemifpherical form. A tree thickly fcattered with this plant, has fomewhat the appearance at a diftance as if overgrown with Ivy. The Mifletoe, however, is of a lighter green than the Ivy; efpecially when full of berries, which are of a light tranfparent Pea-green colour, and about the fize of the common Field Pea; but when full ripe they become paler, taking the appearance of white currants. The pulp is vifcid in the extreme, being of the confiftence of thick gum water. Each berry inclofes one vetch-like feed. In the cyder counties the Mifletoe is a mifchievous intruder upon the Apple Tree; fo much, that were not the Farmers to cut it out every three or four years, or as often as neceffary, it would deftroy the tree. It is very common to fee Crab Trees, efpecially in or near woods, entirely killed by this truly parafitical plant. This is a curious fact in Nature, and affords ample fubject for reflection.

 The

The Misletoe may be said to be a superior order of Plants; for, like the animal creation, it feeds not upon the juices of the earth, but upon those of vegetables. This, added to its supposed medicinal qualities, assisted, probably, in rendering it sacred among the antient Britons; especially when found growing upon the Oak; which tree they also held sacred.

The PROPAGATION of this Plant is supposed to be, naturally, by the Misletoe Thrushes, which delight in its glutinous berries, and which in autumn, the season of their becoming ripe, repair in flights to the places where the Misletoe abounds. It seems to remain unascertained whether the seed be conveyed in the fœces of the bird, or whether, sticking to its beak amongst the glutinous matter, the bird in cleaning its beak wipes it off upon the branch of the tree it happens to perch upon. This last is the more probable supposition; as it has been found, that by striking the seeds upon the clean smooth part of the bark of some or all of the trees abovementioned, this plant may be artificially propagated. We do not learn, however, that the attempt has yet been successful upon the Oak or the Willow. It seems probable that the Bird, in wiping its beak across the branch, ripples the cuticle or outer rind; and this ought perhaps to be copied in attempting artificial propagation. In places where this plant is unknown, the cultivation of it would add a striking variety to shrubery quarters.

VITEX.

LINNEAN Class and Order, *Didynamia Angiospermia*. Each flower contains four males and one female; two of the males being longer than the other two; and the seeds being covered. There are eight SPECIES; one only of which is proper for our Collection.

VITEX *Agnuscastus*: The AGNUSCASTUS, or the TREE OF CHASTITY; *a deciduous shrub*; native of marshy,

marſhy, moiſt places in ſome parts of France, Spain, and Italy.

The TREE OF CHASTITY (being held by the antients as conducive to that amiable virtue) affords two *Varieties:*

The Broad-leaved Chaſte Tree.

The Narrow-leaved Chaſte Tree.

One deſcription will nearly ſerve for both ſorts; though it has been obſerved, that the Narrow-leaved ſort will grow to be the talleſt. The branches are produced from the bottom and ſides of the ſtalk. They are very pliable, and the joints are long. It is difficult to expreſs the colour of the bark. To ſay it is gray is not proper; and to ſay it is brown is not true; it is of a colour between both, though, in different ſoils, the bark of ſome trees will be of a darker colour than others. The leaves are digitated, being compoſed of ſeveral folioles, which ſo unite at their baſe in one common footſtalk as to reſemble an open hand. Theſe folioles are of a dark green colour; and their number is uncertain; being five, ſix, ſeven, and ſometimes eight. They are narrow, and the longeſt grow always in the middle, whilſt the ſhorter occupy the outſides. This character is common to both the ſorts; though it is obſervable, that the folioles of the Broad-leaved ſort are both ſhorter and broader, which occaſions its being ſo called. Their edges are alſo ſerrated, whilſt thoſe of the Narrow-leaved are intire; and in this the moſt important difference of theſe plants conſiſts. The flowers of both ſorts are produced at the ends of the branches, in whorled ſpikes. Theſe ſpikes are pretty long, and their colour is that of a blueiſh purple. They appear in September and October; and are not ſucceeded by ſeeds in England. Each individual flower is inconſiderable; but the whole ſpike makes a good ſhow: and the circumſtances of the flowers being produced late, even often when moſt other flowers are over, as well as being alſo very fragrant, greatly heighten their value. The early froſts often deſtroy the beauty of theſe ſpikes, before and when they are in full blow; ſo that it is no wonder their ornamental fruit ſeldom, if ever, ſucceeds them.

There is a *Variety* of each kind with white flowers.

The PROPAGATION of theſe ſorts is eaſily done, either by layers or cuttings. 1. The young ſhoots being layered, any time in the winter, will have roots by the autumn following; though it will be proper not to take them up until the ſpring, as they ſhoot late in the autumn, and have often their ends deſtroyed by the froſts. When this work is deferred until the ſpring, all the killed ends may be taken off; and all danger from ſevere froſts being over, they will meet with no check in their preparing to ſhoot. The removing of theſe trees in the ſpring, however, is not abſolutely neceſſary; for it may be done any time in the winter though the cutting off the dead ends ſhould be deferred until the latter end of March, when they ſhould be gone over with the knife, and cut down to within an eye or two of the ground, whether planted in nurſery lines, or finally ſet out to ſtand. 2. Plenty of plants may be ſoon raiſed by cuttings. About the middle of March is the beſt time for planting them; and they ſhould be ſet in a ſhady border of good light garden mould. Nothing but weeding, and now and then watering, will be required all ſummer; though, if the place is not naturally well ſheltered, they muſt be defended from black froſts by ſticking plenty of furze buſhes all around them. If this be judiciouſly done, it will take off the keen edge of froſty winds ſufficiently, and will occaſion much leſs trouble and expence than reed hedges, &c. All theſe plants are very hardy; but they require this protection, to preſerve the young ſhoots. Here they may grow until they are fully planted out: and if it be a moiſt, light, rich ſoil, and a well ſheltered ſituation, they will like it the better.

VITIS.

LINNEAN Claſs and Order, *Pentandria Monogynia*: Each flower contains five males and one female. There are

are eight SPECIES; four of which are adapted to ornamental plantations.

1. VITIS *Labrusca:* The WILD VIRGINIA GRAPE; *a climber*; native of many parts of North America.

2. VITIS *Vulpina:* The FOX GRAPE; *a climber*; native of Virginia.

3. VITIS *Laciniosa:* The PARSLEY-LEAVED GRAPE; *a lofty climber*; native of Canada.

4. VITIS *Arborea:* The PEPPER TREE; *a shrub or climber*; native of Virginia and Carolina.

1. The WILD VIRGINIA GRAPE, if desired for its climbing property, should be planted among pretty large trees or shrubs; for, by the assistance of its well-holding tendrils, it will arrive to a great height; and if the shrubs that grow near it be low growing ones, it will entirely overtop them; and in summer, its leaves being large, almost conceal them from the sight. These large ornamental leaves have their edges indented, and are nearly divided into three lobes, though they are of a heart-shaped appearance; and downy on their under side. The flowers are produced in bunches, like the other species of the Vine; and they are succeeded by round, rough-flavoured, black fruit.

2. The FOX GRAPE. The name of this species naturally brings the fable of the fox and grapes to the memory; and it is very common for those who are not skilled in the history and nature of plants, to ask if this species is not possessed of more excellent properties, or produces more desirable fruit, than most of the other sorts of the vine; whereas, alas! this sort is called the Fox Grape from the ill flavour of its fruit, which is like the scent of a fox, and which name the inhabitants of Virginia, where it grows naturally, have given it on that account. It must, like the former, be planted among largish trees; for it will overtop the small ones. The leaves are large, smooth on both sides, of a heart-shaped figure, and their edges are indented. The flowers are produced in the Vine-like bunches; and they are succeeded by black fruit of the above named disagreeable flavour.

3. The PARSLEY-LEAVED GRAPE. The leaves of this sort are finely divided, and at a distance resemble those of parsley, though larger. The stem is very

thick, and the ſhoots are ſtrong; ſo that when it is planted for a climber, the talleſt trees muſt be appropriated for its ſupport; otherwiſe it will be too powerful for trees of lower growth.

4. The PEPPER TREE is a weaker ſhooting plant than any of the others, and affords ſingular beauty from its leaves. Their upper ſurface is of a fine ſhining green colour; their under is paler, and they are compoſed of a multitude of folioles of the moſt elegant and delicate texture. The ſhoots will arrive to a tolerable height by their tendrils, if they have trees near for their ſupport; but they are very liable to be killed down very low in ſevere winters; on which account the plant ſhould be ſtationed at firſt in a well ſheltered place. Every ſpring the Gardener ſhould carefully cut off not only the dead ſhoots, but ſhorten them within an eye or two of the old wood, which will make them ſhoot ſtronger, and the leaves will be larger and finer. The flowers are white, and are produced in bunches from the wings of the ſtalks; but we have never yet perceived any fruit to ſucceed them. The name Pepper Tree is a cant name, and was given it without any meaning by the inhabitants where it grows naturally.

All theſe ſorts are PROPAGATED by cuttings, layers, or ſuckers. 1. The cutting muſt be the bottom of the laſt year's ſhoot; and if there be a bit of the old wood to it, it will be the better. 2. When raiſed from layers, the young branches ſhould be pegged down, and a little ſoil drawn over them. They will ſtrike root, and become good plants by the ſeaſon following. 3. Suckers may be taken from theſe plants, and immediately planted; or may be ſet in the nurſery for a year to gain ſtrength before they are ſet out.

U L E X.

LINNEAN Claſs and Order, *Diadelphia Decandria*: Each flower contains ten males and one female; the males being joined at the baſe in two ſets. There

are

are two SPECIES; one of which is a greenhouse plant; the other is,

U'LEX *Europæ'us:* The FURZE, WHIN, or GORSE; *an evergreen shrub*; native of England, France, and Brabant.

The FURZE is so extremely common in this country, that how ornamental soever it may be in nature, it cannot with much propriety be admitted into our *ornamental* plantations. Its *uses* however are many; as a fuel where wood and coals are scarce; and as hedge wood upon light barren land: its use as horse provender too seems to be fully proved, though not yet established.

HANBURY enumerates the following *Varieties*; some of which, if properly trained, may add a kind of secondary ornament to our grounds and shruberies.

The White-flowered Furze.
The long Narrow-spined Furze.
The Short-spined Furze.
The large French Furze.
The Round-podded Furze.
The Dwarf Furze.

The FURZE is PROPAGATED from seeds sown very shallow, in February or March. See the Article HEDGES.

U L M U S.

LINNEAN Class and Order, *Pentandria Digynia*: Each flower contains five males and two females.— LINNEUS makes only three SPECIES of ULMUS:

1. U'LMUS *Campe'stris:* Leaves double sawed; unequal at the base.

2. U'LMUS *America'na:* Leaves equally sawed; unequal at the base.

3. U'LMUS *Pu'mila:* Leaves equally sawed; equal at the base.

MILLER enumerates six SPECIES:

" 1. U'LMUS *Campe'stris:* Elm with oblong acute-pointed

pointed leaves, which are doubly ſawed in their edges, and unequal at their baſe; called the COMMON ROUGH or BROAD-LEAVED WITCH ELM.

2. U'LMUS *Sca'ber:* Elm with oblong oval leaves, which are unequally ſawed, and have leafy empalements to the flowers; called The WITCH HAZEL, or VERY BROAD-LEAVED ELM; by ſome unſkilful perſons called The ENGLISH ELM.

3. U'LMUS *Sati'va:* Elm with oval acute-pointed leaves, which are double ſawed, and unequal at the bſse; called The SMALL-LEAVED or ENGLISH ELM.

4. U'LMUS *Gla'ber:* Elm with oval ſmooth leaves, which are ſharply ſawed on their edges; called The SMOOTH-LEAVED WITCH ELM.

5. U'LMUS *Holla'ndica:* Elm with oval acute-pointed rough leaves, which are unequally ſawed, and a fungous bark; called The DUTCH ELM.

6. U'LMUS *Mi'nor:* Elm with oblong ſmooth acute-pointed leaves, which are doubly ſawed; called The SMOOTH NARROW-LEAVED ELM, and by ſome The UPRIGHT ELM."

These ſix ſpecies of MILLER are all of them comprehended in the ULMUS *Campeſtris* of LINNEUS; ſo that Miller is ſilent as to LINNEUS's ſecond and third ſpecies; and ſo is HANBURY, who only treats botanically of one ſpecies; namely, the ULMUS *Campeſtris* of LINNEUS: He nevertheleſs enumerates ſeven SORTS:

1. The true Engliſh Elm,
2. The Narrow leaved Corniſh Elm,
3. The Dutch Elm,
4. The Black Worceſterſhire Elm,
5. The Narrow-leaved Wych Elm,
6. The Broad leaved Wych Elm,
7. The Upright Wych Elm.

In another part of his work he ſays, "It would be endleſs, as well as needleſs, to enumerate the ſorts of Elms: I have counted in my time more than twenty, in woods, hedges, &c. that have fell in my way when in queſt of plants." The fact is, no genus of plants whatever is more incomprehenſible to the Botaniſt than the Ulmus; for although we ſee among the cultivated Elms of this country, individuals as different from each other as are ſome individuals of diſtinct genera, yet every man who has attended cloſely to the ſeveral kinds of Elms growing in different parts of the kingdom, muſt have obſerved ſuch a chain of intermediate kinds as renders claſſification extremely difficult; and muſt frequently have met with an individual, which he was puzzled to find a name for. LINNEUS, no doubt, having experienced this, lumped the whole maſs of cultivated Elms in one ſpecies,—The *Ulmus Campeſtris*: and as a BOTANIST he may be right: As PLANTERS, however we muſt beg leave to attempt a ſeparation; and yet we are obliged to confeſs, that we cannot *deſcribe* more than TWO *obviouſly diſtinct* VARIETIES. With reſpect to the ſecond and third ſpecies of LINNEUS, we take it for granted they have not yet been introduced, or are but little known, in this country (if we are wrong, we beg to be ſet right): We therefore proceed to

ULMUS *Campeſtris* The CULTIVATED ELM; *a tall deciduous tree*; found growing more or leſs, in one or other of its Varieties, in hedges about villages, in moſt parts of Europe.

The CULTIVATED ELM. Notwithſtanding the chain of Varieties above mentioned, if we examine the two extremities, we ſhall find two plants very different from each other in their general appearance; and ſufficiently diſtinct in the analyſis to be conſidered, in a work of this nature, as diſtinct ſpecies: The leaf of the one is nearly oval, with an obtuſe lance-like point; that of the other nearly circular, ſaving a narrow ſlender point, growing as it were out of the periphery of the circle. The membrane of the one is groſs and rigid, of the other comparatively thin, delicate, and ſupple: When held againſt a ſtrong light, the former appears opake; the latter, comparatively tranſparent. The nerves of *that* are ſtronger, ſet cloſer, and run more parallel; of *this*, more ſlender, fewer in number, and divide more into branches: *That*, in its general appearance, bears ſome reſemblance to the leaf of the Cheſnut; *this*, a very ſtrong one to that of the Hazel. The branches of the Coarſe-leaved ſort are clean, ſtraight, and ſlender, with a ſilvery bark; thoſe of the Fine-leaved kind more divided, run ſhorter lengths, and

and are covered with a less delicate bark. The general tendency of the latter is more upright, being easily trained to a great length of stem; that of the former is to divide into spreading arms, and when attempted to be trained up with a tall straight stem, generally grows stooping with a nodding head like the panicles of an oat. The Coarse leaved kind matures its seed in this island, and is probably a native; the Fine-leaved sort seldom if ever perfects its seed with us, and is probably an exotic.

We therefore proceed to treat separately of these two sorts; considering the intermediate kinds as subordinate Varieties of these two:

1. The COARSE-LEAVED ELM, or the Chesnut-leaved Elm, or the Broad-leaved Elm, or the Wych Elm, or the North-country Elm.

2. The FINE-LEAVED ELM, or the Hazel-leaved Elm, or the Narrow-leaved Elm, or the South-country Elm.

1. The COARSE-LEAVED ELM will grow to a very great size. Mr. Marsham mentions a Wych Elm by Bradley Church in Suffolk, which in 1754 measured (at five feet high) twenty-five feet five inches and a half, and in 1767 twenty-six feet three inches. The leaves of this species of Elm have been already described to be of an oval figure, with a thick membrane and strong nerves; their size varies with the individuals they grow upon, some trees of this species bearing leaves considerably smaller and *much narrower* than those of some individuals of the Hazel-leaved sort: The common distinction therefore of these two kinds of Elms into *Broad-leaved* and *Narrow-leaved* is altogether improper—their *figures* forbid it: It would be equally proper to distinguish an oval from a circle, by calling the former broad, and the latter narrow.

The outline or general appearance of this tree is sometimes strongly featured, coming near to that of the Oak: In general, however, it is liable to be ragged, rather than irregular, and in point of *ornament* is frequently exceeded by the Lime, the Beech, and its sister tree the Fine-leaved Elm. Its *uses* are many. The whole tribe of Elms have a peculiar excellency by which they stand alone, and are rendered in a great measure

meaſure independent of other woods. The Oak is pre-eminent for durability, the Aſh for toughneſs, the Beech for cloſeneſs of texture and cleanneſs of grain, and the Elm for its tenacity or adheſive quality, being leſs liable to be ſplit than other woods: This renders it ſingularly uſeful for many important purpoſes. The keels of ſhips are now almoſt univerſally laid with Elm, and ſometimes the gunwales, eſpecially of ſhips of war, are made of this wood; it being leſs liable to ſplinter off in action even than Engliſh Oak; as keels made of this wood are leſs apt to ſplit in taking the ground. Another very important uſe of Elm is for naves of wheels of carriages of every kind, whether of uſe or of pleaſure. There is a ſort in Yorkſhire peculiarly adapted to this purpoſe, which goes by the name of the *Nave Elm*; it is of the Coarſe-leaved kind.

The Coarſe-leaved Elm may be PROPAGATED from ſeeds, or by layering. HANBURY ſays, "In order to propagate them by layers, proper ſtools for the purpoſe muſt be firſt obtained; to procure which, let a piece of good ground be double dug, and plant Elms of about four or five feet high over it, at the diſtance of about ten feet: If they make good ſhoots in the firſt year after, they may be cut down early in the ſpring following; if not, they ſhould remain two years before they are headed for ſtools; which ſhould be by cutting them down to within half a foot of the ground. After they are cut down, they ſhould be ſuffered to grow undiſturbed for two years: The ground between the ſtools muſt be dug in the winter, and conſtantly hoed as the weeds ariſe in the ſummer; and at the end of that time, that is two years, the branches growing from theſe ſtools will be fit for layering; which may be performed thus: Excavate a piece of ground wide enough to receive a whole branch, and let the hollow be about half a foot deep; then ſplaſh the branch with a knife, near the body of the ſtool, that its head may be more readily brought into the prepared place: Next, thruſt an hooked ſtick into the ground, to hold it cloſe; take off all the ſuperfluous branches, which croſs and would otherwiſe incommode thoſe that are to be continued After this, cut all the remaining young branches acroſs half through with the knife; turn the edge towards the

the end, ſlitting it about half an inch. When this is done on all the young branches, the mould ſhould be gently put amongſt them, and every one of them ſhould have their ends bent towards the ſtool, that the ſlit may be open. Laſtly, having the whole vacuity filled with its own mould, ſmooth and even, take the end of each twig off that peeps above the ground, down to one eye, and the branch is layed, and will afford you as many plants as there are buds peeping out of the ground. Proceed in like manner to the other branches of the ſame ſtool, then to the next ſtool in order, and ſo on until the whole buſineſs of layering is finiſhed. By the autumn following, theſe layers will have taken root, and many of them will have made a ſhoot of near a yard in length. It is now neceſſary to take them from their ſtools, and plant them in ſome double dug ground in the nurſery. They ſhould be ſet in rows three feet aſunder, and the diſtance allowed them from each other in the rows ought to be a foot and a half. Here they may ſtand till they are planted out where they are to remain, with no farther trouble than digging the ground between the rows every winter, and in the ſummer carefully watching thoſe which ſhoot out two branches at the head, and nipping the weakeſt of them off. After the layers are taken up, the ſtools muſt have all the wounded parts, occaſioned by the former ſplaſhing, taken away; the old branches alſo ſhould be cut off, pretty cloſe to the ſtem; and in the ſpring they will begin to ſhoot out freſh branches again, for a ſecond layering, which will likewiſe be ready to have the ſame operation performed the ſecond year after: and thus may this layering be performed on theſe ſtools every other year. but Nurſerymen who would raiſe great quantities of trees this way, ſhould be provided with two quarters of ſtools, to come in alternately, ſo that from one or other of them they may annually receive a crop." We have given Mr. HANBURY's method in his own words, in order to convey to our leaders in the fulleſt manner *his method of layering*.

His method of PROPAGATING the Elm from ſeeds, we alſo give at length; for the Elm ſtanding next to the Oak at the beginning of his book, he has treated more

more fully of that article than any other (the Oak only excepted), and frequently refers to it in the courſe of his work. He ſays, "Let the ſeeds be gathered the beginning of June, it being the time when they are full ripe. When gathered, ſpread them three or four days to dry; for if they were to be ſown immediately after they were gathered, they would rot. Having been ſpread about that time, and the mould, which ought to be freſh and good, being in readineſs for their reception, mark out your beds four feet wide, and let the alleys between them be a foot and a half or two feet broad. Rake the mould out of the beds until they are about an inch deep; riddle that which came out of the beds into them again, until the bottom of each bed is raiſed half an inch (*i. e.* half filled) with riddled mould; then gently preſs the mould down with the back of the ſpade, and ſow the ſeeds thinly all over it with an even hand, covering them down with fine earth about half an inch deep. When the ſeeds are all ſown this way, the beds ſhould be hooped, and covered with mats, to be ſhaded in that hot ſeaſon of the year; and they ſhould alſo ſometimes be refreſhed with water: Part of the young plants will come up in about a month, or ſooner; the others not till the ſpring following. From the time the ſeeds are ſown to their appearance above ground, whenever rain falls, be careful to uncover the beds, and as ready to cover them again when the ſcorching beams of the ſun break out. About the end of Auguſt, the mats ſhould be wholly taken away, that the plants may be hardened againſt winter: The ſpring following, a freſh breed will preſent themſelves among thoſe that came up the ſummer before. All the ſummer following they ſhould be conſtantly kept free from weeds, and watered as often as dry weather ſhall render it neceſſary; and in October or ſpring they may be planted out in the nurſery, at the diſtance before preſcribed for the layers, and afterwards ſhould be managed like them."

2. The Fine leaved Elm will alſo grow to a great height and conſiderable bulk: We do not however find any tree of this kind upon record. The largeſt Elms we have ſeen of the Fine-leaved ſort, grow in the Vale of Glouceſter. There are ſeveral in the pariſh

parish of Church-down which girt, at five feet high, from ten to twelve feet. But the finest Elm in the Vale stands in the road between Cheltenham and Tewkesbury—within a few hundred yards of the Boddington Oak (See QUERCUS). It is known by the name of PIFFE's ELM; and the turnpike gate, the fence belonging to which is fastened at one end to this tree, takes its name from it; being called "Piffe's Elm 'Pike." The smallest girt of this tree, which falls about five feet high, is at present (1783) exactly sixteen feet. At ten feet high it throws out large arms, which have formerly been lopped, but which now are furnished with tree-like shoots, rising, by estimation, to seventy or eighty feet high, with an extent proportionable, exhibiting all together the grandest tree we have seen; not so much from its present size, as from that fullness of growth and vigour which it now wears. There is an Elm of the Small leaved sort in Hyde Park whose stem is larger than this; but it is hollow, its head much impaired, and is a mere dotard compared with Piffe's Elm; which we mention the rather as it may be a tree in ages to come, and, standing as it does in a well soiled country, may swell out to twice its present size.

The leaves of this species of Elm have been already fully described; it remains only to observe, that notwithstanding we are accustomed to see trees of this sort trimmed up to mere maypoles, or at best with close aspiring heads, yet, if planted singly and suffered to form their own head, they will take an outline equal to that of the Beech or the Linden; and where an immediate object or skreen is wanted, the Elm has two material advantages: it may be removed when of a great size, and its growth is quicker than that of any other tree which is equally ornamental. The *uses* of this species of Elm are similar to those of the Coarse-leaved kind; and in places where bricks are rendered dear by a want of a proper supply of fuel, as in Surrey and Kent, great quantities of this Elm are cut up fot studs and weather boarding for the sides of barns, stables, and even dwelling-houses; and in the Southern counties in general it is much used in carpenter's work of all kinds as a substitute for Oak.

The

The propriety of planting the Elm depends entirely upon the ſoil: It is the height of folly to plant it upon light ſandy land. There is not, generally ſpeaking, a good Elm in the whole county of Norfolk. By the time they arrive at the ſize of a man's waiſt they begin to decay at the heart, and if not taken at the critical time, they preſently become uſeleſs as timber. This is the caſe in all light ſoils; it is in ſtiff ſtrong land which the Elm delights. It is obſervable, however, that here it grows comparatively ſlow. In light land, eſpecially if it be rich, its growth is very rapid; but its wood is light, porous, and of little value, compared with that grown upon ſtrong land; which is of a cloſer ſtronger texture, and, at the heart, will have the colour and almoſt the heavineſs and the hardneſs of iron: On ſuch ſoils the Elm becomes profitable, and is one of the four *Cardinal Trees* which ought in preference to all others to engage the Planter's attention. It will bear a very wet ſituation.

The method of PROPAGATING this ſpecies of Elm is principally by layering (in the manner already deſcribed); the ſeeds not coming to perfection here. HANBURY recommends in very ſtrong terms the grafting what he calls the True Engliſh Elm upon the Wych Elm; which he ſays has a ſtronger and more porous root, and will thrive upon poorer land. His reaſoning, however, ſeems to flow from a theory perhaps ill grounded, rather than from practice. Nevertheleſs, as he ſeems to have taken particular pains in drawing up directions for this operation, we will, for reaſons already aſſigned, tranſcribe them at length. "The ſtocks for the purpoſe ſhould be the Broad-leaved Wych Elm, which muſt be raiſed from the ſeed, and planted out as before. When they have grown two years in the nurſery, they will be of proper ſize to receive the graft; and the laſt week in January is the beſt time for the work. If a large quantity of Elm ſtocks are to be grafted, procure ſix men in readineſs for the purpoſe. The buſineſs of the firſt man is to take the mould from the ſtem of the ſtocks, with a ſpade, down to the root, laying the top of the root bare; the next man is to follow him with a ſharp pruning-knife, cutting off the heads of the ſtocks, and leaving the ſtumps to be grafted only about two inches above the root; the third man is the grafter himſelf, who having his grafts cut about four or five inches in length, all of the young wood, and ſuch as has never bore lateral branches, in a diſh, takes out one of them, and holding it in his left hand, the taper end being from him, with the knife that is in his right he takes off a ſlope about an inch and half or two inches long; and if the grafter be an artiſt, it will be cut as true as if wrought by a plane. This done, he makes a ſmall cut acroſs, nearly at the top of the ſlope, and then proceeds to prepare the ſtock to receive it, which is effected by ſloping off a ſide of it, of the ſame leng h with the ſloped graft, that the parts may fit as near as poſſible He then makes a cut nearly at the top of the ſtock downward, to receive the tongue he had made in the graft; and having properly joined them, he proceeds to the next. After the grafter follows a perſon with baſs matting, cut into proper lengths; and with theſe he ties the grafts pretty cloſe to the ſtock. The fifth man brings the clay, which ſhould have been prepared a week or longer before, and well worked and beaten over, mixed with a fourth part of horſe dung, and ſome chopped hay, in order to make it hang the better together: with this he ſurrounds the graft and the ſtock. Laſtly, the ſixth man comes and cloſes the clay, ſo that there may be no probability of its being waſhed off. Two or three rows being grafted, let an additional hand or two be employed, either in drawing the earth up above the clay, ſo that it may be wholly covered, or digging the ground between the rows, and levelling it ſo that nothing of the performed work may appear, except the tops of the grafts, above ground. The danger of froſt renders this precaution highly neceſſary; for if it ſhould be delayed a night or two, and ſharp froſts ſhould happen, the clay will moſt of it fall off, and thus the work will require to be repeated; whereas, when it is lapped warm in the manner directed, there will be no danger of ſuch an accident. A good workman, with the above mentioned neceſſary aſſiſtance, will graft about fifteen hundred ſtocks in a day. In the ſpring, the buds will ſwell diſcloſe, and ſhoot forth nearly as ſoon as thoſe of the tree from

which

which they were taken. By the latter end of June, they will be ſhot a foot and half, when they ſhould be freed from the clay; the matting ſhould be alſo taken off, and themſelves left to ſport at eaſe with all the vegetative powers. At this time, of thoſe which have put forth two ſhoots, the weakeſt ſhould be taken up, to ſtrengthen the other, and to lighten the head, which would otherwiſe be ſubject to be broken off by high winds. By autumn the ſhoot will have grown about a yard in length; and in the winter dig the ground between the rows. In this place they may remain till they are of a ſize to be planted out for continuance, with no other trouble than what was directed for the layers; namely, keeping them clear of weeds, digging between the rows in the winter; at the ſame time taking off all very large ſide branches; and in the ſummer pinching off ſuch young ſhoots, in the head, as may have a tendency to make the tree become forked. This practice of grafting will be found a valuable improvement of the Engliſh Elm, if we conſider the nature of the Wych Elm, on which it is grafted. Firſt, the Wych Elm will not only grow to the largeſt ſize of all the ſorts, but will grow the faſteſt. However, this is not to be wondered at, if we examine the root, which we ſhall find more fibrous, and the pores larger and in greater numbers than in any of the other Elms. Now, as all roots are of a ſpongy nature, to receive the juices of the earth for the nouriſhment and growth of the tree, that tree muſt neceſſarily grow the faſteſt whoſe root is moſt ſpongy and porous; and therefore the true Engliſh Elm, being ſet upon the root of the Wych, a greater quantity of nutriment is received from the earth for its encreaſe, in proportion as the root of the Wych Elm is more ſpongy and porous than that of its own ſort. Thus the Engliſh Elm, on this baſis, will arrive at timber many years ſooner than thoſe raiſed by layers, and be alſo forced to a greater ſize. If we conſider too that the roots of the Wych Elm will imbibe ſuch juices as are proper for the growth of its own ſorts, timber thus raiſed muſt be better, as the wood of the Wych Elm is ſo excellent in its kind as to anſwer the purpoſes of all the other kinds."

See more of the ELM under WOODLANDS.

ZANTHOXYLUM.

LINNEAN Claſs and Order, *Dioecia Pentandria:* Male flowers containing five parts, and female flowers containing one part; upon diſtinct plants. There are two SPECIES; one of which, with due care, may be enured to this climate:

ZANTHOXYLUM *Clava Herculis* The TOOTHACHE TREE; *a deciduous ſhrub*; native of Jamaica, Carolina, and Pennſylvania.

The TOOTHACHE TREE (ſo called from its bark being ſaid to be efficacious in that complaint) will grow to the height of about twelve feet. The bark is rough, and armed with ſhort thick ſpines. The leaves are its greateſt ornament; for they are pinnated, are of a fine dark green on their upper ſurface, and yellowiſh underneath, and grow without order on the branches. The folioles are ſpear-ſhaped, long; four or five pair are terminated by an odd one, and the whole leaf has much the reſemblance of thoſe of the Maſtich Tree. The flowers come out in looſe panicles, from the ends of the branches; they are ſmall, and of little figure, having no petals, though the coloured ſegments of the calyx have been taken for petals. They are ſucceeded by roundiſh capſules, containing the ſeeds, which hardly ever ripen in England.

There is a *Variety* of this genus, with leaves compoſed of oval, oblong folioles, which have prickly midribs. This difference is permanent from ſeeds. They are numbered in the nurſeries as two diſtinct ſorts; the firſt is called the Lentiſcus-leaved Toothache Tree; the other the Aſh-leaved Toothache Tree.

Theſe Trees are PROPAGATED from the ſeeds, which we receive from abroad; and theſe are ſeldom leſs than two, and often three or four, years before they come up. They muſt be ſown deep, in largiſh pots, filled with a good, light, ſandy compoſt; and after that, the pots may be plunged into ſome natural ſoil, in a ſhady place, and there left undiſturbed, except having conſtant weeding,

weeding, during the next ſummer and winter. The ſpring following they may be taken up and plunged into a hotbed; and this will bring up many of the ſeeds. They muſt be next hardened by degrees; and afterwards plunged into their former ſtation, to remain there until autumn. In the enſuing winter they muſt be preſerved in the greenhouſe, or under a hotbed frame; and in the ſpring they ſhould have a hotbed as before; and then you may expect to ſee the remainder of the whole crop. The ſame management muſt be repeated until the ſpring following, when they muſt be all ſhaken out of the pots, and each be planted in a ſeparate pot. Watering ſhould be given them, to ſettle the mould to the roots; and they ſhould be plunged into a hotbed as before. After this they muſt be hardened to the air, and ſet abroad in a ſhady place. The plants are now raiſed; but they ſhould be treated as greenhouſe plants for two or three years after; when, in ſome ſpring, they may be turned out of their pots, with their mould, into the places where they are deſigned to remain. The places allotted them ſhould be naturally warm and well ſheltered; for although they are tolerably hardy when old, they require protection at firſt; and with this, nothing but the ſevereſt winters can deſtroy them.

A CLASSICAL ARRANGEMENT OF *TREES AND SHRUBS*, AGREEABLY TO THEIR RESPECTIVE HEIGHTS.

IN forming mixed ornamental Plantations, it is neceſſary to arrange the Plants, according to the Heights to which they ſeverally riſe, in a given climature. If low Shrubs be planted promiſcuouſly, among Foreſt Trees, the latter quickly riſe above them; firſt hiding, and at length overgrowing them. On the contrary, if the taller Plants be placed in the inward or central parts of the maſs of planting, and the lower, outwardly towards the margin, all the plants enjoy air and head-room; and form, collectively, a rich bank of foliage.

Perceiving, in practice, the utility of liſting the given plants, agreeably to their reſpective growths (ſee VOL. I. MIN. 8 and 11.) we have cauſed a general Analyſis of the Plants treated of in this Volume to be made, with reſpect to height, and have endeavoured to claſs them, according to their ſeveral growths, in this country.

An accuracy of arrangement cannot reaſonably be expected, in a *firſt attempt* of this nature; the heights to which many exotics riſe, in this iſland, are imperfectly known: nor is a critical exactneſs, in this caſe, requiſite. It is not an even ſurface of foliage, like what we ſee in

Confervatories, we afk for in Grounds. Even if it were, it would be difficult to produce it. Soil, afpect, and the habits of individual plants, ever tend to occafion an inequality of growth, and a defirable variety of furface: a flight error in the Claffification will only tend to increafe this Variety.

To give character and perfpicuity to the arrangement, we have diftributed the Plants in the following CLASSES.

I. TALL TREES,—or fuch as rife, in fuitable foils and fituations, to *fixty feet or upwards.*

II. MIDDLERANK TREES,—or fuch as ufually rife to *between forty and fixty feet.*

III. LOW TREES, of *thirty to forty feet* growth.

IV. VERY TALL SHRUBS; namely, thofe from *twenty to thirty feet.*

V. TALL SHRUBS, as from *twenty down to twelve feet high.*

VI. MIDDLERANK SHRUBS, between *twelve and fix feet in height.*

VII. LOW SHRUBS, between *fix and three feet.*

VIII. VERY LOW SHRUBS; namely, thofe *under three feet growth.*

CLASS THE FIRST.

TALL TREES.

CLASS THE SECOND.

MIDDLERANK TREES.

A'cer Negu'ndo	3	Aſh-leaved Maple
A'cer Platanoi'des	3	Norway Maple
A'cer Sacchari'num	4	Sugar Maple
A'cer Ru'brum	4	Scarlet Maple
Be'tula A'lba	34	Common Birch
Be'tula Ni'gra	35	Black Birch
Be'tula Le'nta	36	Canada Birch
Be'tula A'lnus	37	Alder
Ca'rpinus Be'tulus	51	Common Hornbeam
Ce'ltis Auſtra'lis	58	Southern Celtis
Ce'ltis Occidenta'lis	58	Weſtern Celtis
Cratæ'gus Tormina'lis	96	Wild Service
Fra'xinus America'na	136	American Aſh
Ju'glans Re'gia	170	Common Walnut
Ju'glans Ni'gra	173	Black Virginia Walnut
Ju'glans A'lba	174	White Virginia Walnut
Ju'glans Cine'rea	174	Pennſylvania Walnut
Liriode'ndron Tulipi'ſera	198	Tulip Tree
Pi'nus Pi'nea	254	Stone Pine
Pi'nus Ce'mbra	256	Cembro Pine
Pi'nus Tæ'da	256	Swamp Pine
Pi'nus Ce'drus	258	Cedar of Lebanon
Po'pulus A'lba	274	White Poplar
Po'pulus Balſami'fera	275	Balſam Poplar
Pru'nus A'vium	284	Wild Red Cherry
Pru'nus Ce'raſus	283	Cherry Tree
Pru'nus Ce'raſus	284	Red Flowering Cherry
Py'rus Commu'nis	294	Double-bloſſomed Pear
Py'rus Commu'nis	294	Twice-flowering Pear
Que'rcus Pri'nus	304	Cheſnut-leaved Oak
Que'rcus Phe'llos	303	Willow-leaved Oak
Que'rcus A'lba	305	White Oak
Que'rcus Ce'rris	306	Auſtrian Oak
Robi'nia Pſeudaca'cia	332	Falſe Acacia

Sa'lix

Sa'lix A'lba	354	White Willow
Sa'lix Purpu'rea	355	Purple Willow
Sa'lix Penta'ndria	355	Sweet-ſcented Willow
Sa'lix Hermaphrodi'tica	356	Shining Willow
Sa'lix Babylo'nica	355	Weeping Willow
Sa'lix Tria'ndria	356	Triandrous Willow
Sa'lix Fra'gilis	357	Crack Willow
Sa'lix Amygdali'na	356	Almond-leaved Willow
Sa'lix Haſta'ta	357	Haſtated Willow
Ta'xus Bacca'ta	395	Yew
Ti'lia America'na	404	American Lime

CLASS THE THIRD.

LOW TREES.

A'cer Campe'ſtre	2	Common Maple
A'cer Monſpeſula'num	3	Montpelier Maple
A'cer Cre'ticum	4	Cretan Maple
A'cer Tarta'ricum	4	Tartarian Maple
A'cer Ru'brum	4	Scarlet Maple
Amy'gdalus Commu'nis	13	Common Almond
Amy'gdalus Pe'rſica	14	Peach Tree
Bigno'nia Cata'lpa	39	Catalpa
Ca'rpinus Be'tulus	53	Flowering Hornbeam
Ca'rpinus Be'tulus	53	American Hornbeam
Ca'rpinus O'ſtrya	53	Hop Hornbeam
Ce'rcis Siliqua'ſtrum	61	Common Judas Tree
Ce'rcis Canade'nſis	62	Canadian Judas Tree
Cratæ'gus Oxyaca'ntha	91	Hawthorn
Cratæ'gus A'ria	95	White Leaf
Cupre'ſſus Sempervi'rens	99	Common Cypreſs
Cupre'ſſus Sempervi'rens	99	Upright Cypreſs
Cupreſſus Sempervi'rens	100	Spreading Cypreſs
Cupreſſus Sempervi'rens	100	Small-fruited Cypreſs
Cupre'ſſus Juni'perus	101	African Cypreſs

Cu-

Cupre'ssus Di'sticha	102	Deciduous Cypress
Cy'tisus Labu'rnum	106	Laburnum
Dio'spyros Lo'tus	116	Indian Date Plum
Fra'xirus O'rnus	135	Flowering Ash
Gledi'tsia Triaca'nthus	142	Gleditsia
I'lex Aquifo'lium	162	Common Holly
I'lex Aquifo'lium	165	Smooth-leaved Holly
I'lex Aquifo'lium	165	Box-leaved Holly
I'lex Aquifo'lium	166	Saw-leaved Holly
I'lex Aquifo'lium	166	Hedgehog Holly
I'lex Cassi'ne	167	Dohoon Holly
Liquida'mber Styraci'flua	195	Virginia Liquidamber
Liquida'mber Peregri'num	196	Canada Liquidamber
Magno'lia Grandiflo'ra	213	Evergreen Magnolia
Me'lia Aze'darach	218	Bead Tree
Me'spilus Germa'nica	223	German Medlar
Mo'rus A'lba	227	Silkworm Mulberry
Mo'rus Ni'gra	228	Garden Mulberry
Mo'rus Papyri'fera	228	Paper Mulberry
Mo'rus Ru'bra	229	Virginia Mulberry
Ny'ssa Aqua'tica	235	Serrated Tupelo Tree
Pi'nus Tæ'da	258	Dwarf Pine
Pi'nus Balsa'mea	265	Hemlock Fir
Pi'nus Orienta'lis	265	Oriental Fir
Py'rus Ma'lus	295	Fig Apple
Py'rus Corona'ria	295	Sweet-scented Crab
Pru'nus Pa'dus	281	Common Bird's Cherry
Pru'nus Pa'dus	282	Virginia Padus
Pru'nus Ce'rasus	284	Double-blossomed Cherry
Pru'nus Armeni'aca	288	Apricot Tree
Pru'nus Dome'stica	288	Plum Tree
Pru'nus Dome'stica	285	Cherry Plum Tree
Pru'nus Dome'stica	285	Double-blossomed Plum
Que'rcus Ni'gra	304	Black Oak
Que'rcus Æ'sculus	305	Italian Oak
Que'rcus Su'ber	307	Broad leaved Cork Tree
Que'rcus Su'ber	307	Narrow-leaved Cork Tree
Que'rcus I'iex	308	Ilex
Que'rcus Cocci'fera	309	Kermes Oak
Que'rcus Molu'cca	309	Live Oak
Robi'nia Hi'spida	333	Rose-flowered Acacia
Sa'lix Vitelli'na	355	Golden Willow

Sa'lix

Sa'lix Phylicifo'lia	356	Phylica-leaved Willow
Sa'lix He'lix	357	Rofe Willow
Sa'lix Ca'prea	358	Sallow
Sa'lix Glau'ca	358	Glaucous Willow
Sa'lix Vimina'lis	358	Ozier
So'rbus Aucupa'ria	369	Mountain Sorb
So'rbus Dome'ſtica	370	Cultivated Sorb
So'rbus Hy'brida	371	Mongrel Sorb

CLASS THE FOURTH.

VERY TALL SHRUBS.

A'cer Pennſylva'ticum	5	Mountain Maple
Æ'ſculus Pa'via	9	Scarlet Efculus
Anno'na Trilo'ba	18	Papaw
Arbu'tus Une'do	23	Oblong fruited Arbutus
Arbu'tus Une'do	24	Red-flowered Arbutus
Arbu'tus Une'do	24	Double-bloſſomed Arbutus
Arbu'tus Andra'chne	26	Andrackny
Bu'xus Sempervi'rens	45	Broad-leaved Tree Box
Bu'xus Sempervi'rens	45	Narrow-leaved Tree Box
Co'rnus Ma'ſcula	83	Cornelian Cherry
Co'rylus Avella'na	89	Hazel
Cratæ'gus Oxyaca'ntha	93	Scarlet Hawthorn
Cratæ'gus Oxyaca'ntha	94	Yellow Hawthorn
Cratæ'gus Oxyaca'ntha	94	White Hawthorn
Cratæ'gus Oxyaca'ntha	94	Maple-leaved Hawthorn
Cratæ'gus Oxyaca'ntha	94	Double-bloſſ. Hawthorn
Cratæ'gus Oxyaca'ntha	94	Glaſtonbury Thorn
Cratæ'gus Crus Ga'lli	96	Cockſpur Hawthorn
Dio'ſpyros Virginia'na	117	Piſhamin Plum
Elæa'gnus Anguſtifo'lia	118	Narrow-leaved Eleagnus
Euo'nymus Europæ'us	123	Broad leaved Spindle
Fra'xinus O'rnus	136	Dwarf Aſh
Fra'xinus America'na	136	Manna Aſh
Guilandi'na Dio'ica	148	Canada Nickar Tree

Juni'perus

Juni'perus Commu'n:s	176	Swediſh Juniper
Juni'perus Oxyce'drus	177	Spaniſh Juniper
Juni'perus Virginia'na	177	Virginia Cedar
Juni'perus Bermudia'na	177	Bermudian Cedar
Juni'perus Barbade'nſis	178	Jamaica Cedar
Juni'perus Thuri'fera	178	Spaniſh Cedar
Juni'perus Ly'cia	178	Lycian Cedar
Juni'perus Phœni'cia	178	Phenician Cedar
Lau'rus No'bilis	187	Evergreen Bay
Lau'rus Be'nzoin	190	Benzoin Tree
Lau'rus Sa'ſſafras	190	Saſſafras Tree
Magno'lia Acumina'ta	210	Long-leaved Magnolia
Magno'lia Tripe'tala	210	Umbrella Tree
Ny'ſſa Aqua'tica	234	Entire leaved Tupelo Tree
Piſta'cia Terebi'nthus	267	Common Turpentine Tree
Piſta'cia Ve'ra	268	Piſtacia Nut Tree
Piſta'cia Trifo'lia	268	Three-leaved Piſtacia
Piſta'cia Narbone'nſis	268	Larger-fruited Piſtacia
Py'rus Cydo'nia	296	Quince
Pru'nus Canade'nſis	282	Canada Padus
Pru'nus Inſiti'tia	286	Bullace Tree
Pru'nus Lauro-Ce'raſus	286	Common Laurel
Pru'nus Luſita'nica	287	Portugal Laurel
Rha'mnus Catha'rticus	312	Common Buckthorn
Rha'mnus Fra'ngula	313	Frangula
Rha'mnus Paliu'rus	314	Chriſti Thorn
Rha'mnus Alate'rnus	316	Common Alaternus
Rha'mnus Alate'rnus	317	Broad-leaved Alaternus
Rha'mnus Alate'rnus	317	Jagged-leaved Alaternus
Sambu'cus Ni'gra	361	Common Black Elder
Sambu'cus Ni'gra	361	White-berried Elder
Sambu'cus Ni'gra	362	Green-berried Elder
Sambu'cus Ni'gra	362	Parſley-leaved Elder
Syri'nga Vulga'ris	388	Common Lilac
Syri'nga Vulga'ris	388	Purple Lilac
Syri'nga Vulga'ris	389	Blue Lilac
Syri'nga Vulga'ris	389	White Lilac
Thu'ya Occidenta'lis	398	Common Arbor Vitæ
Thu'ya Orienta'lis	401	Chineſe Arbor Vitæ
Vibu'rnum La'ntana	406	Common Viburnum
Vibu'rnum O'pulus	408	Marſh Elder
Vibu'rnum O'pulus	409	Gelder Roſe

CLASS

CLASS THE FIFTH.

TALL SHRUBS.

Anagy'ris Fœ'tida	15	Anagyris
Ara'lia Spino'ſa	20	Prickly Angelica
Azale'a Nud flo'ra	30	Red Azalea
Bu'xus Sempervi'rens	46	Gold-edged Box
Bu'xus Sempervi'rens	46	Curled-leaved Striped Box
Ca'rpinus Be'tulus	53	Eaſtern Hornbeam
Ce'ltis Orienta'lis	60	Eaſtern Celtis
Chiona'nthus Virgi'nica	64	Snowdrop Tree
Colute'a Arbore'ſcens	81	Common Bladder Senna
Colute'a Arbore'ſcens	82	Red-podded Bladder Senna
Co'rnus Sangui'nea	83	Common Dogwood
Co'rnus Flo'rida	83	Virginian Dogwood
Co'rnus Flo'rida	84	Female Dogwood
Co'rnus Flo'rida	84	Blue-berried Dogwood
Co'rnus Flo'rida	84	White-berried Dogwood
Co'rnus Flo'rida	84	Swamp Dogwood
Cratæ'gus Aza'rolus	95	Azarole
Cratæ'gus Cocci'nea	96	Virginian Azarole
Cupre'ſſus Thyo'ides	100	American Cypreſs
Euo'nymus Europæ'us	122	Common Euonymus
Euo'nymus Europæ'us	122	Narrow leaved Euonymus
Fa'gus Pu'mila	131	Dwarf Cheſnut
Hi'ppophæ Rhamno'ides	155	European Sea Buckthorn
Hi'ppophæ Canade'nſis	156	Canada Sea Buckthorn
Juni'perus Commu'nis	176	Common Juniper
Juni'perus Sabi'na	180	Upright Savin
Lau'rus Æſtiva'lis	189	Deciduous Bay
Ligu'ſtrum Vulga're	193	Privet
Ligu'ſtrum Vulga're	193	Deciduous Privet
Ligu'ſtrum Vulga're	194	Evergreen Privet
Magno'lia Glau'ca	210	Sea-green Magnolia
Me'ſpilus Arbutifo'lia	223	Arbutus leaved Medlar
Philade'lphus Inodo'rus	242	Carolina Syringa
Phillyre'a Me'dia	243	Smooth-leaved Phillyrea
Phillyre'a Me'dia	243	Privet-leaved Phillyrea
Phillyre'a Me'dia	244	Olive-leaved Phillyrea
Phillyre'a Latifo'lia	244	Broad-leaved Phillyrea

Phillyre'a

Phillyre'a Angustifo'lia	244	Narrow-leaved Phillyrea
Ptele'a Trifolia'ta	291	Ptelea
Pru'nus Maha'leb	283	Perfumed Cherry
Pru'nus Spino'sa	286	Sloe Tree
Py'rus Ma'lus	294	Paradise Apple
Py'rus Corona'ria	295	Sub-evergreen Crab
Rha'mnus Alpi'nus	314	Alpine Frangula
Rha'mnus Fra'ngula	314	Smooth-leaved Frangula
Rha'mnus Oleo'ides	318	Olive-leaved Buckthorn
Rha'mnus Infecto'rius	318	Narrow-leaved Buckthorn
Rhus Coria'ria	325	Tanners' Sumach
Rhus Ty'phinum	326	Virginia Sumach
Rhus Ty'phinum	326	Stag's Horn Sumach
Rhus Gla'brum	327	Smooth Sumach
Rhus Gla'brum	327	New England Sumach
Rhus Gla'brum	327	Carolina Sumach
Rhus Gla'brum	327	Canada Sumach
Rhus Coppa'llinum	328	Canada Lentisc. Sumach
Rhus Co'tinus	330	Venetian Sumach
Sambu'cus Racemo'sa	363	Mountain Elder
Spa'rtium Ju'nceum	373	Common Spanish Broom
Spa'rtium Ju'nceum	373	Double Spanish Broom
Sty'rax Officina'lis	386	Styrax
Ta'marix Ga'llica	392	French Tamarisk
Vibu'rnum Denta'tum	407	Saw-leaved Viburuum
Vibu'rnum Nu'dum	408	Entire-leaved Viburnum
Vibu'rnum Prunifo'lium	408	Plum-leaved Viburnum
Vibu'rnum Acerifo'lium	410	Maple leaved Viburnum
Zantho'xylum Cla'va H.	434	Toothache Tree

CLASS THE SIXTH.

MIDDLERANK SHRUBS.

Amo'rpha Frutico'sa	11	Amorpha
Be'rberis Vulgaris	31	Common Berbery
Be'rberis Cre'tica	32	Box-leaved Berbery
Bupleu'rum Frutico'sum	43	Ethiopian Hartwort
Colute'a Arbore'scens	81	Pocock's Bladder Senna

Cratæ'gus

CLASS

CLASS THE SEVENTH.

LOW SHRUBS.

Geni'sta

Ro'ſa Eglante'ria	338	Double Sweet Briar
Ro'ſa Eglante'ria	338	Double Bluſh Sweet Briar
Ro'ſa Eglante'ria	338	Yellow Sweet Briar
Ro'ſa Caroli'na	340	Pennſylvania Roſe
Ro'ſa Villo'ſa	340	Apple Roſe
Ro'ſa Centifo'lia	340	Hundred-leaved Roſe
Ro'ſa Centifo'lia	340	Provence Roſe
Ro'ſa Centifo'lia	341	Moſs Provence
Ro'ſa Centifo'lia	341	Dutch Hundred-leav. Roſe
Ro'ſa Ga'llica	342	Gallican Roſe
Ro'ſa Sempervi'rens	342	Muſk Roſe
Ro'ſa Sempervi'rens	342	Single Muſk Roſe
Ro'ſa Sempervi'rens	343	Double Muſk Roſe
Ro'ſa Penduli'na	343	Pendulous-fruited Roſe
Ro'ſa A'lba	343	White Roſe
Ru'ſcus Hypoglo'ſſum	350	Alexandrian Laurel
Sa'lſola Frutico'ſa	359	Stonecrop Tree
Spa'rtium Scoparium	373	Starry Broom
Spiræ'a Salicifo'lia	377	Common Spirea Frutex
Spiræ'a Tomento'ſa	378	Red flowering Spirea
Spiræ'a Hypericifo'lia	378	Hypericum Spirea
Spiræ'a Crena'ta	379	Spaniſh Spirea
Spiræ'a Sorbifo'lia	380	Service-leaved Spirea
Vibu'rnum Ti'nus	412	Lauruſtinus
Vibu'rnum Ti'nus	413	Hairy-leaved Lauruſtinus
Vibu'rnum Ti'nus	413	Shining-leav. Lauruſtinus

CLASS THE EIGHTH.

VERY LOW SHRUBS.

Andro'meda Panicula'ta	16	Canada Andromeda
Andro'meda Maria'na	17	Maryland Andromeda
Be'tula Na'na	36	Dwarf Birch
Bu'xus Sempervi'rens	47	Dwarf Box
Da'phne Meze'reum	110	Mezereon
Da'phne Gni'dium	112	Flax-leaved Daphne
Da'phne Cneo'rum	112	Spear-leaved Daphne

Da'phne

Da'phne Tartonrdira	112	Tarton Raire
Da'phne Alpi'na	112	Alpine Chamelea
Da'phne Thymelæ'a	112	Milkwort-leaved Daphne
Da'phne Villo'ſa	113	Hairy-leaved Daphne
Da'phne Laure'ola	114	Evergreen Daphne
Geni'ſta Tincto'ria	140	Dyers' Broom
Geni'ſta A'nglica	140	Dwarf Engliſh Broom
Juni'perus Sabi'na	179	Spreading Savin
Ono'nis Fruticoſa	236	Reſt Harrow
Philade'lphus Corona'rius	242	Dwarf Mock Orange
Rhodode'ndron Ferrug.	320	Ferrug. Dwarf Roſe Bay
Rhodode'ndron Hirſu'tum	321	Hairy Dwarf Roſe Bay
Rhodode'ndron Chamæci'ſ.	321	Chameciſtus
Rhodode'ndron Dau'ricum	321	Daurian Dwarf Roſe Bay
Robi'nia Py'gmæa	334	Dwarf Aſpalathus
Ro'ſa Pimpinelliſo'lia	336	Burnet Roſe
Ro'ſa Spinoſi'ſſima	336	Scotch Roſe
Ru'bus Odora'tus	347	Virginia Raſpberry
Ru'ſcus Aculea'tus	349	Common Butchers Broom
Ru'ſcus Hypophy'llum	349	Broad-leaved Ruſcus
Ru'ſcus Hypoglo'ſſum	350	Hypogloſſum

☞ *It is proper to be remarked, that The* ALPHABET OF PLANTS *is printed agreeably to the firſt Edition. All that is* there *intended, by noting the heights of the Species, at the heads of the Genera, is to give ſome general idea of their reſpective growths, and, moſt eſpecially, as to whether they are* TREES *or* SHRUBS,*—without the trouble of referring to their ſeveral deſcriptions.* Here, *a greater degree of accuracy has been attempted.*

APPENDIX.

CLIMBERS.

Bigno'nia Sempervi'rens	40	Evergreen Bignonia
Bigno'nia U'nguis	41	Claw Bignonia
Bigno'nia Capreola'ta	42	Tendril Bignonia
Bigno'nia Ra'dicans	42	Scarlet Trumpet Flower
Cela'strus Sca'ndens	56	Climbing Staff Tree
Cle'matis Vitice'lla	73	Virgin's Bower
Cle'matis Vio'rna	74	Virginia Climber
Cle'matis Cri'spa	74	Carolina Climber
Cle'matis Orienta'lis	75	Oriental Climber
Cle'matis Vita'lba	75	Traveller's Joy
Cle'matis Cirrho'sa	76	Evergreen Clematis
Cle'matis Fla'mmula	76	Creeping Clematis
Cle'matis Virginia'na	76	Sweet-scented Clematis
Cyna'nchum Acu'tum	103	Acute-leaved Cynanchum
Cyna'nchum Monsp.	104	Round-leaved Cynanchum
Cyna'nchum Subero'sum	104	Carolina Cynanchum
Glyci'ne Frute'scens	144	Carolina Kidney Bean
Glyci'ne A'pios	145	Ash-leaved Milk Vetch
Glyci'ne Tomento'sa	146	Climbing Rest Harrow
Glyci'ne Como'sa	146	Maryland Kidney Bean
Glyci'ne Mono'ica	147	Virginian Glycine
He'dera He'lix	150	Common Ivy
He'dera He'lix	152	Yellow berried Ivy
He'dera He'lix	152	Gold-striped Ivy
He'dera He'lix	152	Silver-striped Ivy
He'dera Quinquefo'lia	152	Deciduous Ivy
Ja'sminum Officina'le	160	Common Jasmine
Ja'sminum Fru'ticans	161	Yellow Jasmine
Ja'sminum Hu'mile	161	Italian Jasmine
Lonice'ra Caprifo'lium	200	Italian Honeysuckle

Loni-

Lonicéra Periclýmenum	201	Engliſh Honeyſuckle
Lonicéra Periclýmenum	202	Oak-leaved Honeyſuckle
Lonicéra Periclýmenum	202	Red Dutch Honeyſuckle
Lonicéra Periclýmenum	202	MidſummerHoneyſuckle
Lonicéra Periclýmenum	202	German Honeyſuckle
Lonicéra Periclýmenum	202	Long-blowingHoneyſuckle
Lonicéra Periclýmenum	202	Evergreen Honeyſuckle
Lonicéra Sempervírens	202	Trumpet Honeyſuckle
Meniſpérmum Canadénſe	221	Canada Moonſeed
MeniſpérmumVirgínicum	221	Virginia Moonſeed
Meniſpérmum Carolínum	221	Carolina Moonſeed
Méſpilus Pyracántha	225	Evergreen Thorn
Paſſiflóra Cærúlea	237	Paſſion Flower
Períploca Grǽca	239	Periploca
Smílax Áſpera	365	Italian Smilax
Smílax Excélſa	365	Oriental Smilax
Smílax Sarſaparílla	365	Peruvian Smilax
Smílax Rotundifólia	366	Canada Smilax
Smílax Laurifólia	366	Laurel-leaved Smilax
Smílax Tamóides	366	Briony leaved Smilax
Smílax Lanceoláta	366	Lance-leaved Smilax
Smílax Herbácea	366	Ivy-leaved Smilax
Solánum Dulcamára	367	Woody Nightſhade
Támus Communis	394	Common Black Briony
Támus Crética	394	Cretan Black Briony
Vítis Labrúſca	421	Wild Virginia Grape
Vítis Vulpína	421	Fox Grape
Vítis Lacinióſa	421	Parſley-leaved Grape
Vítis Arbórea	422	Pepper Tree

CREEPERS.

Ly'cium Ba'rbarum	208	Broad-leaved Boxthorn
Ly'cium Ba'rbarum	209	Narrow-leaved Boxthorn
Ru'bus Frutico'ſus	345	Double-bloſſomed Bramble
Ru'bus Frutico'ſus	346	Thornleſs Bramble
Ru'bus Frutico'ſus	347	Variegated Bramble
Ru'bus Hi'ſpidus	347	American Bramble
Ru'bus Cœ'ſius	347	Dewberry Buſh
Vi'nca Ma'jor	414	Large Green Periwinkle
Vi'nca Mi'nor	415	Comm. Green Periwinkle
Vi'nca Lu'tea	416	Yellow Periwinkle

PARASITICAL PLANT.

Vi'ſcum A'lbum	417	Miſletoe

INDEX

INDEX

TO THE

ENGLISH NAMES,

AND OTHER

NON-LINNEAN TERMS,

IN THE

ALPHABET OF PLANTS.

Chamæ-

INDEX.

D.

E.

F.

G.

H.

J.

K.

Labu'r-

Sages,

INDEX.

END OF THE SECOND VOLUME.

WORKS ON RURAL ECONOMY,

Written by the ſame Author,

AND

To be had of the PUBLISHERS of theſe Volumes;

MINUTES OF AGRICULTURE,

made on a Farm of 300 Acres of various Soils, near *Croydon*, SURREY. To which is added, A DIGEST; wherein the Minutes are ſyſtemized and amplified; and illuſtrated by Drawings: the whole being publiſhed as a Sketch of the Actual Buſineſs of a Farm; as Hints to the inexperienced Agriculturiſt; and as an Overture to SCIENTIFIC AGRICULTURE.

ALSO,

EXPERIMENTS AND OBSERVATIONS CONCERNING AGRICULTURE AND THE WEATHER.

☞ *The above are ſold in One Volume Quarto. Or the latter may be had ſeparate.*

ALSO,

(In two Volumes Octavo,)

THE RURAL ECONOMY OF NORFOLK;

comprising the MANAGEMENT of LANDED ESTATES, and the PRESENT PRACTICE of HUSBANDRY, in that County.

ALSO,

(In Two Volumes Octavo,)

THE RURAL ECONOMY

OF

YORKSHIRE;

comprising the MANAGEMENT of LANDED ESTATES, and the PRESENT PRACTICE of HUSBANDRY, in the agricultural Districts of that County.

ALSO,

(In Two Volumes Octavo,)

THE RURAL ECONOMY

OF

GLOUCESTERSHIRE;

including its DAIRY: together with the Dairy Management of NORTH WILTSHIRE; and the Management of ORCHARDS and FRUIT LIQUOR, in HEREFORDSHIRE.

ALSO,

(In Two Volumes Octavo),

THE RURAL ECONOMY

OF

THE MIDLAND COUNTIES;

including the Management of LIVESTOCK, in LEICESTERSHIRE, and its Environs; together with Minutes on AGRICULTURE and PLANTING, in the District of the Midland Station.

*** *For some Account of the general Design of which the Four last Works form Parts, see the Address prefixed to the* RURAL ECONOMY *of* NORFOLK: *also page 340 of the first Volume of the present Work.*

Printed in the United States
By Bookmasters